OCCUPATIONAL BLOOD-BORNE INFECTIONS: RISK AND MANAGEMENT

Occupational Blood-borne Infections: Risk and Management

Edited by

C.H. Collins

Department of Microbiology
Imperial College of Medicine
at The National Heart and Lung Institute
London
UK

and

D.A. Kennedy

Cranfield Biomedical Centre
Cranfield University
Cranfield
Bedfordshire
UK

CAB INTERNATIONAL

CAB INTERNATIONAL
Wallingford
Oxon OX10 8DE
UK

Tel: +44 (0)1491 832111
Fax: +44 (0)1491 833508
E-mail: cabi@cabi.org

CAB INTERNATIONAL
198 Madison Avenue
New York, NY 10016-4314
USA

Tel: +1 212 726 6490
Fax: +1 212 686 7993
E-mail: cabi-nao@cabi.org

A catalogue record for this book is available from the British Library, London, UK

Library of Congress Cataloging-in-Publication Data
Occupational blood-borne infections: risk and management/edited by C.H. Collins and
  D.A. Kennedy.
    p.   cm.
  Includes index.
  ISBN 0-85198-626-9 (alk. paper)
  1. Bloodborne infections—Prevention.   2. Medical personnel—Diseases—
  Prevention.   I. Collins. C.H. (Christopher Herbert)
  [DNLM: 1. Infection Control.   2. Disease Transmission, Patient-to-Professional—
  prevention & control.   3. Blood-Borne Pathogens.   4. Blood Specimen
  Collection.   5. Occupational Exposure—prevention & control.
  6. Risk Management.   WA 110 015 1997]
RA642.B56023   1997
614.4'4—dc21
DNLM/DLC
for Library of Congress   97-4393
                                                                              CIP

ISBN 0 85198 626 9

Typeset in Stempel Garamond by AMA Graphics Ltd, Preston, UK
Printed and bound in the UK at the University Press, Cambridge

CONTENTS

CONTRIBUTORS

Dr Dominique Abiteboul, MD. *Médicin de Travail, Institut National de Recherche et Sécurité, 30 rue Olivier Noyer, 75680 Paris, Cedex 14, France*

Dr J.A.J. Barbara, PhD, FIBiol. *Head of Microbiology, North London Blood Transfusion Centre, Colindale Avenue, London NW9 5BG, UK*

Ms Melanie Bentley, BS. *International Health Care Worker, Research and Resource Center, Health Sciences Center, University of Virginia, Charlottes-ville, VA 22908, USA*

Dr Elizabeth Bouvet, MD. *Professeur d'Université – Patricien Hospitalier, Groupe Hospitalier Bichat-Claude-Bernard, 46 rue Henri Huchard, 75877 Paris, Cedex 18, France*

Dr C.H. Collins, MBE, MPhil, DSc, FRCPath, FIBiol. *Senior Visiting Research Fellow, Department of Microbiology, Imperial College of Medicine at the National Heart and Lung Institute, Dovehouse Street, London SW3 6LY, UK*

Mrs Glenys Griffiths, BSc, RGN, SCM. *Formerly Senior Nurse, Infection Con-trol, Mayday University Hospital, Thornton Heath, London, UK (Present address: 2, Friezingham Cottages, Friezingham Lane, Rolvenden Lane, Near Cranbrook, Kent TN17 4PT, UK)*

Mr P.N. Hoffman BSc. *Clinical Scientist, Hospital Hygiene Unit, Hospital and Respiratory Infection Division, Central Public Health Laboratory, Colindale Avenue, London NW9 5HT, UK*

Dr Deborah Hunt, DrPH. *Director, Biological Safety, Duke University Medical Center, Durham, NC 27710, USA*

Dr Janine Jagger, MPH, PhD. *Director, International Health Care Worker, Research and Resource Center, Health Sciences Center, University of Virginia, Charlottesville, VA 22908, USA*

Professor D.J. Jeffries, BSc, MB BS, FRCPath. *Head of Department of Virology and Clinical Director, St Bartholomew's and the Royal London School of Medicine and Dentistry, Bartholomew Close, London EC1A 7BE, UK*

Dr D.A. Kennedy, PhD, MIBiol, FIBMS, F10511. *Visiting Fellow, Cranfield Biomedical Centre, Cranfield University, Cranfield, Bedfordshire MK43 0AL, UK*

Dr C.C. Kibbler, MA, MRCP, MRCPath. *Consultant in Medical Microbiology, Department of Medical Microbiology, Royal Free Hospital Trust, Pond Street, London NW3 2QG, UK*

Dr A. Kitchen, PhD, FIBMS, *Lead Scientist, Transfusion Microbiology, National Blood Transfusion Service, London and S.E. Zone. Microbiology Consultant to the National Blood Authority (now International Product Manager – Blood Transfusion, Murex Biotech, Dartford, Kent DA1 5LR, UK)*

Dr Elizabeth McCloy, BSc, FRCP, FFOM. *Formerly Chief Executive, The Civil Service Occupational Health and Safety Service (OHSA), Edinburgh, UK (Present address: The Orchard Cottage, Friday Street, Ockley, Dorking, Surrey RH5 5TE, UK)*

Dr D.A. Morgan, MEd, MPhil, PhD, FIBiol. *Head of Science, British Medical Association, BMA House, Tavistock Square, London WC1H 9JP, UK*

Professor Crispian M. Scully, MD, PhD, MDS, FDSRCPS, FFDRCSI, FRCPath. *Dean, Eastman Dental Institute for Oral Health Care Science, 256 Gray's Inn Road, London WC1X 8LD, UK*

Dr H.A. Waldron, PhD, MD, FRCP, FFOM. *Department of Occupational Health, St Mary's Hospital, Praed Street, London W2 1NY, UK*

PREFACE

The transmission of viral, bacterial and endoparasitic infection by the blood-to-blood route has a long history but assumed great importance in the 1970s when an increasing number of health care workers became infected with hepatitis B, and later, with the onset of the AIDS epidemic.

The mechanisms of transmission and the precautions which might be taken to avoid infection were the subject of many investigations and reports, and this book is an attempt to bring the more important factors between two covers.

The introduction of 'Universal Precautions' was a welcome step but it soon became apparent that these were not truly 'universal'. Those that were designed to reduce the risks in any one field of health care were not necessarily applicable to others or in non health-care occupations. For the central part of this book, the reduction of occupational risks, we therefore invited practitioners and investigators in several different fields and disciplines to offer their own opinions on the ways in which blood-to-blood transmission may be avoided or at least minimized. This they have done, and the result is that we have perspectives from different professional backgounds, all based on a huge wealth of personal experience. Necessarily there is an overlap; some precautions are obviously common to all fields, but there are many differences in opinions and practices, some of which may be adapted to one another.

Around this central material there are chapters on the microbiology and epidemiology of blood-borne infections; the general principles of prevention, prophylaxis and treatment; and finally reviews of the Univeral Precautions as set out by the United States Occupational Health and Safety Agency, and the evolution of Standard Precautions.

C.H. Collins
D.A. Kennedy

Acknowledgements

The editors are grateful for assistance and advice, not only from the contributors, but also from the following: Mr N. Jennings, Mr M.P. Garden and Mr P. Woods (Medical Devices Agency), Dr R. Slade (King's College, London), The Chief Constable of Kent, The Chief Fire Officer, Kent Fire Brigade, and the Forensic Science Service.

Mention of proprietory materials in this volume does not imply specific endorsement by the editors and contributors. Addresses of manufacturers and suppliers may be found in Trades Directories.

CHAPTER 1
Viral agents of blood-borne infections

D.J. Jeffries

INTRODUCTION

A wide range of viruses that infect man and other animals is present in the blood (viraemia) at some stage of the infection. If blood is transferred to another individual, at that critical stage, and the contact person is not immune to the virus, infection may result. Such a transfer may occur during transfusion of blood and of blood products, exposure of cuts, abrasions, weeping eczema, etc. to the blood of the contact individual, or as a result of percutaneous inoculation with a contaminated needle or other sharp instrument. If the rate of exposure to the blood of others is high, as in the case of injecting drug users sharing inadequately decontaminated needles and syringes, the risks of acquiring blood-borne viruses rise accordingly. As most virus diseases are of short duration with a low level of viraemia, and there is control of the infection by the immune system, the possibility of blood transfer and acquisition of infection during the short period of viraemia is unlikely and diseases such as measles, mumps, chickenpox, etc. are not normally blood-borne. In some severe virus diseases, e.g. Lassa fever, Marburg, Ebola, Congo-Crimean haemorrhagic fever, a high level of viraemia may persist for a few days and this fact, together with the haemorrhagic nature of the illnesses, may expose contacts and health-care staff to a high risk of blood-borne infection. Some virus infections are known to persist in the host and those that produce chronic viraemia with the continuous or intermittent presence of infectious virus offer a continuing risk of transmission of infection.

Three viruses that induce long-term carrier states are recognized as the major risk to health-care workers, rescue workers and others in the community whose occupation or lifestyle brings them into contact with human blood. These viruses are human immunodeficiency virus type 1 (HIV-1), hepatitis B virus and hepatitis C virus. Human immunodeficiency virus type 2(HIV-2) is a rare infection at present in most industrialized countries and, although transmission to health-care workers has not been reported, as its transmission routes are similar to those of HIV-1, occupational transmission is a theoretical possibility. Long-term carriers of human lymphotropic virus type 1 (HTLV-1, the cause of adult T cell leukaemia/lymphoma and tropical spastic paraparesis) and the related virus HTLV-2 are identified in the population of all countries. No occupational transmission of these viruses has been reported to date although as

Occupational Blood-borne Infections (eds C.H. Collins and D.A. Kennedy)

with the other blood-borne viruses (including HIV-2) they can be transmitted by blood transfusion. The newly discovered human hepatitis virus, hepatitis G, is thought to be a member of the same family of viruses as hepatitis C (Flaviviridae) and also produces a chronic carrier state. As yet, diagnostic tests are not commercially available and it is too early to know whether this virus will be associated with occupational infection.

In addition to the nature of the virus itself, the state of health of the source individual may influence the likelihood of transmission of blood-borne agents. Thus, the common childhood infection erythema infectiosum due to parvovirus B19 is normally of short duration and is unlikely to transmit horizontally by the blood-borne route. If the individual is immunocompromised, however, the level of viraemia may be much higher and its duration prolonged. The risks of blood-borne transmission to contacts may then become higher.

Laboratory workers may be exposed to blood-borne viruses in animal hosts other than the human. In many cases viruses, particularly in distant species, are unable to cross the species barrier and infect humans. Viruses that can transfer from one species to another cause what are known as zoonotic infections and this is most likely to occur between closely related species. As part of the worldwide research into HIV infection, some workers use monkeys infected with simian immunodeficiency virus (SIV) as a convenient model for studying pathogenesis, antiviral therapy and vaccine development. SIV, which is viraemic in the monkey host, has caused human infections.

It should be remembered that viruses identified as blood-borne may be present in other parts of the body and may also be detectable in other body fluids. As examples, HIV replication occurs predominantly in the lymphoid system and the brain and the hepatitis viruses replicate in the liver. Thus, any unfixed tissue should be regarded as potentially infectious and, in health care, a number of other body fluids from persistently infected individuals are considered likely to carry blood-borne viruses. These fluids are cerebrospinal, pleural, pericardial, synovial and amniotic fluids, breast milk, semen and female genital tract secretions and any other fluid that is visibly blood-stained.

In this chapter, each of the common and less common blood-borne viruses is described in turn, to provide a basis for presenting details of prevention and control in other sections of the book.

HEPATITIS B VIRUS (HBV)

Hepatitis B virus (HBV) is a member of the Hepadnavirus family of DNA viruses (Hepadnaviridae). The complete virus particle (Dane particle) is a 42 nm spherical structure with a lipoprotein envelope composed of hepatitis B surface antigen (HBsAg).

During the course of acute infection and in individuals who are carriers of HBV infection a surplus of HBsAg is produced by the liver and is present in circulation. This non-infectious coat material is recognized as spheres (20 nm) and rods and its associated antigens are detected in the major screening tests for acute and chronic HBV infection (HBsAg screening). The appearance of HBV and associated HBsAg particles may be visualized by electron microscopy (Fig. 1.1) and the structure of the virus is presented schematically in Fig. 1.2.

An internal core protein (HBcAg) surrounds the viral DNA and a subunit of HBcAg, HBeAg, is present in the plasma during active viral replication. HBeAg has been recognized as a marker of potential infectivity and it is detectable during acute hepatitis and in individuals with a highly infectious carrier state. In addition, the association of HBeAg with viral replication indicates the likelihood of active liver damage.

The mean incubation period of acute HBV infection is 75 days (range 45–180 days). As with all types of viral hepatitis, the degree of illness is variable and is clearly influenced by the individual host's response. If the immune response is immature (as in neonates and infants) or compromised (e.g. due to immunosuppressive drugs or illness, such as malignancy or HIV infection), asymptomatic infection is common, and the person is likely to develop a persistent infection. It is clear from this observation that in addition to its role in controlling acute HBV infection, the immune response is responsible for causing the acute illness. Some carriers of HBV infection, who regain competence of their immune system, either from relief of immunosuppressive activity or by therapy (interferon-alpha), eliminate the virus from circulation and, during the process, develop acute hepatitis. In the immunocompetent, the onset of jaundice is usually preceded by a prodromal illness which may last up to 2 weeks and is

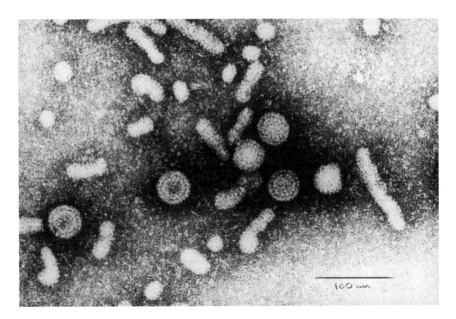

Fig. 1.1.   Hepatitis B virus particles viewed by electron microscopy with negative staining.

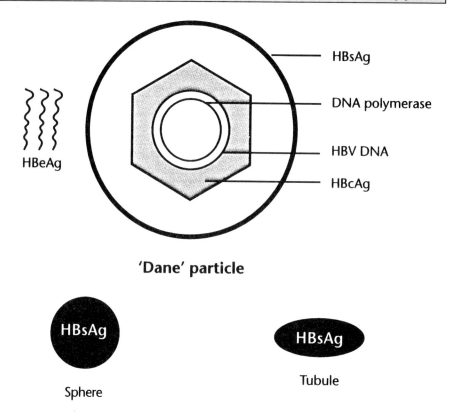

Fig. 1.2.   Diagram to illustrate the structure of hepatitis B virus and the subunit particles of HBsAg.

characterized by nausea, vomiting, diarrhoea, anorexia, headaches, malaise and a distaste for cigarettes. Fever is usually present and there may be upper abdominal discomfort. In addition, a serum sickness-like syndrome may be seen, consisting of a rash (urticarial or maculopapular) and a polyarthritis affecting small joints. The prodromal illness is followed by jaundice and, at this stage, the systemic symptoms often improve. As the jaundice deepens, the urine becomes dark and the stools are pale due to intrahepatic cholestasis. In the majority of patients the illness is over within 3–6 weeks. In rare cases the disease may be very severe with fulminant hepatitis, liver coma and death occurring usually without the appearance of jaundice.

Persistent infections with the development of a long-term carrier state occurs in at least 90% of infected neonates, 23% of children aged 1–10 years and 5% or less of adults.

The antigen and antibody responses in acute HBV infection are shown schematically in Fig. 1.3. Following the appearance of HBsAg and HBeAg in the blood, the first antibody to be detectable is anti-HBcAg. At first this antibody is mainly of the IgM class and this then switches to IgG; anti-HBcAg

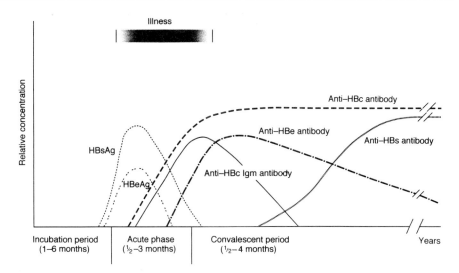

Fig. 1.3. Time course of serological events during, and following, uncomplicated hepatitis B.

IgG then persists, and is detectable as an indicator of past infection, for the life of the individual. Development of anti-HBeAg antibody leads to the elimination of HBeAg from the circulation and this correlates with a reduction or removal of infectivity. The last antibody to develop, anti-HBsAg, is associated with protection against reinfection and it is this antibody which is produced by HBV immunization of health-care workers and others who receive vaccine.

The serological profile of a carrier of HBV is presented in Fig. 1.4. In this particular example, the high infectivity state with persistence of HBeAg and abnormal liver function (ALT–alanine aminotransferase) is seen to be present for 5 years. After this, with the elimination of HBeAg from plasma, restoration of the ALT to normal and the appearance of anti-HBeAg, the individual has become a 'low-infectivity' HBV carrier. This is characterized by the persistence of HBsAg and the presence of anti-HBeAg antibody. In this state, although there may be a small risk of transmission of HBV by blood transfusion or from mother to baby during labour, the risk from percutaneous inoculation is very low. In addition to the risks of horizontal and vertical transmission of infection in the HBeAg positive carrier state, the continued presence of viral replication is associated in some individuals with progressive liver damage (chronic active hepatitis and cirrhosis and an increased risk of primary liver cancer.

In clinical and virological practice, the detection of HBeAg, as a marker of potential infectivity, has served us well. In recent years, however, the appearance of pre-core mutants of HBV has been documented. These mutant viruses, which do not produce circulating HBeAg, may appear during chronic infection or as a result of interferon therapy and may be transmitted to others. Fortunately, they are rare in most countries at present. Their presence can be detected

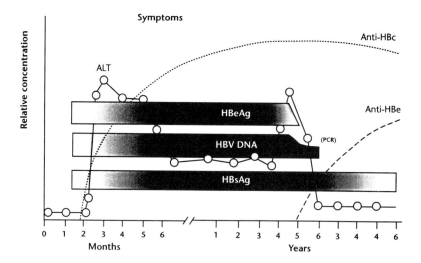

Fig. 1.4. Time course of serological events in hepatitis B virus infection leading to long-term carrier status (details in text).

by testing for HBV DNA by dot-blot hybridization or polymerase chain reaction (PCR) in HBsAg positive individuals who are negative for HBeAg.

HEPATITIS D VIRUS (HDV)

Previously known as the 'delta agent', HDV is a defective virus, or virusoid, which requires the presence of helper activity of HBV to allow it to replicate. Thus, HDV infection only occurs in carriers of HBV or in people who acquire the two viruses simultaneously. HDV is a blood-borne virus with transmission routes similar to those of HBV. The virus is particularly prevalent in the Amazon basin, Equatorial Africa, the Middle East and Mediterranean basin and in Asiatic parts of the former Soviet Union.

The particle of HDV is 36 nm in diameter and the circular single-stranded RNA genome is enclosed by a nucleocapsid antigen, the HDV antigen. The virus particle is encapsulated by HBsAg.

The mean incubation period of HDV infection is 35 days. Acute hepatitis caused by HDV is usually severe and carriers with persistent HBV and HDV infection usually develop rapidly progressive disease and cirrhosis at an earlier stage than those with HBV alone. The serological events occurring in HDV infection occurring simultaneously with HBV are shown in Fig. 1.5 and as a superinfection of a HBV carrier in Fig. 1.6.

Anti-HBs   ·········   Anti-δ —··—

Anti-HBc (IgM and IgG) — — —   ALT   - - - - ·

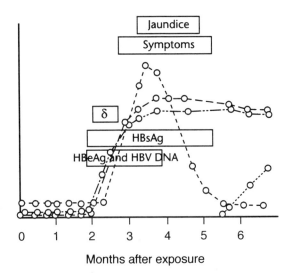

Fig. 1.5.   Serological events following double infection with hepatitis D virus and hepatitis B virus. δ = Antigen.

Anti-HBc (IgG) — — —   Anti-δ —··—

ALT   - - - - ·

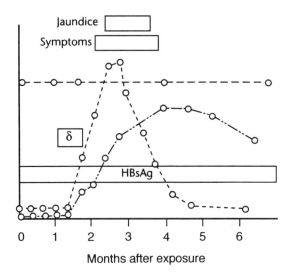

Fig. 1.6.   Serological events following HDV infection in a HBV carrier. δ = Antigen.

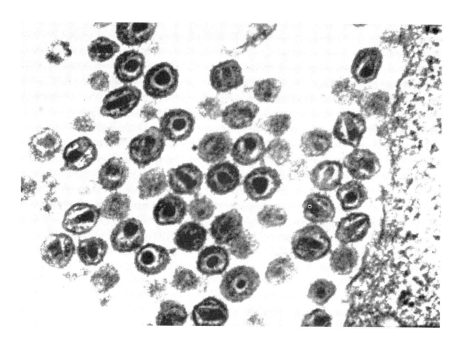

Fig. 1.8. Human immunodeficiency virus particles viewed by electron microscopy.

The major cellular receptor for HIV, the CD4 antigen, is expressed on helper T-lymphocytes and cells of the antigen-presenting series including follicular dendritic cells and macrophages. Recent studies indicate the involvement of other surface receptors in the binding and entry process. After binding, the virus is thought to enter the target cell by virus-to-cell fusion involving interaction of gp41 with the host cell membrane. Following uncoating, the viral RNA is reverse-transcribed to DNA and the complementary DNA copy of the virus genome is integrated into the host cell DNA where it is described as the provirus. Although it was originally thought that, following infection, the virus remained relatively inactive or latent for several months or years, it is now recognized that virus replication and consequent destruction of T helper cells occurs from early in the course of the infection. Following an early release of high levels of virus into the bloodstream at the time of seroconversion, the level of viraemia normally declines to a low or moderately low level until the later stages of infection (Fig. 1.9). During the early months and years of the infection, high levels of virus production and immune cell damage are occurring in the lymphoreticular system. In the later stages, reduction in numbers and functional capacity of T helper cells and other cells of the immune system leads to the onset of opportunistic diseases and this is accompanied by an increase in the level of virus in the plasma and in blood cells. There is now evidence to indicate that the level of viraemia in the first year of infection (months 6–12) correlates with the likely period of freedom from disease and time to death from AIDS.

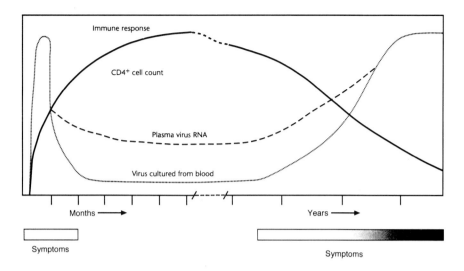

Immune response

CD4+ cell count

Plasma virus RNA

Virus cultured from blood

Months ———▶   Years ———▶

Symptoms   Symptoms

Fig. 1.9. Relative relationships between viraemia and CD4 count in human immunodeficiency virus infection.

Thus, the higher the level of virus in the plasma, the shorter is the likely period from infection to AIDS and death. Other prognostic factors including the genetic profile of the infected individual, the nature of the virus strain involved and the severity of the acute, seroconversion illness have also been described. In some individuals, neuroglial cell infection may lead to progressive encephalopathy and resulting dementia. The exact mechanism(s) of cellular damage by HIV has not been clearly defined. Latest research highlights the importance of high level viral replication of virus in target cells as a major cause of damage and the process of apoptosis due to circulating cytotoxic factors appears to be a significant effect. Other possible mechanisms including cell fusion and autoimmune destruction of cells may play a role in cellular damage but their contribution is not clearly elucidated.

The serological responses to HIV infection are illustrated in Fig. 1.10. Most individuals develop specific antibody within 3 months of infection and during seroconversion there may be an acute illness resembling glandular fever (infectious mononucleosis). This illness commonly presents with fever, muco-cutaneous ulceration, a maculopapular rash, lymphadenopathy and pharyngitis. Following the resolution of this acute illness there is usually a long symptom-free period (months or years) although later in the course of the infection generalized lymph node enlargement may be present (persistent generalized lymphadenopathy). Non-specific illnesses, including fever, night sweats and lymphadenopathy, are associated with progressive immune dysfunction and, when AIDS develops, it is characterized by the appearance of opportunistic infections and tumours. The reader is referred to specialized texts on HIV and AIDS for details of AIDS-defining conditions.

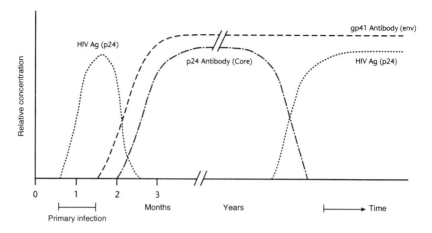

Fig. 1.10. Schematic representation of serological responses to human immuno-deficiency virus type 1.

Infectious virus is present in the blood and certain other body fluids (see Introduction) at all stages of infection. The presence of high levels of viraemia at the time of seroconversion and in the later stages of infection suggests that infectivity is highest at these times, and this is supported by epidemiological studies. Epidemiological evidence indicates that the most important vehicles of infection are blood, semen and female genital tract secretions. Thus, worldwide, most infections have been transmitted by sexual intercourse and blood transfer, either by transfusion or by use of contaminated syringes and needles. Infection of babies from infected mothers has been attributed to transplacental infection, exposure to blood and genital tract secretions during delivery, and to breast feeding.

VIRAL HAEMORRHAGIC FEVERS (VHF)

Viral haemorrhagic fevers are potentially serious and life-threatening diseases which are caused by a number of different viruses. Their existence is dependent on the presence of animal reservoirs of infection and/or the availability of suitable vectors for transmission such as mosquitos and ticks. Environmental conditions in the UK and other industrialized countries do not support the natural existence of these viruses, but the popularity of travel to remote and exotic locations, together with the speed of modern air travel, means that there is a constant danger that travellers will arrive during the incubation period or with active disease due to VHF.

Currently there are four viral causes of VHF which are of concern in the UK because they are known to be capable of person-to-person spread. These are Lassa fever virus, Marburg virus, Ebola virus and Crimean/Congo haemorrhagic fever virus.

Lassa Fever Virus

This virus belongs to the arenavirus family (Arenaviridae) and was first recognized as a cause of human disease in 1969 when a missionary nurse working in a small hospital in Lassa, North Eastern Nigeria, developed a febrile illness with pharyngitis and a haemorrhagic rash, from which she subsequently died. Two nurses who cared for her became infected and one died. The other nurse was flown to New York, where the virus was isolated. One of the virologists working with infected cell cultures contracted the disease but recovered. Another laboratory worker became infected and died. Laboratory precautions against infection were said to be minimal.

Since 1969, at least 12 patients who had contracted Lassa fever in Africa have been transferred to Europe or North America. These patients have been nursed in isolation facilities and there has been no evidence of nosocomial or occupational transmission. There is evidence from serological studies that the virus is endemic and widespread in West Africa and also that asymptomatic and mild infection is common. The potential danger in industrialized countries is presented by the individual returning from Africa with undiagnosed fever. Most of these patients will have active malaria or other infections but care must be taken with their management and with specimen handling until a specific diagnosis has been made.

The virus may be present in the serum and throat of patients during the first 2 weeks of their illness and it may be excreted in their urine for many weeks. The incubation period of Lassa fever is usually 7–10 days (range 3–17 days). The onset of disease is gradual with fever, malaise, myalgia, headache and sore throat. The tonsils and pharynx may be inflamed with exudates and ulceration. The fever is accompanied by profound lethargy or prostration and lasts an average of 16 days. As the disease progresses, there may be facial oedema, pleural effusion and ascites and the most severe cases develop haemorrhage into the skin and mucous membranes. The haemorrhagic manifestations are likely to lead to death.

Lassa fever virus is transmitted naturally by exposure to the urine of the multimammate rat (*Mastomys natalensis*) which enters buildings in rural areas. Person-to-person spread has been reported in West Africa as a result of inoculation injury involving blood, intimate personal contact or by exposure to pharyngeal secretions.

Marburg Virus

This virus is a member of the filovirus family (Filoviridae) and was first reported as a cause of severe VHF in 1967. The virus was present in a consignment of African green monkeys (*Cercopithecus aethiops*) that had been caught in Uganda and had been shipped via Heathrow airport in London to vaccine plants in Marburg and Frankfurt in Germany and Belgrade in the former Yugoslavia. There were 25 primary infections of laboratory workers and six secondary cases resulting from person-to-person spread. Seven of the 25 primary cases died.

Marburg disease appeared again in 1975 when a young Australian male, who had hitch-hiked through Zimbabwe, died in a Johannesburg hospital. His travelling companion and a nurse who cared for him became infected but both recovered. Further cases occurred in European tourists visiting Africa in 1980, 1982 and 1987.

The incubation period of Marburg disease is usually 7 days (extremes 4–16 days). The illness starts with a rapid rise in body temperature accompanied by severe headache, backache and myalgia. Gastrointestinal disturbances, including vomiting and diarrhoea usually follow and this may lead to dehydration. Many patients develop a measles-like rash after 3–8 days and, as with severe measles, this may be followed by desquamation. Many patients bleed spontaneously and this is associated with severe thrombocytopenia; renal failure is common in fatal cases and as the disease progresses there is extreme lethargy and deterioration in mental function. In non-fatal attacks the fever usually subsides in 10–20 days but the period of subsequent debility is prolonged.

The natural reservoir of Marburg virus is unknown. Person-to-person contact may lead to transmission, mainly through direct contact with infected blood. There is no epidemiological evidence of aerosol transmission although the possibility of airborne transmission in laboratory animals with high titre virus has not been excluded.

Ebola Virus

Ebola virus is also a member of the Filoviridae although it shows very little evidence of common antigenicity with Marburg virus. The first known cases occurred in simultaneous outbreaks in northern Zaire and southern Sudan in 1976. Several hundred cases occurred. The infection spread nosocomially and in one hospital of a total of 230 staff, 76 were infected with 41 deaths. In Zaire there were 318 cases with 280 deaths and the mortality was 89% in 149 secondary cases. Secondary infection arose as a result of re-use of unsterilized syringes and other equipment and by direct contact with body fluids.

Sporadic cases of Ebola virus infection occurred in 1977 and 1994. Further epidemics occurred in the Sudan (34 cases in 1979) and Zaire (more than 300 cases in 1995). In the latter outbreak more than 50% of those affected were hospital or home-based carers of patients.

The clinical features of Ebola infection are similar to those of Marburg disease although it tends to be more severe and have a higher fatality rate. As with Marburg virus, the natural reservoir is unknown.

Crimean/Congo Haemorrhagic Fever Virus (CCHF)

First recognized with an outbreak of more than 200 cases of haemorrhagic fever in the Crimea in 1944–1945, the virus was subsequently isolated from blood samples and from ticks (*Hyalomma marginatum marginatum*) and classified as a member of the bunyavirus family (Bunyaviridae). Many of the initial cases were helping with the harvest which would have exposed them to ticks. Identification of the disease and its causative virus led to the realization that it had previously occurred in outbreaks in other areas of the former Soviet Union and since then it has been reported from areas bordering the Black and Caspian Seas, Bulgaria and the former Yugoslavia.

The identity of this virus with a virus isolated from a febrile child in Zaire in 1956, was reported in 1969. The virus is widespread in East and West Africa and more recently its presence (and occasionally the disease) has been reported from Dubai, Iraq, South Africa, Greece, Albania, Turkey and the Indian sub-continent.

CCHF has an incubation period of 7–12 days and following the onset of fever, headache, myalgia and irritability there is usually nausea, vomiting and abdominal pain. Fever usually resolves by crisis within 8 days. The conjunctivae and pharynx are inflamed and a fine petechial rash usually appears on the trunk and then generalizes. Patients are often depressed and somnolent. Haemorrhagic manifestations including skin haemorrhages, haematemesis and melaena usually appear on the 4th or 5th day. The mortality rate may be as high as 30–50% although the severity of the disease varies in different geographical regions. Involvement of the central nervous system with neck stiffness, excitation and coma usually carries a poor prognosis.

CCHF virus is spread naturally by tick bites and hospital infection has resulted from direct exposure to blood samples.

CONCLUSIONS

The risks of infection by blood-borne viruses range from the relatively common and ubiquitous (HIV, HBV and HCV) to the rare and exotic (VHF viruses). Infection may be present in any individual and there may be no obvious risk factors or clinical signs to raise the index of suspicion of risk to health-care workers, rescue workers or others.

Specific protection with a safe, potent vaccine exists for HBV. The logical approach to protection with the other viruses includes awareness of potential risks, avoidance of contact with blood and other body fluids, careful and appropriate conduct of invasive procedures including care with sharp instruments and needles and prompt reporting of accidents and exposure incidents as soon as possible, so that appropriate prophylaxis (when available) can be administered.

The possibility of exotic infections (including VHF) in travellers should be constantly in mind although patient management should not be prejudiced by the fear of rare blood-borne viruses which can be easily controlled by good technique and the adoption of appropriate protective measures.

RECOMMENDED READING

Advisory Committee on Dangerous Pathogens (1995) *Protection against Blood-borne Infections in the Workplace: HIV and Hepatitis.* HMSO, London.

Advisory Committee on Dangerous Pathogens (1997) *Management and Control of Viral Haemorrhagic Fevers.* HMSO, London.

Barre-Sinoussi, F., Chermann, J.C., Rey, F. *et al.* (1983) Isolation of a T-lymphotropic retrovirus from a patient at risk for acquired immunodeficiency syndrome (AIDS). *Science* 220, 868–71.

Bowen, E.T.W., Platt, G.S. and Lloyd, G. (1977) Viral haemorrhagic fever in southern Sudan and northern Zaire. *Lancet* 1, 571–3.

Choo, Q.L., Kuo, G., Weiner, A.J. *et al.* (1989) Isolation of a cDNA clone derived from a bloodborne non A non B viral hepatitis genome. *Science* 244, 359–62.

Department of Health (1986) *Memorandum on The Control of Viral Haemorrhagic Fevers.* HMSO, London.

Fields, B.N., Knipe, D.M. and Howley, P.M. (eds) (1996) *Fields Virology,* 3rd edn. Lippencott-Raven, Hagerstown, MD.

Frame, J.D., Baldwin, J.M., Gocke, D.J. and Troep, J.M. (1970) Lassa Fever, a new virus disease of man from West Africa. I. Clinical description and pathological findings. *American Journal of Tropical Medicine and Hygiene* 19, 670–6.

Harrison, T.J. and Zuckerman, A.J. (1995) Viral hepatitis. In: Jeffries, D.J. and De Clercq, E. (eds) *Antiviral Chemotherapy.* John Wiley & Sons, Chichester, pp. 415–40.

Locarnini, S.A. and Cunningham, A.L. (1995) Clinical treatment of viral hepatitis. In: Jeffries, D.J. and De Clercq, E. (eds) *Antiviral Chemotherapy.* John Wiley & Sons, Chichester, pp. 441–530.

McCormick, J.B. and Fisher-Hoch, S.P. (1990) Viral haemorrhagic fevers. In: Warren, K.S. and Mahmoud, A.A.F. (eds) *Tropical and Geographical Medicine*, 2nd edn. McGraw-Hill, New York, pp. 700–727.

Rizzetto, M., Canese, M.G., Arico, S. *et al.* (1977) Immunofluorescence detection of a new antigen – antibody system (δ-anti δ) associated to hepatitis B virus in liver and serum of HBsAg carriers. *Gut* 18, 997–1003.

Simpson, D., Knight, E.M., Courtois, G. *et al.* (1967) Congo virus: a hitherto unde-scribed virus occurring in Africa. I. Human isolations – clinical notes. *East African Medical Journal* 44, 87–92.

Smith, C.E.G., Simpson, D.I.H. and Bowen, E.T.W. (1967) Fatal human disease from vervet monkeys. *Lancet* 2, 1119–21.

CHAPTER 2
Bacterial and other agents of blood-borne infections

C.H. Collins

In occupational medicine blood-borne bacterial and endoparasitic infections are not as important as those associated with viruses. Bacteria and protozoa known to have caused occupational infections in health-care workers as a result of percutaneous or mucous membrane exposure to blood, and those that have been involved in nosocomial transmission of infection by tattooing and blood transfusions, are listed in Table 2.1. Although this book is not directly concerned with the latter, infected transfusion blood offers a potential for occupational infections among staff who handle it (see Kitchen and Barbara, Chapter 13 this volume).

There are fewer reports of incidents of transmission of these agents than of viruses and the occupational risks are on a much smaller scale. Nevertheless,

Table 2.1. Nosocomially-transmitted blood-borne bacteria and endoparasites.

---

1. Bacteria and protozoa known to cause occupational infections in health care workers from percutaneous or mucous membrane exposure to blood
Borrelia spp. (relapsing fever)
Treponema pallidum (syphilis)
Mycobacterium leprae (leprosy)
Mycobacterium tuberculosis
Rickettsia rickettsii (Rocky Mountain spotted fever)
Plasmodium spp. (malaria)

2. Bacterial and protozoa known to cause nosocomial transmission of infection through blood transfusions or tattooing
Brucella spp. (undulant/Mediterranean fever)
Treponema pallidum (syphilis)
Mycobacterium leprae (leprosy)
Babesia microti
Plasmodium spp. (malaria)
Trypanosoma brucei gambiense (African trypanosomiasis)
Trypanosoma cruzi (Chagas' disease)
Leishmania spp.

From Hunt (1995).

during bacteraemia and parasitaemia quite large numbers of the agents may be present in the circulating blood. Some transmissions do occur, and there is a considerable potential for others among health-care workers, especially those who work in laboratories. Even if blood contains relatively few of these organisms, some laboratory procedures may concentrate them into much larger numbers. Thus we may distinguish between *primary* transmission, where the agent was transmitted directly from blood to blood, and *secondary* transmission, where it was first deliberately concentrated by laboratory culture – in either artificial media or in the gut of laboratory-bred arthropods, and then transmitted, either by the inhalation of infected aerosols during subculturing and other manipulations (see below), or by accidental injection or contact (see Kennedy and Collins, Chapter 11, this volume).

BACTERIAL AGENTS

Bacteraemia is not an uncommon condition, and the list of organisms responsible, and which have been isolated from blood, usually by culture of venous blood (Hickey and Shanson, 1993), is quite long (Table 2.2).

Two other bacteria must be added to this list: *Mycobacterium tuberculosis* and *M. avium-intracellulare* complex are frequently present in the blood of patients with AIDS and are isolated from blood cultures.

There is thus a high potential for transmission in the laboratory, especially where laboratory manipulations with blood cultures are not carried out in microbiological safety cabinets.

Reports of occupational blood-borne transmissions, however, as indicated above, are scanty, and seem to be confined to only a few agents: *Brucella* spp., *Borrelia* spp., *Mycobacterium leprae*, *M. tuberculosis*, *Rickettsia rickettsii* and *Treponema pallidum*

Table 2.2.   Bacteria that have been isolated from blood cultures in the UK.

Acinetobacter	*Mycobacterium avium-intracellulare*
Bacillus	*Mycobacterium tuberculosis*
Bacteroides	*Neisseria meningitidis*
Candida	*Pasteurella multocida*
Clostridium	*Proteus*
Corynebacterium	*Pseudomonas aeruginosa*
Cryptococcus neoformans	*Salmonella typhi*
Enterobacter	*Salmonella* other species
Escherichia coli	*Shigella dysenteriae*
Haemophilus influenzae	*Staphylococcus aureus*
Klebsiella	*Streptococcus pyogenes*
Listeria	*Xanthomonas maltophila*
Morganella	

Adapted from Hickey and Shanson (1993).

Brucella Spp.

Brucellas are small, Gram-negative, non-sporing coccobacilli, 0.6–1.5 μm in length and 0.5–0.7 μm in width. They may be grown in laboratory culture media but are fastidious in their nutritional requirements.

There are three species. *Brucella abortus,* responsible for undulant fever in humans and contagious abortus in cattle, is the one most frequently encountered. *Brucella melitensis* is the agent of Mediterranean or Malta fever and *B. suis* causes disease in pigs but both are now rare in developed countries. In most cases of undulant fever the agent is ingested in milk or, in occupational infections inhaled in dust or in aerosols generated during calving etc. The number of laboratory-acquired infections, however, is very high: surveys of laboratory-acquired brucellosis (Pike, 1976; Collins, 1993; see also US Department of Labor, 1989) suggest that most infections were acquired by the air-borne route from cultures; one may postulate that many were derived indirectly from blood, and that at least some when the aerosols were generated and inhaled, or through needlestick injuries, during subculture from the blood culture bottle, i.e. secondary transmission. It has been shown that in acute brucellosis 70–90% of blood cultures will contain brucella organisms (Arnow *et al.*, 1984; Gotuzzo *et al.*, 1986; US Department of Labor, 1989). For example, Batchelor *et al.* (1992) reported a laboratory-acquired infection that followed a misidentification as a result of which Containment Level 3 requirements were not observed.

Although brucellosis is now uncommon – and therefore may not be suspected by clinicians – blood-borne infections acquired in laboratories must be regarded as a serious potential hazard.

Borrelia Spp.

Borrelias are large spirochaetes, 10–20 μm long by 0.2–0.3 μm in width. They are difficult to cultivate *in vitro*. The agent of African relapsing fever is *Borrelia duttoni*, transmitted from wild rodents and pigs by ticks. European relapsing fever, caused by *B. recurrentis,* is transmitted from human to human by the louse. The organisms are present in the circulating blood in the acute stage of the disease. Direct transmission to humans may therefore occur, either by contact or needlestick injuries. In laboratories both species are propagated in arthropods and secondary transmission is therefore possible. The US Department of Labor (1989) recognizes borrelias as agents of blood-borne disease. Pike (1976) lists 45 cases of laboratory-acquired infection although it is not clear how many of these were the result of blood-to-blood contact (see also Felsenfeld, 1971, and Collins, 1993). There is a potential for laboratory-acquired infections when blood is examined for the organisms by dark-field microscopy, or inoculated into rats for diagnosis.

A third species, *B. burgdorferi*, the agent of Lyme disease, is tick-borne, and has been grown in blood cultures. There is therefore a potential for secondary transmission.

Mycobacterium leprae

Mycobacterium leprae is an acid-fast bacillus, 1–8 μm long by 0.3–0.5 μm in width. It has never been cultured in artificial media and is propagated in the laboratory in the footpads of mice and in armadillos. The transmission of leprosy by blood (or tissue) seems most unlikely, apart from the fact that the incidence of the disease in developed countries is low. There are two reports, however: Marchoux (1934) describes one such case in a country where leprosy is virtually non-existent, and Parritt and Olsen (1947) report two cases associated with tattooing. The former infection may be the laboratory-acquired infection mentioned by Pike (1976).

Mycobacterium tuberculosis

Mycobacterium tuberculosis is an acid-fast bacillus, 1–4 μm long by 0.3–0.6 μm in width. It may be present in the blood of HIV-positive patients with tuberculosis and Kramer *et al.* (1993) reported a primary cutaneous infection in a nurse who sustained a needlestick injury while attending to a patient with AIDS and undiagnosed tuberculosis. It is not infrequently recovered in blood cultures from HIV-positive patients with tuberculosis and is therefore a potential source of infection to laboratory workers, either by contact or the inhalation of aerosols during microbiological manipulations (see also Schluger *et al.*, 1994, who found evidence of *M. tuberculosis* DNA in the lymphocyte fraction of peripheral blood).

Rickettsia rickettsii

Rickettsias are very small, obligate intracellular bacteria, 0.5–1.0 μm in length by 0.2–0.4 μm in width. They have not been cultured *in vitro*. They are transmitted between vertebrate hosts by lice, fleas, mites and ticks and are the agents of, among others, typhus and Rocky Mountain spotted (RMS) fevers. The organisms are present in the circulating blood and blood-to-blood and secondary transmission from blood (during laboratory manipulations) is possible. Some of the 63 cases of laboratory-acquired infection noted by Pike (1976) (see also Collins, 1993) may well have been the result of needlestick injuries during clinical investigations, as, possibly, was that of RMS fever reported by Sexton *et al.* (1975).

Treponema pallidum

The spirochaetes, which are 4–14 μm in length by 0.2 μm in width are difficult to culture *in vitro*. They are present in large numbers in the blood in the secondary (haematogenous) stage of syphilis, and intermittently in untreated cases. Occupational transmission has occurred by needlestick injuries (reviewed by Collins and Kennedy, 1987) and tattooing (Stokes *et al.*, 1945). Laboratory-acquired infections (non-venereal, e.g. cutaneous), some of them from needlestick injuries during patient procedures are reported by Pike (1976) and Collins (1993). Infected blood, collected for transfusion in the absence of pretesting, is a potential hazard.

PROTOZOA

Blood-borne parasites include members of the following genera: *Plasmodium, Trypanosoma, Leishmania* and *Babesia* (Table 2.1). All are normally arthropod-borne, but blood-to-blood infections may occur as a result of direct (primary) contact or percutaneous exposure to infected blood at a time when there are large numbers of the parasites present.

Plasmodium Spp.

Malaria parasites have two alternate hosts. Sexual reproduction takes place in the gut of female *Anopheles* mosquitoes and the resulting sporozoites are transmitted to humans from the insect's salivary glands during blood sucking. In the human, asexual reproduction takes place in the liver and then in the peripheral blood. The mosquito is infected when it takes a blood meal from the human during the haematogenous phase. There are four species of malarial parasites: *Plasmodium vivax* and *P. ovale* (agent of benign tertian malaria), *P. falciparum* (malignant tertian malaria), and *P. malariae* (quartan malaria).

Cases of transmission of malarial parasites by blood transfusion are now rare. Gordon (1941) reported a case that resulted from the use of stored blood. The Malarial Surveillance Report (1971) included nine cases, and others have been reviewed by Chojnacki *et al.* (1968) and Dover and Schultz (1971). There is thus a potential for occupational infection from this source (see Guerrero *et al.*, 1983).

Cases of occupational infections with malarial parasites, usually *P. falciparum*, include several arising from needlestick injuries during the withdrawal of blood from patients. There was no evidence that any of these patients had had blood transfusions or had travelled in malarial regions.

Cannon *et al.* (1972) reported two cases of malaria associated with needlestick injuries, one infected with *P. falciparum*, the other with *P. vivax*. Bruce-Chwatt (1982), Carriere *et al.* (1993) and Haworth and Cook (1995) reported cases of *P. falciparum* transmission. In the case reported by Bending and Maurice (1980) a student pricked himself while making a blood smear.

An unusual case was discussed by Varma (1982). A pregnant woman with no history of exposure developed falciparum malaria. Blood had been taken from her some 30 minutes after blood had been collected from a malarial patient. A re-used blood-contaminated needle was probably responsible for the transmission.

No obvious mechanism of transmission was implicated in the cases described by Brumpt (1949) and by Petithory and Lebeau (1977), although a contaminated syringe might have been involved in the first of these.

Some infections were probably the result of blood-to-blood contact as there was no history of needlestick injury or any other exposure. Burne (1970) reported a case where blood had been withdrawn without mishap a short time after the operator had pared his finger nails, injuring the 'quick', which bled. In the case discussed by Bouree and Fouquet (1978) a nurse acquired falciparum malaria from a child patient but no obvious mode of transmission was discovered. Borsch *et al.* (1982) reported transmission of *P. vivax* to a nurse who did not wear gloves and had scratches on her finger tips.

Autopsies on malarial patients are not without hazards: a prosector suffered a finger prick and then developed the disease (Holm, 1924).

The transmission of malaria among intravenous drug abusers is well documented from 1926 onwards: see Wenyon (1926), Netter (1929), Biggam (1929), Dover (1971), Rosenblatt and Marsh (1971), Friedman *et al.* (1973) and Baker and Crawford (1978). Chung *et al.* (1940) found malarial parasites in syringes used for the intravenous injection of heroin, and Garcia *et al.* (1986) found *P. vivax* in the blood of drug abusers.

Finally, an interesting case of malaria acquired by self-inoculation was reported by the Centers for Disease Control (CDC, 1991). A patient with Lyme disease attempted malaria therapy, against medical advice, using an illicitly obtained suspension of *P. vivax*.

Trypanosoma Spp.

Trypanosomes are flagellated protozoa and normally have two hosts, a mammal and an arthropod, both being required for a full life cycle.

South American trypanosomiasis – Chagas' disease – is caused by *Trypanosoma cruzi*. The animal host is the armadillo and the arthropod host assassin (Reduviid) bugs. The parasite is transmitted to humans when the faeces of infected reduviid bugs are rubbed into the skin after a bite, or into an abrasion. Laboratory-acquired disease is by needlestick injury, either from patient blood (Pizzi *et al.*, 1963; Hanson *et al.*, 1974; Brenner, 1984) or animal blood (Hofflin *et al.*, 1987; Herwaldt and Juranek, 1995). Chagas' disease is endemic in Latin America and offers a potential occupational hazard, e.g. to blood transfusion staff (Bruce-Chwatt, 1972; Hunt, 1995).

African trypanosomiasis – sleeping sickness – is caused by *Trypanosoma brucei* ssps. *gambiense* and *rhodesiense*). The vertebrate hosts are cattle and antelopes and the arthropod host is the tsetse fly (*Glossina* spp.). Humans become infected when bitten by the fly.

A case of direct needlestick injury transmission from patient blood is reported by Emeribe (1988), and from a cultured strain in the same way by Receveur and Vincendeau (1993). Robertson *et al.* (1980) report a human infection acquired from experimentally-infected rat blood. A transfusion-induced infection was reported by Hira and Hussein (1979).

Leishmania Spp.

Leishmanias are protozoa related to the trypanosomes. The vertebrate hosts are dogs and rodents and the arthropod hosts are sandflies (*Phlebotomus* spp.). Both are required for a full life cycle.

The three species, *Leishmania donovani* (agent of visceral leishmanias, 'Kala Azar'), *L. tropica* (agent of cutaneous leishmaniasis) and *L. braziliensis* (agent of American, nasopharyngeal leishmaniasis) are transmitted from animals to humans by the bite of the fly. The six laboratory-associated infections described by Herwaldt and Juranek (1995) were acquired from the blood of animals in which the parasites were being passaged (Sampaio *et al.*, 1983; Sadick *et al.*, 1984; Evans and Pearson, 1988). The agent may also be present in blood collected for transfusion (Bruce-Chwatt, 1972).

Babesia microti

Babesias are amoeboid parasites of the Class Piroplasmea that are found in vertebrate animal erythrocytes. *Babesia microti* may be transmitted from cattle to humans by the bite of *Ixodes* ticks. During the parasitaemic phase of the disease the parasites are present in up to 80% of peripheral erythrocytes and may therefore be transmissible. Apart from transmission during blood transfusion (Wittner *et al.*, 1981; Grabowski *et al.*, 1982; Smith *et al.*, 1986; Mintz *et al.*, 1991), there seem to be no reports of occupational transmission, although there is obviously a potential for it (Herwaldt and Juranek, 1995).

FUNGI

Cryptococcus

Transmission of fungal pathogens by blood is rare, but Glaser and Garden (1985) reported a case where cryptococci but not HIV were transmitted from a patient infected with both.

REFERENCES

Arnow, P.M., Smaron, M. and Ormiste, V. (1984) Brucellosis in a group of travellers in Spain. *Journal of the American Medical Association* 251, 505–507.

Baker, J.E. and Crawford, G.P.M. (1978) Malaria: a new facet of heroin addiction in Australia. *Medical Journal of Australia* 2, 427–428.

Batchelor, B.I., Brindle, R.J., Gilks, G.F. *et al.* (1992) Biochemical mis-identification of *Brucella melitensis* and subsequent laboratory-acquired infection. *Journal of Hospital Infection* 22, 159–162.

Bending, M.R. and Maurice, P.D.L. (1980) Malaria: a laboratory risk. *Postgraduate Medical Journal* 56, 344–345.

Biggam, A.G. (1929) Malignant malaria associated with the administration of heroin intravenously. *Transactions of the Society of Tropical Medicine and Hygiene* 23, 147–153.

Borsch, G., Odendalhl, J., Sabin, G. *et al.* (1982) Malaria transmission from patient to nurse. *Lancet* ii, 1212.

Bourree, P. and Fouquet, E. (1978) Paludisme: contamination directe interhumain. *Nouvelle Presse* 7, 1865.

Brenner, Z. (1984) Laboratory-acquired Chagas disease: an endemic disease among parasitologists. In: Morel, C.M. (eds) *Genes and Antigens of Parasites: a Laboratory Manual*, 3rd edn. Cruz, Rio de Janeiro.

Bruce-Chwatt, L.J. (1972) Blood transfusion in tropical disease. *Tropical Diseases Bulletin* 69, 825–862.

Bruce-Chwatt, L.J. (1982) Imported malaria: an uninvited guest. *British Medical Bulletin* 38, 179–185.

Brumpt, L-C. (1949) Paludisme autochthone ou paludisme accidental. Contribution a la pathologie par la seringue. *Bulletin et Memoires de la Société medicale des hôpitaux de Paris* 65, 392–397.

Burne, J.C. (1970) Malaria by accidental inoculation. *Lancet* ii, 936.

Cannon, N.J., Walker, S.P and Dismukes, W.E. (1972) Malaria acquired by accidental needle puncture. *Journal of the American Medical Association* 222, 1425.

Carriere, J., Dalry, A., Hilmansdottir, I. *et al.* (1993) Transmission de *Plasmodium falciparum* à la suite d'une piqûre accidentelle. *Presse Médicale* 22, 1707.

CDC (1991) Update: Self-induced malaria associated with Lyme Disease – Texas. *Morbidity and Mortality Weekly Reports* 40, 465.

Chojnacki, M.C., Brazisky, J.H. and Barrett, O. (1968) Transfusion induced falciparum malaria. *New England Journal of Medicine* 279, 984–985.

Chung, H.L., Liu, W.Y., Wang, C.W. *et al.* (1940) Transmission of malaria by drug addicts in Peiping: demonstration of malarial parasites in syringes used for intravenous injection of heroin. *Chinese Medical Journal* 57, 32–38.

Collins, C.H. (1993) *Laboratory-acquired Infections*, 3rd edn. Butterworth-Heinemann, Oxford.

Collins, C.H. and Kennedy, D.A. (1987) A Review: microbiological hazards of occupational needlestick and 'sharps' injuries. *Journal of Applied Bacteriology* 62, 385–402.

Dover, A.S. (1971) Malaria in a heroin user. *Journal of the American Medical Association* 225, 1987.

Dover, A.S. and Schultz, M.G. (1971) Transfusion induced malaria. *Transfusion* 11, 353–357.

Emeribe, A.O. (1988) Gambiense trypanosomiasis acquired from needle scratch. *Lancet* i, 470.

Evans, T.G. and Pearson, R.D. (1988) Clinical and immunological response following accidental inoculation of *Leishmania donovani*. *Transactions of the Royal Society of Tropical Medicine and Hygiene* 82, 854–856.

Felsenfeld O. (1971) *Borrelia: Strains, Vectors, Human and Animal Borreliosis.* H. Green, St Louis.

Friedman, C.T.H., Dover, A.S., Roberts, R.R. *et al.* (1973) A malaria epidemic among heroin users. *American Journal of Tropical Medicine and Hygiene* 22, 302–307.

Garcia, J.J.G., Arnalich, F., Pena, J.M. *et al.* (1986) Outbreak of *Plasmodium vivax* malaria among heroin users in Spain. *Transactions of the Royal Society of Tropical Medicine and Hygiene* 80, 549–552.

Glaser, J.B. and Gardner, A. (1985) Inoculation of cryptococcus without transmission of the acquired immune deficiency syndrome. *New England Journal of Medicine* 312, 266.

Gordon, E.F. (1941) Accidental transmission of malaria through administration of stored blood. *Journal of the American Medical Association* 116, 1200.

Gotuzzo, E., Carrillo, C., Guerra, J. *et al.* (1986) An evaluation of diagnostic methods for brucellosis – the value of bone marrow culture. *Journal of Infectious Disease* 153, 122–125.

Grabowski, E.F., Giardina, P.J.V., Goldberg, D. *et al.* (1982) Babesiosis transmitted by a transfusion of frozen-thawed blood. *Annals of Internal Medicine* 96, 466–467.

Guererro, I.A., Weniger, B. and Schuktze, M.G. (1983) Transfusion malaria in the United States 1972–1981. *Annals of Internal Medicine* 99, 221–226.

Hanson, W.L., Devlin, R.F. and Roberson, E.L. (1974) Immunoglobulin levels in a laboratory-acquired case of Chagas' disease. *Journal of Parasitology* 60, 302–309.

Haworth, F.L.M. and Cook, G.C. (1995) Needlestick malaria. *Lancet* 346, 1361.

Herwaldt, B.L. and Juranek, D.D. (1995) Protozoa and helminths. In: Fleming, D.O., Richardson, J.H., Tulis, J.I. and Vesley, D. (eds) *Laboratory Safety: Principles and Practice*, 2nd edn. American Society for Microbiology, Washington.

Hickey, M.M. and Shanson, D.E. (1993) Septicaemia in patients with and without AIDS at Westminster Hospital, London. *Journal of Infection* 27, 243–250.

Hira, P.R and Hussein, S.P. (1979) Some transfusion-induced parasitic diseases in Zambia. *Journal of Hygiene, Epidemiology, Microbiology and Immunology* 4, 436–444.

Hofflin, J.M., Sadler, R.H., Araujo, F.G. *et al.* (1987) Laboratory-acquired Chagas' disease. *Transactions of the Royal Society for Tropical Medicine and Hygiene* 81, 437–440.

Holm, K. (1924) Ueber einen Falle von Infection mit Malaria tropica an der Leiche. *Klinische Wochenschrift* 3, 1633–1634.

Hunt, D.L. (1995) Human immunodeficiency virus type 1 and other blood-borne pathogens. In: Fleming, D.O., Richardson, J.H., Tulis, J.I. and Vesley, D. (eds) *Laboratory Safety. Principles and Practice*, 2nd edn. American Society of Microbiology, Washington DC.

Kramer, F., Sasse, S.A., Sims, J.C. *et al.* (1993) Primary cutaneous tuberculosis after a needlestick injury from a patient with AIDS and undiagnosed tuberculosis. *Annals of Internal Medicine* 119, 594–595.

Malarial Surveillance Annual Report (1971) Public Health Service publication 78–8152. Government Printing Office, Washington DC.

Marchoux, P.E. (1934) Un cas d'inoculational accidentelle du bacille de Hansen en pays non-lepreux. *International Journal of Leprosy* 2, 1–7.

Mintz, E.D., Anderson, J.F., Cable, R.G. *et al.* (1991) Transfusion-transmitted babesiosis: a case report from a new endemic area. *Transfusion* 31, 365–368.

Netter, L. (1929) Un cas de paludisme accidental. *Bulletin de la Société de pathologie exotique filiales* 22, 318.

Parritt, R.J. and Olsen, R.E. (1947) Two simultaneous cases of leprosy developing in tattoos. *American Journal of Pathology* 23, 815–817.

Petithory, J. and Lebeau, G. (1977) Contamination probable de laboratoire par *Plasmodium falciparum*. *Bulletin de la Société de pathologie exotique filiales* 70, 371–375.

Pike, R.M. (1976) Laboratory-associated infections: summary and analysis of 3,921 cases. *Health Laboratory Science* 13, 105–114.

Pizzi, T., Niedman, G. and Jarpa, A. (1963) Communicación de tres casos de enfermedad de Chagas aguda producidos por infecciones de laboratorio. *Bolletino Chilean della Parasitologia* 18, 32–36.

Receveur, M.C. and Vincendeau, P. (1993) Laboratory-acquired gambiense trypanosomiasis. *New England Journal of Medicine* 329, 209–210.

Robertson, D.H.H., Pickens, S., Lawson, J.H. *et al.* (1980) An accidental laboratory infection with African trypanosomes of a defined stock. I. The course of the infection. *Journal of Infection* 2, 105–112.

Rosenblatt, J.E. and Marsh, V.H. (1971) Induced malaria in narcotic addicts. *Lancet* ii, 189–190.

HEPATITIS VIRUSES

Currently, there are at least six viruses that can cause hepatitis: hepatitis A, B, C, D, E and G. Although the clinical signs and symptoms caused by these viruses are similar, each virus is a distinct entity. Epidemiologically, the viruses can be divided into two groups: the hepatitis A and E viruses are transmitted primarily by the faecal–oral route – and are considered only briefly in this chapter – while the hepatitis B, C, D, and G viruses are transmitted by direct contact with blood or body fluids (Table 3.1). The biology of these viruses is discussed by Jeffries (Chapter 1 this volume).

The first recorded cases of blood-borne hepatitis were those associated with the inoculation of shipyard workers in Bremen, Germany in 1833 with glycerinated human lymph to prevent smallpox (Lurman, 1855). Outbreaks were repeatedly observed throughout the nineteenth and early part of the twentieth centuries. Many of these were due to the re-use of needles and syringes, e.g. during the intravenous arsenical treatment of syphilis patients (Bigger and Dubi, 1943; MacCallum, 1945), or to the administration of human plasma for immunoprophylaxis in measles and mumps (Propert, 1938; Beeson *et al.*, 1944). Sporadic cases of hepatitis after blood transfusions were also recognized (Beeson, 1943; Morgan and Williamson, 1943), indicating that the aetiologic agent was probably present in blood and transmitted through direct inoculation.

Laboratory-acquired infections were recognized soon afterwards when, in 1949, a laboratory worker in a blood bank was infected with 'serum hepatitis' (Leibowitz *et al.*, 1949). The first study that quantified the higher incidence of hepatitis in health-care workers was published by Byrne (1966) who estimated an annual attack rate of 69 cases of hepatitis per 100,000 health-care workers when compared with 15 cases per 100,000 in the general population. Although the definitive agent(s) for these cases cannot be determined at present, we must assume that most of them were due to transmission of hepatitis B, C, D or G which are now known to be spread by the blood-borne route.

Table 3.1. Hepatitis viruses and their modes of transmission.

Virus	Mode of transmission
Hepatitis A virus ('infectious')	Faecal–oral, rarely blood transfusion, clotting factors
Hepatitis B virus ('serum')	Parenteral, sexual, perinatal
Hepatitis C virus ('non-A, non-B')	Parenteral, rarely sexual, perinatal (HIV + mother)
Hepatitis D virus ('delta virus')	Parenteral, hepatitis B virus dependent
Hepatitis E virus	Faecal–oral
Hepatitis G virus	Parenteral, sexual??

Hepatitis A and E

Types A and E hepatitis viruses are causes of acute hepatitis and are spread by the faecal–oral route (commonly referred to as 'infectious hepatitis'). Both types A and E are self-limiting diseases and do not lead to chronic infection. Although not classified as a blood-borne agent, hepatitis A has rarely been transmitted through blood transfusion and clotting factors during the brief viraemic phase of the disease (Noble *et al.*, 1984; Centers for Disease Control and Prevention: CDC, 1996a). Health-care workers are not generally considered at risk for infection with hepatitis E; however, a hepatitis attack rate of 42% was demonstrated in the medical staff working in Somalian refugee camps during a hepatitis E outbreak (CDC, 1987). The exact mode of transmission in this setting is unknown, although we assume that the route was faecal–oral.

Hepatitis B

After the discovery of the Australia antigen (hepatitis B surface antigen or HBsAg) by Blumberg *et al.* (1965), hepatitis B virus was identified as the causative agent for 'serum hepatitis' and distinguished from 'infectious hepatitis'. The development of serological tests for hepatitis B virus (HBV) led to the recognition that it is the major cause, worldwide, of acute and chronic hepatitis, cirrhosis, and hepatocellular carcinoma. The virus is an important cause of mortality in East Asia and in sub-Saharan Africa (London, 1981), and of liver disease in industrialized countries, primarily in high-risk groups such as intravenous drug users, homosexual men, and prostitutes (CDC, 1988). Other groups at substantial risk of infection include sexual partners of HBV carriers, heterosexuals with multiple partners, haemodialysis patients, institutionalized patients, and health-care workers with occupational exposure to blood (CDC, 1991d) (Table 3.2).

Hepatitis B virus is responsible for 43% of acute hepatitis cases reported in the United States (Alter and Mast, 1994). The CDC estimated that 200,000 to 300,000 primary HBV infections occurred annually during the past decade (CDC, 1991d). Of these, more than 10,000 persons were hospitalized with hepatitis B each year, and approximately 250 died with complications of the disease. Unlike hepatitis A and E viruses, HBV can cause chronic infection. Approximately 5% of those infected fail to resolve their infection and become chronically infected (HBsAg carriers). The prevalence of the carrier state is even higher in 'high risk' populations such as intravenous drug users (7%), haemophiliacs (7%), renal dialysis patients (2–15%), and male homosexuals (6%) (Hoofnagle *et al.*, 1978). In highly endemic areas of the world where transmission occurs at an early age, carrier rates can be higher than 10% of the population. It is estimated that there are 170 million chronic carriers in the world today (Szmuness, 1978). Not only do carriers have a 15–25% rate of mortality due to either cirrhosis or hepatocellular carcinoma, but they are the major reservoir of HBV transmission (Alter and Mast, 1994).

Since 1989 heterosexual activity has been the most frequently reported risk factor in HBV infection in the United States, accounting for approximately

Table 3.2. Comparative epidemiology of HBV, HCV, HGV and HIV.

Virus	Blood donor seroprevalence (%)	High-prevalence areas	High-risk groups and seroprevalence (%)
HBV [1,2,3]	3–5	East Asia, sub-Sahara Africa	Immigrants from high-endemicity areas (70–85); IVDA (60–80); homosexual men (35–80); household contacts (30–60); haemodialysis patients (20–80); health-care workers (3–35)
HCV [4]	0.3–1.5	Japan, south USA, Mediterranean Europe and Africa	IVDA (60–90); haemophilia patients (80–100)
HGV [5]	1.7	?	IVDA (33); haemophilia patients (18); post-transfusion (17)
HIV [6,7]	0.04	Africa, USA, Caribbean	Geographical variability in: IVDA (50–80); homosexual men (10–70); haemophilia patients (70 – type A, 35 – type B); prostitutes (0–69)

1. London (1981).
2. CDC (1988).
3. Alter and Mast (1994).
4. van der Poel (1994).
5. Linnen *et al.* (1996).
6. Chamberland and Curran (1990).
7. WHO (1995).

32% of reported cases. Hepatitis B infection rates are high in population groups exposed to many different sexual partners, such as female prostitutes (Papaevanglion *et al.*, 1975) and/or with a history of other sexually transmitted diseases (Alter *et al.*, 1988). Between 1980 and 1985 homosexual activity accounted for 20% of cases of HBV; since then the number of cases in this risk group has decreased to the current level of about 11% since 1988 (Alter and Mast, 1994).

Direct percutaneous inoculation of the virus is an important mode of transmission in the United States, Western Europe, and other technologically advanced countries. Transmission can occur via contaminated blood or blood products, haemodialysis, or other medical procedures. However, the shift from paid to volunteer donors and the testing for HBsAg in all blood products have reduced post-transfusion hepatitis B rates in the past two decades. Currently, from the two million transfusions each year, the American Red Cross reports fewer than 100 cases of post-transfusion hepatitis B. This is comparable with the frequency of hepatitis B cases in the population at large (6.3 cases per 100,000), which might indicate coincidental infection (Dodd, 1995).

The proportion of reported hepatitis B cases reported to the Centers for Disease Control and Prevention resulting from injecting drug use increased during the 1980s. In 1987, it accounted for 28% of the reported cases, making it the most commonly identified risk behaviour in the United States that year (CDC, 1988). Since 1989 the number of cases due to injection drug users has declined by 40%, possibly as a result of safer needle-using practices (Alter and Mast, 1994).

One of the most important routes of transmission in endemic areas of the world is that from an infected mother to her child. Neonatal transmission has been clearly documented both from chronic carrier mothers and mothers with acute hepatitis B (Schweitzer et al., 1973). Such infected infants commonly develop persistent or chronic infections. It has been estimated that in Taiwan, 40–50% of HBsAg carriers were infected perinatally (Okada et al., 1976). Infants may be infected in utero, by exposure to infected maternal blood during delivery, or in the first few months of life, probably by direct contact with infected mothers or siblings. Clusters of carriers in families have been documented, although the exact means of viral acquisition is not clear (Ohbayashi et al., 1972).

Among the reported cases of HBV infections in the United States, the largest proportion (approximately 40%) is that with no identifiable risk factor (Alter et al., 1992). Whether these cases are acquired by non-parenteral routes or inapparent parenteral routes is unknown. As a HBV carrier may not have any symptoms of disease, direct contact may not be recognized. Hepatitis B virus is commonly transmitted by routes other than those that are overtly parenteral. Transmission can occur from environmental surfaces that might contact mucous membranes or open skin breaks, such as by toothbrushes (CDC, 1977), eating utensils, razors (Gocke, 1974; Pattison et al., 1974), by hospital equipment such as endoscopes (Morris et al., 1975), or by indirect exposures to laboratory equipment (Lauer et al., 1979; see Kennedy, Chapter 6, this volume). Hepatitis B has been transmitted by the bite of an infected patient (CDC, 1974; Cancio-Bello et al., 1982) and by eye contact with blood containing HBV (Kew, 1973; Bond et al., 1982).

Hepatitis B in the health-care setting

Most studies indicate an increase in the prevalence of HBV-infected, hospitalized patients over the past two decades from 0.05–1.5% in the 1970s (Cherubin et al., 1972; Linnemann et al., 1977) to 3–6% in the late 1980s (Handsfield et al., 1987; Gordin et al., 1990; Louie et al., 1992). Possible explanations may include: (i) an increase of HBV infection in the general population in the 1980s (Alter et al., 1990a); (ii) an increase in immigrants from South-East Asia where prevalence rates range between 11 and 16% (CDC, 1991b); and (iii) an increase in the number of immunocompromised patients who are likely to develop chronic HBV carriage, such as those with AIDS, renal disease, or malignancies (Szmuness et al., 1981). Hadler et al. (1991) found that HIV-positive males developed chronic hepatitis B at a higher rate (21%) than HIV-negative males (7%).

Health-care personnel have been known to be at greater risk for HBV infection than the general population (Lewis *et al.*, 1973; Maynard, 1978). In the 1970s and early 1980s, epidemics of hepatitis B were common in patients and hospital staff, especially in haemodialysis units (London *et al.*, 1969; Garibaldi *et al.*, 1973; Snydman *et al.*, 1976) and clinical laboratories (Williams *et al.*, 1974; Levy *et al.*, 1977). The incidence of clinical cases of hepatitis B in health-care workers before the availability of the hepatitis B vaccine (i.e. before 1982) was reported to be between 50 and 120 per 100,000 (Schneider, 1979; Hansen *et al.*, 1981), much higher than that of the general population of < 10 cases per 100,000 (CDC, 1991d). In a seroepidemiological survey Dienstag and Ryan (1982) found a significantly higher prevalence of serologic markers for HBV infection in health-care workers than in volunteer blood donors. They found the markers were a function of contact with blood, with a seroprevalence of 15% or more in occupations associated with the emergency ward, laboratory, blood bank, intravenous team, and the surgical house officers.

Since 1982, changes in infection control practices, including the recommendation, in 1987, of Universal Precautions (CDC, 1987), and the availability of the hepatitis B vaccine to health-care workers have undoubtedly been responsible for a decline in the numbers of occupational hepatitis B infections. The CDC Hepatitis Surveillance Study (Alter *et al.*, 1990a) found a decline in the proportion of reported hepatitis B cases associated with health-care workers from 4% in 1982 to 1% in 1988. Several reports from university hospitals confirm the decline after initiation of the hepatitis B vaccination programme. Gerberding (1994) found an increase in the HBV vaccine acceptance rate from 20% in 1984 to 80% by 1990 at San Francisco General Hospital, with no reported occupational, acute hepatitis during that time period. Another hospital reported a decline in clinical hepatitis B in health-care workers from 82 cases per 100,000 health-care workers in 1980–1984 to no cases between 1985 and 1989 (Lanphear *et al.*, 1993). At Duke University the number of cases of clinical hepatitis B in hospital employees has declined from two cases per 1000 employees in 1979 to none since 1992, with a vaccine acceptance rate of 90% (T.Z. Weddle, Durham, North Carolina, 1995, personal communication).

Transmission of hepatitis B from health-care workers to patients has also been documented in at least 30 clusters involving over 350 patients (CDC, 1991c; Harpaz *et al.*, 1996). A combination of risk factors is associated with these outbreaks. Of the health-care workers whose hepatitis B e antigen (HBeAg) was tested (18/21), all were HBeAg positive, indicating higher levels of circulating virus and higher infectivity. This finding is consistent with studies that have documented the risk of transmission after a needlestick from an HBeAg-positive patient is higher (30%) than from an HBeAg-negative patient (6%) (Alter *et al.*, 1976; Seeff *et al.*, 1978).

In most of the reported clusters of nosocomial patient infections, transmission occurred before the increased awareness of health-care workers about transmission of blood-borne pathogens and before the implementation of

Universal Precautions (see Hunt, Chapter 19, and Kibbler, Chapter 20 this volume). Many of the infected health-care workers did not wear gloves or use other infection control practices that are routinely used in the health-care setting today. Other potential means of transmission included unintentional injuries to the infected health-care worker during invasive procedures, such as needlestick injuries while 'blind suturing' (manipulating needles in a body cavity without being able to see them). The latest report involved patients who were potentially infected when the surgeon tied wire sutures during cardiac surgeries (Harpaz *et al.*, 1996). Most cases have involved cardiovascular surgeons (11 cases), followed by dentists or oral surgeons (9), obstetricians or gynaecologists (5), a cardiopulmonary bypass-pump technician (1), a general practitioner (1), and an inhalation therapist (1) (CDC, 1991c). Although such transmissions are rare, they are totally preventable if health-care workers take advantage of the hepatitis B vaccine.

Hepatitis C

In 1989 Choo *et al.* discovered the hepatitis C virus (HCV) genome and developed a serological test for the agent. Since then, HCV has been found to be the primary agent of parenterally-transmitted non-A, non-B (NANB) hepatitis, and a major cause of acute and chronic hepatitis throughout the world (Alter, 1993; Ebeling, 1994; van der Poel, 1994). In the United States, 21% of reported acute hepatitis can be attributed to HCV (Alter and Mast, 1994). Likewise, HCV is responsible for between 10 and 25% of hospitalized cases with acute hepatitis in Europe (Ebeling, 1994).

Acute hepatitis C is generally a mild event. By extrapolation from transfusion-related hepatitis C infections, it can be estimated that only about 25% of acute HCV infections present with a clinical illness that is indistinguishable from other viral hepatitis infections (Alter, 1995). The most important aspect of the HCV, however, is its ability to persist in the host and cause chronic hepatitis infection. Chronic hepatitis has been documented in over 60% of HCV-infected patients, with evidence of the virus (HCV-RNA) in more than 90%. Of those with chronic infection, chronic active hepatitis or cirrhosis has been found in 29–76% after several years of infection (Alter *et al.*, 1992), and in 88% in one study that followed infected patients for up to 18 years (Seeff *et al.*, 1992). The persistence of a chronic state is an indication that no neutralizing antibodies are induced with hepatitis C infection. In fact, the primary mechanism of persistence seems to reside in the virus' ability to mutate under immune pressure, existing as a series of immunologically distinct strains. The virus can also exist in the liver in a quiescent state, escaping the immune response (Alter, 1995).

Hepatitis C virus has also been confirmed as an independent risk factor for hepatocellular carcinoma (HCC) (Simonetti *et al.*, 1992). In one study from Japan 21 patients with HCC were all found to be anti-HCV positive, with a mean interval of 30 years from acute infection to clinical recognition of HCC (Kiyosawa *et al.*, 1990). Multiple studies have shown that the prevalence of anti-HCV among patients with HCC ranges from 13 to 74%. Case-control studies have estimated a five- to seven-fold higher risk for HCC in anti-HCV-positive patients compared with anti-HCV-negative patients (Yu *et al.*, 1990; Di Bisceglie *et al.*, 1991; Nalpas *et al.*, 1991; Tanaka *et al.*, 1991; Simonetti *et al.*, 1992).

The prevalence of HCV in the population is highly variable. The highest rates (60–90%) are found in groups exposed to large or repeated direct percutaneous exposures, such as injecting drug users and haemophilia patients (Alter, 1993) (see Table 3.2). Epidemiological studies show that intravenous drug addicts form the largest known risk group for HCV. It has been estimated that two-thirds of addicts are anti-HCV seropositive within 2 years of regular use of IV drugs, increasing to close to 100% seropositivity after 8 years (Bell *et al.*, 1990). Since most drug users become chronic HCV carriers, they can transmit the virus after more than 10 years of abstinence (Kolho and Krusius, 1992), forming an important reservoir for contaminated blood units, particularly from paid blood donations. In spite of the high prevalence of infection, cases of acute hepatitis C disease among injection drug users increased through 1989, but then declined more than 50%, possibly related to safer needle-using practices (Alter, *et al.*, 1990b; Alter, 1993).

Haemophiliacs were a risk group with a high rate of infection in the early 1980s. Since 1985 inactivation procedures, such as heat treatment and solvent/detergent mixture, and donor screening for high-risk groups, have decreased the risk of transmission of HCV by blood products to this group (Makris *et al.*, 1990). However, the risk has not been eliminated, as evidenced by the recent transmission of HCV in intravenous immunoglobulin preparations, even after detergent treatment (Schneider and Geha, 1994). Infection rates are highest in haemophiliacs when large-pool, commercial and non-inactivated products have been used, and when large amounts of products are used (i.e. > 100,000 units of cryoprecipitate per year) (Makris *et al.*, 1990).

Moderate prevalence rates are found in those with smaller but repeated direct or inapparent percutaneous exposures, such as long-term cancer survivors (20%) (Locasciulli *et al.*, 1993), bone marrow transplant recipients (29%) (Locasciulli *et al.*, 1991), and haemodialysis patients (Alter *et al.*, 1990b). Seroprevalence rates in haemodialysis patients range from 2 to 10% in northern Europe (van der Poel *et al.*, 1991; Kolho *et al.*, 1993), 28 to 45% in southern Europe (Chiaramonte *et al.*, 1991; Beccari *et al.*, 1993), and from 20 to 43% in the United States (Alter *et al.*, 1990b; De Medina *et al.*, 1992). A high rate of infection is also seen in renal transplant recipients; however, the rate is proportional to the duration of haemodialysis before transplantation (Niu *et al.*, 1993). Studies that have demonstrated high seropositivity rates in haemodialysis patients independent of blood transfusions, and documented outbreaks of hepatitis C in dialysis units suggest that HCV can be transmitted nosocomially (Niu *et al.*, 1992; Niu *et al.*, 1993).

Lower rates are found in those with inapparent parenteral or mucosal exposures, such as sexual transmission. Conflicting data have been collected about the magnitude or even existence of a sexual mode of transmission. Before the discovery of HCV and subsequent serological tests, a case-control study found that persons with NANB hepatitis were more likely than the general population to have had multiple sexual partners or sexual contact with a person with a history of hepatitis (Alter *et al.*, 1982). In contrast, there are many studies that, after correcting for intravenous drug use as a risk, suggest either that sexual transmission of HCV does not occur or that it occurs in very low frequency. For example, studies of male homosexuals (Osmond *et al.*, 1993) and female prostitutes (Lissen *et al.*, 1993) demonstrate that the seroprevalence of HCV infection in these groups is between 4 and 8% (Alter, 1995). Likewise, in studies of heterosexual partners of HCV-infected patients, low rates of transmission have been found: 4% in European studies (Esteban *et al.*, 1989; Brackmann *et al.*, 1993; Bresters *et al.*, 1993), and 1.4% in a National Institute of Health study of HCV-positive donors (Conry-Cantilena *et al.*, 1992). Although transmission of HCV by sexual activity seems to be low, the number of persons engaging in sex and the frequency of this occurrence could provide a reservoir for transmission.

As with rates for sexual transmission of HCV, data for vertical transmission are also conflicting. Most studies show low rates of 0–5% transmission from mother to infant (Wejstal *et al.*, 1992; Alvarez *et al.*, 1992; Lam *et al.*, 1993). However, Giovannini *et al.* (1990) showed HCV infection in 11 of 25 infants (44%) born to anti-HCV-positive mothers. All 11 mothers of the infected infants were co-infected with HIV, supporting the suggestion that HIV infection may facilitate non-parenteral transmission of HCV. It has now been shown that co-infection with HIV facilitates the replication of HCV, and that higher titres of HCV-RNA in mothers has resulted in higher rates of transmission to infants (Ohto *et al.*, 1994). In fact, studies have consistently shown that no infants were infected when infected mothers exhibited low titres ($< 10^6$ infectious doses per ml) (Lin *et al.*, 1994).

The lowest prevalence rates are found in those groups that have no high-risk characteristics, such as volunteer blood donors. Worldwide, the rates of HCV infection among volunteer blood donors range from 0.3 to 1.5%; rates in the United States, United Kingdom and Northern Europe are in the 0.5–0.7% range (Contreras *et al.*, 1991; van der Poel *et al.*, 1991; Kolho and Krusius, 1992), with higher rates of 1.3–1.5% in Italy, Saudi Arabia and Japan (Sirchia *et al*, 1989; Watanabe *et al.*, 1990; Saeed *et al.*, 1991).

Consistent with the low prevalence rates of HCV in blood donors, the risk of HCV acquisition from blood products is also low. Over time, the rates of post-transfusion HCV infection have declined as safety measures to protect the blood supply have been implemented. Before surrogate testing (serum alanine transaminase levels; antibodies to hepatitis B core antigen) was routinely performed on blood units (i.e. before 1986), the per unit risk of HCV infection was 0.45%. After surrogate testing, the risk dropped to 0.19% per unit. In 1989, antibody testing for HCV became available, and the risk of infection dropped to approximately 0.06% per unit (Dodd, 1995). Further reductions can be expected when additional tests for HCV antigens become available to detect seronegative HCV carriers.

Table 3.4. Transmissions of hepatitis B, hepatitis C and human immunodeficiency viruses from health-care workers to patients.

	HBV[1,2]	HCV[3]	HIV[4]
Health-care workers (total)	30	1	1
Dentists/oral surgeons	9		1
Cardiothoracic surgeons	11	1	
Obstetricians/gynaecologists	5		
Other	5		
Patients infected	> 350	5	6
Risk factor	HBeAg+HCW; prior to Universal Precautions; suturing	Wire suturing	?

1. Harpaz *et al.* (1996).
2. CDC (1991c).
3. Esteban *et al.* (1996).
4. CDC (1993).

disease or increasing the risk for chronic liver disease (Hadler *et al.,* 1984; Bonino *et al.,* 1987).

The prevalence of HDV in hepatitis B carriers is low (1.4–8%) in the general population, but high in populations who experience repeated percutaneous exposures or blood products, such as injecting drug users (20–53%) or haemophilia patients (48–80%) (Alter and Hadler, 1993). Routes of transmission other than percutaneous, such as sexual and perinatal, are rare. Because HDV testing is not routinely done, and co-infection with HBV is clinically indistinguishable from hepatitis B alone, the true incidence of HDV infection and the risk to health-care workers is not known. However, transmission of hepatitis D to health-care workers by needlestick and non-parenteral exposure in a haemodialysis unit has been reported (Lettau *et al.,* 1986). Prevention of hepatitis B infection by vaccination also provides protection against subsequent infection with the dependent hepatitis D virus.

Hepatitis G and GB Agents

Recent advances in molecular technologies have led to the isolation of additional viral agents associated with hepatitis. In 1995, a new flavivirus was cloned from a patient with chronic hepatitis who was also infected with HCV (Kim *et al.,* 1995). The new virus, designated as hepatitis G virus (HGV), is associated with both acute and chronic hepatitis, with viraemia persisting up to 9 years in patients with hepatitis (Linnen *et al.,* 1996).

Risk factors for HGV acquisition are the same as those for HCV, including transfusion, injection drug use, and multiple sexual partners. In fact, seroepidemiological studies from the United States, Europe, South America and Australia have shown that 33% of intravenous drug users, 18% of haemophiliacs, 17% of patients with post-transfusion non-A–E hepatitis, and 13% of patients with chronic non-A–C hepatitis have serological evidence of infection with

HGV. The hepatitis G virus can exist as a co-infecting agent with either HBV or HCV. In these studies 10% of patients with chronic HBV infection and 19% of patients with chronic HCV were co-infected with HGV. In the United States, 1.7% of blood donors (13/769) were positive for HGV-RNA (Linnen et al., 1996). Although no seroepidemiological studies have been done regarding health-care workers, the similar frequency and modes of transmission of HGV and HCV in the community imply that this new blood-borne agent has the potential for occupational transmission as well.

Another group of flaviviruses has been isolated with three distinct agents, designated GBV-A, GBV-B, and GBV-C (Simons et al., 1995). The GBV-C agent causes acute and chronic hepatitis, and may be a West African variant of HGV (Bowden et al., 1996). No information is available at this time about the epidemiological characteristics of these viral agents. However, the initial source of serum for the isolation of the GB agent was obtained almost 30 years ago from a Chicago surgeon (initials G.B.) who developed acute hepatitis from an unknown source (Deinhardt et al., 1967).

HUMAN IMMUNODEFICIENCY VIRUS TYPE 1 (HIV-1)

Evidence for human infection with HIV, the causative agent of acquired immuno-deficiency syndrome (AIDS), can be found as early as 1959 in Zaire, Africa (Nahmias et al., 1986). As of December, 1995, 1,291,810 AIDS cases were reported to the World Health Organization (WHO) Global Programme on AIDS from 193 countries, representing a 23% increase from the previous year (WHO, 1995). The Americas account for the most number of reported AIDS cases (659,662), followed by Africa (442,735), Europe (154,103) and Asia (28,630). European rates of AIDS per 100,000 population include Spain at 16, Italy and France with 9.3, the United Kingdom with 2.7 and Germany with 2.6.

In the United States alone, 513,485 AIDS cases have been reported (CDC, 1995g). The numbers of cases substantially increased in 1993 and early 1994 because of a case definition change that now includes a broader range of AIDS indicator diseases and conditions (CDC, 1992). The USA nationwide rate has increased from 3.5 per 100,000 population in 1985 to 30 per 100,000 in 1994, with 84% of cases residing in large metropolitan areas (>500,000 population) (CDC, 1995a). In 1993, AIDS became the leading cause of death in persons aged 25–44 years of age in the USA, and accounted for 18% of all deaths (CDC, 1996b).

Epidemiological information gathered since the early 1980s indicates that the modes of transmission of HIV-1 have remained the same. The HIV agent is transmitted through sexual contact, percutaneous or mucous membrane expo-sure to blood, birth or breast feeding from an infected mother, or transfusion of HIV-contaminated blood. However, trends of HIV infection for certain popu-lations reflect the evolution of the epidemic.

In Africa, the Caribbean, and some areas of South America, heterosexual and perinatal transmission and HIV-infected blood are the primary modes of transmission. In Western Europe, North America, Australia, New Zealand, and parts of South America, homosexual/bisexual men and intravenous drug users remain the predominantly affected groups (Piot *et al.*, 1988). In the USA, however, AIDS cases attributed to homosexual or bisexual behaviour have decreased from 65% of cases in 1984 to 43% in 1994. Conversely, cases due to heterosexual contact increased from 1.2% in 1984 to 10.3% in 1994 in the USA (CDC, 1986, 1995b) (Table 3.5).

The findings demonstrate a disproportionate increase in US women and racial/ethnic minorities. Women accounted for only 7% of AIDS cases in 1984 (CDC, 1986), 10% in 1987 (Chamberland and Curran, 1990), and increased to 18% in 1994 (CDC, 1995c). The largest percentage of women with AIDS reported injecting drug use as a risk factor (41%), followed by heterosexual contact with a partner at risk for or known to have HIV infection or AIDS (38%). Over three-quarters of cases among women in 1994 occurred among blacks and Hispanics; rates per 100,000 population for white women were 3.8, black women, 62.7, and Hispanic women, 26 (CDC, 1995c).

Overall, reported US AIDS cases in the white population decreased from 63% of cases in 1984 to 41% in 1994 (CDC, 1986, 1995b). Meanwhile, the cases in the black population increased from 26% to 39% in the same decade, and the Hispanic AIDS population increased from 14% to 19%.

Table 3.5. Changing epidemiology of adults with AIDS in the United States within one decade.

Risk group	1984[a] (% of reported AIDS)	1994[b] (% of reported AIDS)
Transmission group		
Homosexual	65	43
IVDA	17	27
Homosexual + IVDA	9	5
Heterosexual	1.2	10
Transfusion recipient	1.2	1
Haemophilia/coagulation disorder	0.8	0.6
No identified risk	5.4	12
Race/ethnicity		
Whites	63	41
Black	26	39
Hispanic	14	19
Other race/ethnic	1.4	1
Sex		
Male	93	82
Female	7	18

[a]CDC (1986).
[b]CDC (1995b).

Percentages of AIDS cases related to intravenous drug use increased from 17% in 1984 to 27% in 1994 (CDC, 1986, 1995b). Cases in children associated with perinatal HIV transmission have also continued to rise. In 1994, 92% of children with AIDS acquired it from mothers at risk for HIV infection (CDC, 1995b).

The annual incidence of AIDS cases associated with blood transfusions and therapeutics for haemophilia has stabilized since the serological screening of blood donations and heat treatment of clotting factors was initiated in 1985. The percentage of AIDS patients associated with transfusion of blood or blood products has dropped from 3% in 1987 to 1% in 1994 (Chamberland and Curran, 1990; CDC, 1995b). Since 1985 35 cases of AIDS have been associated with blood donations from donors in the antibody negative 'window period' of their HIV infection. Test kits for detection of HIV-1 p24 antigen are expected to be licensed soon to detect the contaminated units during this period (CDC, 1996d). The US Food and Drug Administration (FDA) has issued guidelines recommending its use within 3 months of licensure (FDA, 1995). The new test is expected to reduce the current risk of transfusion-transmitted HIV (1:450:000–1:660,000 per unit) to 1:562,000–1:825,000 per unit (Lackritz et al., 1995). Likewise, the percentage of AIDS cases due to treatment of haemophilia patients with contaminated blood products dropped slightly from 1% in 1987 to 0.6% in 1994 (Chamberland and Curran, 1990; CDC, 1995b).

In 1994, 11.9% of AIDS cases were assigned to a 'no identified risk' (NIR) category (CDC, 1995b), most representing cases still under investigation. Generally, about 20% of NIR cases have died before follow-up is completed; other case patients refuse to provide information, or are lost to follow-up. Historically, upon investigation, 83% of the NIRs are classified into an identified risk category, providing an overall NIR rate of 3% over the years (CDC, 1991a). Upon investigation, 40% of the NIRs report a history of sexually transmitted diseases, and 33% of men give a history of sexual contact with a prostitute (Castro et al., 1988).

HIV in the Health-care Setting

National surveillance data of health-care workers show that there is no higher risk for developing AIDS in those working in health-care than for those in the general public. Approximately 5.3% of reported AIDS cases with a known work history were employed in the health-care setting (Chamberland et al., 1988), with 95% reporting recognized non-occupational risk factors. After surveillance investigations, only 1.4% of health-care workers with AIDS are classified with NIR. Demographically, they are more similar to other AIDS cases than to health-care workers in general: 73% of the NIR health-care workers are male (vs. 23% of all US health-care workers) (Bureau of Labor Statistics, 1988). Also, the only occupation that is over-represented among the NIR cases is that of maintenance workers (20% for NIR cases vs. 6% for cases with identified risk) (Chamberland et al., 1988). Those occupations in the health-care setting that have historically been represented as high risk for other blood-borne pathogens, i.e. surgeons and laboratory technologists, are not represented at a higher rate.

Seroprevalence studies conducted in health-care settings across the country support the premise that the risk of HIV-1 transmission in this workplace is relatively low. These studies have examined 7595 US and European health-care workers and found nine seropositives (0.12%) in workers with no identified social risk factors (Hirsch *et al.*, 1985; Shanson *et al.*, 1985; Weiss *et al.*, 1985; Boland *et al.*, 1986; Ebbesen *et al.*, 1986; Gilmore *et al.*, 1986; Harper *et al.*, 1986; Lubick and Schaeffer, 1986; Gerberding *et al.*, 1987; Klein *et al.*, 1988; Marcus, 1988; Weiss *et al.*, 1988; Gerberding, 1994). A recent serosurvey of hospital-based surgeons in 21 hospitals in moderate to high AIDS incidence areas across the USA was conducted by the CDC Serosurvey Study Group (Panlilio *et al.*, 1995). This study also found a low prevalence of 0.14% (one seropositive in 740 surgeons with no community risk identified). This same finding is demonstrated by seroprevalence studies in Kinshasa, Zaire, a region where community prevalence of HIV is high (6–8%), infection control resources are limited, and needles and syringes are usually washed by hand, sterilized, then re-used (Mann *et al.*, 1986). These studies reaffirm that seropositivity in hospital workers is no higher than that of the community, and that transmission appears to be low.

Although the risk of occupational HIV transmission appears to be low, case reports of health-care workers infected with the virus through occupational exposure have been documented. In December, 1995, 49 health-care workers experienced seroconversions after exposures to HIV-infected patients (CDC, 1996f) (Table 3.6). The primary mode of transmission in these cases is percutaneous injury, accounting for 42 (86%) of the cases. Five infections (10%) occurred after mucocutaneous exposures, one infection (2%) after both percutaneous and mucocutaneous, and one infection (2%) after an undefined mode of transmission. Forty-four (90%) of the exposures that resulted in infection were to blood, one exposure was to visibly bloody fluid, one to an unspecified fluid, and three to concentrated virus in the research laboratory. In addition, 102 possible occupational HIV transmissions have been reported from health-care workers who have been investigated and found to have no other social risk factors for HIV infection, but who have experienced non-documented occupational exposures to blood, body fluids, or laboratory levels of HIV. Most of the documented as well as possible occupational infections occurred in nurses (19 documented, 24 possible), followed by clinical laboratory workers (15 documented, 15 possible).

A recent case-control study described risk factors associated with occupational HIV infection after percutaneous exposures in cases reported from national surveillance systems in the United States, France, and the United Kingdom (CDC, 1995f). The findings indicate an increased risk for occupational infection following a percutaneous exposure if it involved a larger quantity of blood, such as a device visibly contaminated with the patient's blood, or a procedure that involved a large-gauge, hollow-bore needle, particularly if used for vascular access. This is consistent with laboratory studies that have

Table 3.6. Occupational HIV infection risk for health care workers.

Exposure type	Documented[1,2] (through 12/95)	Risk from single HIV+ exposure[3,4]
Percutaneous	42	0.3
	Risk factors identified: Large gauge hollow needle Deep exposure Visible blood on device Vascular access Patient source with terminal illness Lack of ZDV prophylaxis	
Mucocutaneous	5	0.1
Both per- and mucocutaneous	1	
Unknown	1	
Skin	0	<0.1

1. CDC (1996f).
2. CDC (1995f).
3. Tokars *et al.* (1993).
4. Gerberding (1995).

indicated that less blood is transferred by suture needles (solid bore) than by phlebotomy needles (hollow-bore) of similar diameter (Mast and Gerberding, 1991; Bennett and Howard, 1994).

Another factor associated with increased risk of occupational infection was exposure to a source patient in the terminal stage of illness. This factor, as well as those above, may have a direct association with the amount or dose of virus present at the time of exposure. Saag *et al.* (1991) evaluated the plasma viraemia levels in patients infected with HIV and found none of the asymptomatic adults, 12% of adults with AIDS-related complex, and 93% of AIDS patients had cell-free infectious virus in their plasma. In fact, the titre of HIV in the blood of patients with AIDS diagnoses may be as many as 100 times that in persons with asymptomatic HIV infection (Ho *et al.*, 1989).

Finally, analysis of data in the case-control study suggested that use of Zidovudine (ZDV) post-exposure might be protective for health-care workers (CDC, 1995f). The risk for HIV infection in this study was reduced approximately 79% in health-care workers who were given ZDV prophylactically following exposure. Because of these results, the CDC recently recommended the use of antiviral agents for post-exposure treatment after high-risk exposure to HIV-infected patients (CDC, 1996e). (See also Morgan, Chapter 8, Waldron, Chapter 15 and McCloy, Chapter 16, this volume.)

The best direct measure of risk of HIV transmission due to a single type of occupational exposure is found in prospective cohort studies that document an

HIV exposure event with follow-up serological monitoring of the exposed health-care worker. The hierarchy of measured risk for each type of occupational exposure is consistent with the percentages of documented cases of HIV infected health-care workers reported to the CDC (Table 3.6). For example, the risk of infection due to a single percutaneous injury (i.e. needlestick) is calculated to be 0.3% (Tokars *et al.*, 1993), much higher than the calculated risk of infection due to either mucous membrane exposure (0.1%) or skin exposure (<0.1%) (Gerberding, 1995). It is speculated that risks from high risk percutaneous exposures such as those identified in the recent study (i.e. source patient with AIDS or higher dose of blood or virus) may actually be higher.

The risk of nosocomial transmission of HIV from a health-care worker to a patient appears to be much less than that of an occupational infection. In one dental practice, six patients were infected with the same strain of HIV as the practising dentist. No specific procedures were identified as the mode of transmission (Table 3.4). However, no other transmissions have been documented after testing 20,136 patients of 58 infected health-care workers (CDC, 1993). On the basis of these pooled results, the maximum risk of transmission of HIV from an infected health-care worker to a patient is 0.03%.

HUMAN IMMUNODEFICIENCY VIRUS TYPE 2 (HIV-2)

In 1986, a second human retrovirus capable of causing AIDS was isolated from patients of West African origin (Kanki *et al.*, 1986). Originally named HTLV-IV and later renamed HIV-2, the virus is endemic to western Africa where it is the dominant HIV. The population of this region has a high rate of HIV-2 infection (8.9% of adults in Guinea-Bissau), but a low rate of AIDS (Paulsen *et al.*, 1989), suggesting that the ability of the virus to cause disease is less efficient than HIV-1. Although found in high rates in this region, it has limited distribution in other areas of the world. It has been reported in Europe and Canada, and 62 cases have been reported in the United States (CDC, 1995e). At least 48 (77%) of the US cases were born in, had travelled to, and/or had a sex partner from western Africa.

In June, 1992, the US Food and Drug Administration recommended screening all blood and blood products with the combined HIV-1/HIV-2 enzyme immunoassays (EIA) (George *et al.*, 1990; CDC, 1992). Since that time three cases of HIV-2 have been detected among the 74 million blood donations in US blood centres (CDC, 1995e).

HIV-2 seems to be transmitted in the same way as HIV-1. To date, no occupational infections have been documented, although there is documentation of parenteral transmission through intravenous drugs and blood transfusions (Dufoort *et al.*, 1988).

HUMAN T-LYMPHOTROPIC VIRUS TYPES I AND II (HTLV-I, HTLV-II)

The HTLV-I and -II viruses are classified as oncornaviruses and are associated with malignancies such as leukaemia and lymphoma. The HTLV-I was the first human retrovirus to be isolated (Poiesz *et al.*, 1980), and is a cause of adult T-cell leukaemia-lymphoma, and tropical spastic paraparesis (Gessain *et al.*, 1985). HTLV-I is found endemically in Japan, Africa, the Caribbean basin, and the south-eastern United States (Hinuma *et al.*, 1982; Blayney *et al.*, 1983; Wong-Staal and Gallo, 1985). The rates of seroprevalence vary from 20% of adults in endemic areas of Japan to 1.4 per 10,000 in screened blood donations in the United States (CDC, 1990).

HTLV-I is transmitted as other blood-borne pathogens, i.e. sexual contact, perinatally, and through contaminated blood. Percutaneous exposure by contaminated needles has been documented as a risk in drug abusers where prevalence rates range from 18 to 49% (Robert-Guroff *et al.*, 1986; Lee *et al.*, 1989, 1990). One occupational transmission has been reported in a health-care worker from Japan who was caring for a patient with adult T-cell leukaemia-lymphoma (Katoaka *et al.*, 1990).

In 1982, a related virus, HTLV-II, was isolated from a patient with a T-cell variant of hairy cell leukaemia (Kalyanaraman *et al.*, 1982). Because of the lack of a specific screening test for HTLV-II, the epidemiology is not defined. However, one study identified 52% of HTLV-I/II seropositive blood donors as infected with HTLV-II when analysed by a specific DNA amplification test. These donors were more associated with a risk factor of IV drug use than the HTLV-I seropositives (Lee *et al.*, 1991).

RELATIVE RISK FOR OCCUPATIONAL INFECTIONS WITH BLOOD-BORNE PATHOGENS

The three most common blood-borne pathogens found in the health-care setting (i.e. HBV, HCV and HIV) present with different risks for transmission. Since the recognition of the 'new' epidemic of HIV in 1981, the potential transmission of this blood-borne pathogen has been compared with the historical problems associated with occupational hepatitis B infections, the model agent for nosocomial and occupational transmission in the health-care setting. In the early 1980s, the CDC estimated that 12,000 health-care workers became occupationally infected with HBV each year, resulting in more than 250 deaths (CDC, 1989). In contrast, the total number of reported occupationally-acquired HIV infections in 15 years is 151 (49 documented, 102 possible), an average of approximately 10 per year.

The risk of HBV infection from a single percutaneous exposure (i.e. needle-stick injury) to hepatitis B surface antigen positive blood (6–30%) (CDC, 1985) is much higher than the risk of HIV infection from a similar exposure to HIV-positive blood (0.3%). As previously mentioned, the risk of HCV infection from a single percutaneous exposure is approximately 2.5% (0.75–10%), a risk between that of HBV and HIV (Table 3.3). These risks correlate directly with average titres of the viruses found in infected sources. Viral titres of hepatitis B during the acute stage of disease can reach 10^2–10^8 viral particles per ml of blood (Shikata *et al.*, 1977; Tabor *et al.*, 1983), hepatitis C titres generally reach 10–10^6 viral particles per ml (Tabor *et al.*, 1983), and HIV titres average around 10–10^3 viral particles per ml (Ho *et al.*, 1989).

Although HBV transmission rates are higher than other blood-borne pathogens in the occupational setting, it is generally a self-limiting disease in the general population (only 5% develop chronicity) and is entirely preventable with vaccination for health-care workers at risk. The chronic nature of HCV and HIV disease increases the reservoir of infected sources in the population and the potential for exposures.

OTHER BLOOD-BORNE PATHOGENS

Several other infectious agents may not be included under the category of classical 'blood-borne' pathogens, but may warrant mention because of the potential for occupational transmission because: (i) the agents may be found in high titres during the septic phases of their disease processes; (ii) rates of community infections may be increasing, and, thus, provide a larger reservoir for exposure; or (iii) occupational infections may not be recognized or diagnosed because of the unusual route of exposure. The risk of infection in the occupational setting has not been quantified for these infectious agents; however, documented cases of occupational parenteral transmission or transmission through blood transfusions has been recorded in the literature (see Collins, Chapter 2 and Kitchen and Barbara, Chapter 13, this volume).

Many of the agents are found in high quantities during the septic phase of disease. For example, the numbers of human parvovirus B19 virions can reach up to 10^{10} per ml during a brief viraemic stage (Barbara and Contreras, 1990). Although it is generally transmitted through contact with infectious respiratory secretions or vertically from mother to fetus, parvovirus B19 has been transmitted through blood transfusions and use of blood products. In fact, up to 90% of recipients of factor VIII are likely to be seropositive for parvovirus B19.

Treponema pallidum (the causative agent of syphilis) is found in highest numbers during the secondary haematogenous stage of the disease, but may also be found in blood if syphilis is left untreated. During 1990, the rate for primary and secondary syphilis in the USA (20.3 per 100,000 population) was the highest since the 1940s (CDC, 1996c). Outbreaks have been linked epidemiologically to increased crack-cocaine use and HIV infection. The synergistic effect of syphilis with HIV infection poses increasing challenges for prevention programmes, particularly in inner cities. The increasing numbers of patients with syphilis and HIV increases the possibilities for occupational exposures and infection. Nosocomial transmission of syphilis has been documented through needlestick injuries, blood transfusions, and tattooing (Stokes *et al.,* 1945; Pike, 1976; Collins and Kennedy, 1987).

Several other agents are not classically considered to be 'blood-borne' pathogens, and are primarily transmitted through community exposures such as sexual contact (*Treponema*, cytomegalovirus) or vectors (*Rickettsia rickettsii* (Rocky Mountain spotted fever), *Plasmodium* (malaria) or Colorado tick fever virus). Reports in the literature indicate that transmission can occur in the occupational setting during disease stages when agents may be present in high numbers, particularly after parenteral exposures. Unless exposures are noted, health-care workers might be misdiagnosed because they do not meet the epidemiological pattern for their disease process.

Table 3.7 is a partial list of those agents that have been reported in the literature to have caused occupational or post-transfusion infections through blood exposure. Some of these agents are reviewed in Chapters 1 and 2. The increasing prevalence of recognized and new infectious agents, and their changing epidemiology, should be a reminder to all health-care providers to use consistent, Universal Precautions when in contact with any blood or body fluids.

Table 3.7. Blood-borne pathogens other than HBV, HCV and HIV documented as the sources of occupational infection with blood or body fluid, or transfusion-associated infections.

Viruses
HTLV-1 (Kataoka *et al.,* 1990)
Viral haemorrhagic fever viruses
   Lassa (Frame *et al.,* 1970)
   Marburg (Smith *et al.,* 1982)
   Ebola (CDC, 1995d)
   Crimea-Congo (Burney *et al.,* 1980)
Colorado tick fever (Monath, 1990)
Cytomegalovirus (Barbara and Contreras, 1990)
Parvovirus B19 (Cohen *et al.,* 1988)

Bacteria
Treponema pallidum (Pike, 1976)
Borrelia (Felsenfeld, 1971)
Mycobacterium leprae (Parrit and Olsen, 1947)
Brucella (Pike, 1976)

Parasites
Babesia microti (Smith *et al.,* 1986)
Plasmodium malariae (Borsch *et al.,* 1982)
Trypanosoma brucei gambiense (Hira and Hussein, 1979)
Trypanosoma cruzi (Bruce-Chwatt, 1972)
Leishmania (Bruce-Chwatt, 1972)

Rickettsia
Rickettsia rickettsii (Sexton *et al.,* 1975)

REFERENCES

Alter, H.J. (1995) To c or not to c: these are the questions. *Blood* 85, 1681–1695.

Alter, H.J., Seeff, L.B. and Kaplan, P.M. (1976) Type b hepatitis: the infectivity of blood positive for e antigen and DNA polymerase after accidental needlestick exposure. *New England Journal of Medicine* 295, 909–913.

Alter, M.J. (1993) The detection, transmission, and outcome of hepatitis C virus infection. *Infectious Agents and Disease* 2, 155–166.

Alter, M.J. and Hadler, S.C. (1993) Delta hepatitis and infection in North America. In: Hadziyannis, S.J., Taylor, J.M. and Bonino, F. (eds) *Hepatitis Delta Virus: Molecular Biology, Pathogenesis, and Clinical Aspects.* Wiley-Liss, New York, p.243.

Alter, M.J. and Mast, E.E. (1994) The epidemiology of viral hepatitis in the United States. *Gastroenterology Clinics of North America* 23, 437–455.

Alter, M.J., Gerety, R.J., Smallwood, L. *et al.* (1982) Sporadic non-A, non-B hepatitis: frequency and epidemiology in an urban United States population. *Journal of Infectious Disease* 145, 886–893.

Alter, M.J., Ahtone, J. and Weisfuse, I. (1988) Hepatitis B virus transmission between heterosexuals. *Journal of the American Medical Association* 256, 1307.

Alter, M.J., Hadler, S.C. and Margolis, H.S. (1990a) The changing epidemiology of hepatitis B in the United States: need for alternative vaccination strategies. *Journal of the American Medical Association* 263, 1218–1222.

Alter, M.J., Hadler, S.C., Judson, F.N. *et al.* (1990b) Risk factors for acute non-A, non-B hepatitis in the United States and association with hepatitis C virus infection. *Journal of the American Medical Association* 264, 2231–2235.

Alter, M.J., Margolis, H.S., Krawczynski, K. *et al.* (1992) The natural history of community-acquired hepatitis C in the United States. *New England Journal of Medicine* 327, 1899–1912.

Alvarez, L.P., Hernandez-Sampelayo, T., Casado, C. *et al.* (1992) Mother-to-infant transmission of HIV and hepatitis C virus infections in children born to HIV-seronegative mothers. *AIDS* 6, 427–430.

Anon. (1984) Needlestick transmission of HTLV-III from a patient infected in Africa. *Lancet* ii, 1376–1377.

Barbara, J.A.J. and Contreras, M. (1990) Infectious complications of blood transfusions: viruses. *British Medical Journal* 300, 450–453.

Beccari, M., Sorgato, G., Rizzolo, L. *et al.* (1993) Anti-HCV reactivity in dialysis patients: discrepancies between first and second generation tests. *Nephron* 63, 238.

Beeson, P.B. (1943) Jaundice occurring one to four months after transfusion of blood or plasma. *Journal of the American Medical Association* 121, 1332.

Beeson, P.B., Chesney, G. and McFarlan, A.M. (1944) Hepatitis following injection of mumps convalescent plasma. *Lancet* i, 814.

Bell, J., Batey, R.G., Farrell, G.C. *et al.* (1990) Hepatitis C virus in intravenous drug users. *Medical Journal of Australia* 153, 274–276.

Bennett, N.T. and Howard., R.J. (1994) How much blood is inoculated in a needlestick injury from suture needles? *Journal of the American College of Surgeons* 178, 107–110.

Bigger, J.W. and Dubi, S.D. (1943) Jaundice in syphilitics under treatment. *Lancet* i, 457.

Blayney, D.W., Blattner, W.A., Robert-Guroff, M. *et al.* (1983) The human T-cell leukemia-lymphoma virus in the southeastern United States. *Journal of the American Medical Association* 250, 1048–1052.

Blumberg, B.S., Alter, H.J. and Visnich, S. (1965) A 'new' antigen in leukemia sera. *Journal of the American Medical Association* 191, 541.

Boland, M., Deresztes, J., Evans, P. *et al.* (1986) HIV seroprevalence among nurses caring for children with AIDS/ARC. Abstract THP.212. 3rd International Conference on AIDS, Washington, D.C.

Bond, W.W., Petersen, N.J., Favero, M.S. *et al.* (1982) Transmission of type B viral hepatitis via eye inoculation of a chimpanzee. *Journal of Clinical Microbiology* 15, 533–534.

Bonino, F., Negro, F. and Baldi, M. (1987) The natural history of chronic delta hepatitis. In: Rizzetto, M., Gerin, J.L. and Purcell, R.H. (eds) *Hepatitis Delta Virus and its Infection*. Alan R. Liss, New York, p.145.

Borsch, G., Odendahl, J., Sabin, G. *et al.* (1982) Malaria transmission from patient to nurse. *Lancet* ii, 1212.

Bowden, D.S., Moaven, L.D. and Locarnini, S.A. (1996) New hepatitis viruses: are there enough letters in the alphabet? *Medical Journal of Australia* 164, 87–89.

Brackmann, S.A., Gerritzen, A., Oldenburg, J. *et al.* (1993) Search for intrafamilial transmission of hepatitis C virus in hemophilia patients. *Blood* 81, 1077–1082.

Bresters, D., Mauser-Bunschoten, E.P. Reesink, H.W. *et al.* (1993) Sexual transmission of hepatitis C virus. *Lancet* 342, 210.

Bruce-Chwatt, I.J. (1972) Blood transfusion and tropical disease. *Tropical Diseases Bulletin* 69, 825–862.

Bureau of Labor Statistics (1988) Employment and earnings. US Department of Labor, Bureau of Labor Statistics, Washington, D.C.

Burney, M.I., Ghafoor, A., Saleen, M. *et al.* (1980) Nosocomial outbreak of viral hemorrhagic fever caused by Crimean hemorrhagic fever – Congo virus in Pakistan, January, 1976. *American Journal of Tropical Medicine and Hygiene* 29, 941–947.

Byrne, E.B. (1966) Viral hepatitis: an occupational hazard of medical personnel. Experience of the Yale-New Haven Hospital, 1952–1965. *Journal of the American Medical Association* 195, 362–364.

Cancio-Bello, T.P., de Medina, M., Shorey, J. *et al.* (1982) An institutional outbreak of hepatitis B related to a human biting carrier. *Journal of Infectious Disease* 146, 652–656.

Castro, K.G., Lifson, A.R., White, C.R. *et al.* (1988) Investigations of AIDS patients with no previously identified risk factors. *Journal of the American Medical Association* 259, 1338–1342.

Causse, X., Germanaud, J. and Barthes, J.P. (1993) Prevalence of anti-hepatitis C virus (HCV) antibodies in health care workers. *International Symposium on Viral Hepatitis and Liver Disease*, Tokyo, p. 227 (abstract).

CDC (1974) Hepatitis transmitted by a human bite. *Morbidity and Mortality Weekly Reports* 23, 24.

CDC (1977) Hepatitis B transmission modes: evidence against enteric transmission. *Hepatitis Surveillance Report*. No. 41. (September), 20–22.

CDC (1985) Recommendations for protection against viral hepatitis. *Morbidity and Mortality Weekly Reports* 34, 313–335.

CDC (1986) Update: Acquired immunodeficiency syndrome – United States. *Morbidity and Mortality Weekly Reports* 35, 17–21.

CDC (1987) Recommendations for prevention of HIV transmission in health care settings. *Morbidity and Mortality Weekly Reports* 36 (suppl 2), 3S–18S.

CDC (1988) Changing patterns of groups at high risk for hepatitis B in the United States. *Morbidity and Mortality Weekly Reports* 37, 429–437.

CDC (1989) Guidelines for prevention of transmission of human immunodeficiency virus and hepatitis B virus to health-care and public safety workers. *Morbidity and Mortality Weekly Reports* 38 (Suppl 6), 3–31.

CDC (1990) Human T-lymphotropic virus type 1 screening in volunteer blood donors – United States, 1989. *Morbidity and Mortality Weekly Reports* 39, 915–924.

CDC (1991a) The HIV/AIDS epidemic: the first ten years. *Morbidity and Mortality Weekly Reports* 40, 357–369.

CDC (1991b) Screening for hepatitis B virus infection among refugees arriving in the United States, 1979–1991. *Morbidity and Mortality Weekly Reports* 40, 784–786.

CDC (1991c) Recommendations for preventing transmission of human immuno-deficiency virus and hepatitis b virus to patients during exposure-prone invasive procedures. *Morbidity and Mortality Weekly Reports* 40 (RR-8), 1–9.

CDC (1991d) Hepatitis B virus: A comprehensive strategy for eliminating transmission in the United States through universal childhood vaccination: Recommendations of the Immunization Practices Advisory Committee (ACIP). *Morbidity and Mortality Weekly Reports* 40 (RR-13), 1.

CDC (1992) 1993 Revised classification system for HIV infection and expanded surveillance case definition for AIDS among adolescents and adults. *Morbidity and Mortality Weekly Reports* 41 (no. RR-17), 1–19.

CDC (1993) Update: Investigations of persons treated by HIV-infected health-care workers-United States. *Morbidity and Mortality Weekly Reports* 42, 329–337.

CDC (1995a) Summary of Notifiable Diseases, United States – 1994. *Morbidity and Mortality Weekly Reports* 43 (53), 1–80.

CDC (1995b) Update: acquired immunodeficiency syndrome – United States, 1994. *Morbidity and Mortality Weekly Reports* 44, 64–67.

CDC (1995c) Update: AIDS among women – United States, 1994. *Morbidity and Mortality Weekly Reports* 44, 81–84.

CDC (1995d) Outbreak of Ebola viral hemorrhagic fever – Zaire, 1995. *Morbidity and Mortality Weekly Reports* 44, 381–382.

CDC (1995e) Update: HIV-2 infection among blood and plasma donors – United States, June 1992–June 1995. *Morbidity and Mortality Weekly Reports* 44, 603–606.

CDC (1995f) Case-control study of HIV seroconversion in health care workers after percutaneous exposure to HIV-infected blood – France, United Kingdom, and United States, January 1988 – August 1994. *Morbidity and Mortality Weekly Reports* 44, 929–933.

CDC (1995g) *HIV/AIDS Surveillance Report.* December 31, 1995.

CDC (1996a) Hepatitis a among persons with hemophilia who received clotting factor concentrate – United States. September – December, 1995. *Morbidity and Mortality Weekly Reports* 45, 29–32.

CDC (1996b) Update: mortality attributable to HIV infection among persons aged 25–44 years – United States, 1994. *Morbidity and Mortality Weekly Reports* 45, 121–125.

CDC (1996c) Outbreak of primary and secondary syphilis – Baltimore City, Maryland, 1995. *Morbidity and Mortality Weekly Reports* 45, 166–169.

CDC (1996d) Persistent lack of detectable HIV-1 antibody in a person with HIV infection – Utah, 1995. *Morbidity and Mortality Weekly Reports* 45, 181–185.

CDC (1996e) Update: Provisional public health service recommendations for chemo-prophylaxis after occupational exposure to HIV. *Morbidity and Mortality Weekly Reports* 45, 468–472.

CDC (1996f) Health care workers' with documented and possible occupationally acquired AIDS/HIV infection – United States through December, 1995. *HIV/ AIDS Surveillance Report 1995*, 7, Table 16.

Chamberland, M.E. and Curran, J.W. (1990) Epidemiology and prevention of AIDS and HIV infection. In: Mandell, G.L., Douglas, R.G. and Bennett, J.E. (eds). *Principles and Practice of Infectious Diseases*, 3rd edn. Churchill Livingstone, New York, pp. 1029–1046.

Chamberland, M., Conley, L., Lifson, A. *et al*. (1988) AIDS in health care workers: a surveillance report. Abstract 9020, *4th International Conference on AIDS*, Stockholm, Sweden.

Cherubin, C.E., Purcell, R.H. and Lander, J.J. (1972) Acquisition of antibody to hepatitis B antigen in three socioeconomically different medical populations. *Lancet* ii, 149–151.

Chiaramonte, M., Stoffolini, T., Caporaso, N. *et al*. (1991) Hepatitis C virus infection in Italy: A multicentric sero-epidemiological study. A report from the HCV study group of the Italian Association for the Study of the Liver. *Italian Journal of Gastroenterology* 23, 555–558.

Choo, Q.L., Kuo, G., Weiner, A.J. *et al*. (1989) Isolation of a cDNA clone derived from a blood-borne non-A, non-B viral hepatitis genome. *Science* 244, 359–362.

Cohen, B.J., Courouce, A.M., Schwarz, T.F. *et al*. (1988) Laboratory infection with parvovirus B19. *Journal of Clinical Pathology* 41, 1027–1028.

Collins, H. and Kennedy, D.A. (1987) A review: microbiological hazards of needlestick and 'sharps' injuries. *Journal of Applied Bacteriology* 62, 385–402.

Conry-Cantilena, C., Viladomiu, L., Melpolder, J.C. *et al*. (1992) Risk factors for hepatitis C virus infection in a US urban blood donor population. *Transfusion* 32, 465–467.

Contreras, M., Barbara, J.A.J. and Anderson, C.C. (1991) Low incidence of non-A, non-B post-transfusion hepatitis in London confirmed by hepatitis C virus serology. *Lancet* 337, 753–757.

Deinhardt, F., Holmes, A.W., Capps, R.B. *et al*. (1967) Studies on the transmission of disease of human viral hepatitis to marmoset monkeys. I. Transmission of disease, serial passage and description of liver lesions. *Journal of Experimental Medicine* 125, 673–687.

De Luca, M., Ascione, A., Vacca, C. *et al*. (1992) Are health care workers really at risk of HCV infection? *Lancet* 339, 1364–1365.

De Medina, M., Ortiz, C., Krenc, C. *et al*. (1992) Improved detection of antibodies to hepatitis C virus in dialysis patients using a second generation enzyme immunoassay. *American Journal of Kidney Disease* 20, 589–591.

Di Bisceglie, A.M., Order, S.E., Klein, J.L. *et al*. (1991) The role of chronic viral hepatitis in hepatocellular carcinoma in the United States. *American Journal of Gastroenterology* 86, 335–338.

Dienstag, J.L. and Ryan, D.M. (1982) Occupational exposure to hepatitis B virus in hospital personnel: infection or immunization? *American Journal of Epidemiology* 115, 26–39.

Dodd, L.G., McBride, J.H., Gitnick, G.L. *et al*. (1991) Prevalence of non-A, non-B hepatitis/hepatitis C virus antibody in laboratory quality-assurance sera. *Clinical Chemistry* 1991, 797–803.

Dodd, R.Y. (1995) Transfusion-transmitted hepatitis virus infection. *Hematology and Oncology Clinics of North America* 9, 137–154.

Dufoort, G., Courouce, A.-M., Ancelle-Park, R. *et al.* (1988) No clinical signs 14 years after HIV-2 transmission after blood transfusion. *Lancet* ii, 510.

Ebbesen, P., Melbye, M., Scheutz, F. *et al.* (1986) Lack of antibodies to HTLV-III/LAV in Danish dentists. *Journal of the American Medical Association* 256, 2199.

Ebeling, F. (1994) Importance of hepatitis C virus infection in Europe and North America. *Current Studies on Hematology and Blood Transfusion* 61, 164–181.

Esteban, J.I., Gomez, J., Martell, M. *et al.*(1996) Transmission of hepatitis C virus by a cardiac surgeon. *New England Journal of Medicine* 334, 555–560.

Esteban, J.I., Esteban, R., Viladomiu, L. *et al.* (1989) Hepatitis C virus antibodies among risk groups in Spain. *Lancet* ii, 294.

Felsenfeld, O. (1971) *Borrelia: Strains, Vectors, Human and Animal Borreliosis.* Warren H. Green, Inc., St. Louis.

FDA (1995) Recommendations for donor screening with a license test for HIV-1 antigen (memorandum to all registered blood and plasma establishments). Rockville, Maryland: US Department of Health and Human Services, Public Health Service, Food and Drug Administration, Center for Biologics Evaluation and Research.

Frame, J.D., Baldwin, Jr., J.M., Gocke, D.J. *et al.* (1970) Lassa fever, a new virus disease of man from West Africa: I. Clinical description and pathological findings. *American Journal of Tropical Medicine and Hygiene* 19, 670–676.

Garibaldi, R.A., Forrest, J.N. and Bryan, J.A. (1973) Hemodialysis-associated hepatitis. *Journal of the American Medical Association* 225, 384–389.

George, J.R., Forrest, J.N. and Phillips, A. (1990) Efficacies of US Food and Drug Administration-licensed HIV-1 screening enzyme immunoassays for detecting antibodies to HIV-2. *Aids* 4, 321–326.

Gerberding, J.L. (1994) Incidence and prevalence of human immunodeficiency virus, hepatitis B virus, hepatitis C virus, and cytomegalovirus among health care personnel at risk for blood exposure: final report from a longitudinal study. *Journal of Infectious Disease* 170, 1410–1417.

Gerberding, J.L. (1995) Management of occupational exposures to blood-borne viruses. *New England Journal of Medicine* 332, 444–451.

Gerberding, J.L., Bryant-LeBlanc, C.E., Nelson, K. *et al.* (1987) Risk of transmitting the human immunodeficiency virus, cytomegalovirus, and hepatitis B virus to health care workers exposed to patients with AIDS and AIDS-related conditions. *Journal of Infectious Disease* 156, 1–8.

Gessain, A., Barin, F., Vernant, J.C. *et al.* (1985) Antibodies to human T-lymphotropic virus type-1 in patients with tropical spastic paralysis. *Lancet* ii, 407–409.

Gilmore, N., Ballachey, M.L. and O'Shaughnessy, M. (1986) HTLV-III/LAV serologic survey of health care workers in a Canadian teaching hospital. Abstract 200. *2nd International Conference on AIDS,* Paris, France.

Giovannini, M., Tagger, A., Ribero, M.L. *et al.* (1990) Maternal infant transmission of hepatitis C virus and HIV infections: A possible interaction. *Lancet* 335, 1166.

Gocke, D.J. (1974) Type b hepatitis – good news and bad news. *New England Journal of Medicine* 291, 1409.

Gordin, F.M., Gibert, C., Hawley, H.P. *et al.* (1990) Prevalence of human immunodeficiency virus and hepatitis B virus in unselected hospital admissions: implications for mandatory testing and universal precautions. *Journal of Infectious Disease* 161, 14–17.

Hadler, S.C., De Monzon, M. and Ponzetto, A. (1984) Delta virus infection and severe hepatitis: an epidemic in Yupca Indians of Venezuela. *Annals of Internal Medicine* 100, 339–344.

Hadler, S.C., Judson, F.N., O'Malley, P.M. et al. (1991) Outcome of hepatitis B virus infection in homosexual men and its relationship to prior human immunodeficiency virus infection. Journal of Infectious Disease 163, 454–459.

Handsfield, H.H., Cummings, M.J. and Swenson, J.D. (1987) Prevalence of antibody to human immunodeficiency virus and hepatitis B surface antigen in blood samples submitted to a hospital laboratory. Implications for handling specimens. Journal of the American Medical Association 258, 3395–3397.

Hansen, J.P., Falconer, J.A., Hamilton, J.D. et al. (1981) Hepatitis B in a medical center. Journal of Occupational Medicine 23, 338–342.

Harpaz, R., Vonseidlein, L., Averhoff, F.M. et al. (1996) Transmission of hepatitis b virus to multiple patients from a surgeon without evidence of inadequate infection control. New England Journal of Medicine 334, 549–554.

Harper, S., Flynn, N., VanHorne, J. et al. (1986) Absence of HIV antibody among dental professionals, surgeons, and household contacts exposed to persons with HIV infection. Abstract THP.215. 3rd International Conference on AIDS, Washington, D.C.

Hayashi, J., Kishihara, Y., Kouzaburo, Y. et al. (1995) Transmission of hepatitis C virus by health care workers in a rural area of Japan. American Journal of Gastroenterology 90, 794–799.

Herbert, A-M., Walker, D.M., Davies, K.J. et al. (1992) Occupationally acquired hepatitis C virus infection. Lancet 339, 305.

Hinuma, Y., Komoda, H., Chosa, T. et al. (1982) Antibodies to acute T-cell leukemia virus-associated antigens in sera from patients with ATL and controls in Japan: a nation-wide seroepidemiologic study. International Journal of Cancer 29, 631–635.

Hira, P.R. and Hussein, S.F. (1979) Some transfusion-induced parasitic infections in Zambia. Journal of Hygiene, Epidemiology, Microbiology and Immunology 4, 436–444.

Hirsch, M.S., Wormser, G.P., Schooley, R.T. et al. (1985) Risk of nosocomial infection with human T-cell lymphotropic virus III (HTLV-III). New England Journal of Medicine 312, 1–4.

Ho, D.D., Moudgil, T. and Alam, M. (1989) Quantitation of human immunodeficiency virus type 1 in the blood of infected persons. New England Journal of Medicine 321, 1621–1625.

Hoofnagle, J.H., Seeff, L.B. and Bales, Z.B. (1978) Serologic responses to type B hepatitis. In: Vyas, G.N., Cohen, S.N. and Schmid, R. (eds) Viral Hepatitis. Franklin Institute Press, Philadelphia. pp. 219–244.

Juanes, R.J., Lalago, E. and Arrazola, M.P. (1993) HCV risk for health care workers. International Symposium on Viral Hepatitis and Liver Disease, Tokyo. p.228 (abstract).

Kalyanaraman, V.S., Sarngadharan, M.G., Robert-Guroff, M. et al. (1982) A new subtype of human T-cell leukemia virus (HTLV-II) associated with a T-cell variant of hairy cell leukemia. Science 218, 571–573.

Kanki, P., Barin, F., M Boup, S. et al. (1986) New human T-lymphotropic retrovirus related to simian T-lymphotropic virus type III (STLV-III). Science 232, 238–243.

Kataoka, R., Takehara, N. and Iwahara, Y. (1990) Transmission of HTLV-I by blood transfusion and its prevention by passive immunization in rabbits. Blood 76, 1657–1661.

Kelen, G.D., Green, G.B., Purcell, R.H. et al. (1992) Hepatitis B and hepatitis C in emergency department patients. New England Journal of Medicine 326, 1399–1404.

Kew, M.C. (1973) Possible transmission of serum (Australia-antigen positive) hepatitis via the conjunctiva. *Infection and Immunity* 7, 823–824.

Kim, J.P., Linnen, J. and Wages, J. (1995) Hepatitis G virus (HGV), a new hepatitis virus associated with human hepatitis. *Journal of Hepatology* 23, Suppl 1, 78.

Kiyosawa, K., Sodeyama, T., Tanaka, E. *et al.* (1990) Interrelationship of blood transfusion, non-A, non-B hepatitis and hepatocellular carcinoma: analysis by detection of antibody to hepatitis C virus. *Hepatology* 12, 671–675.

Kiyosawa, K., Tanaka, E., Sodeyama, T. *et al.* (1994) South Kuso Hepatitis Study Group: transmission of hepatitis C in an isolated area in Japan: community-acquired infection. *Gastroenterology* 106, 1596.

Klein, R.S., Phelan, J.A., Freeman, K. *et al.* (1988) Low occupational risk of human immunodeficiency virus infection among dental professionals. *New England Journal of Medicine* 318, 86–90.

Klein, R.S., Freeman, K., Taylor, P.E. *et al.* (1991) Occupational risk for hepatitis C virus infection among New York City dentists. *Lancet* 338, 1539–1542.

Kolho, E.K. and Krusius, T. (1992) Risk factors for hepatitis C virus antibody positivity in blood donors in a low-risk country. *Vox Sanguinis* 63, 192–197.

Kolho, E.K., Oksanen, K., Honkanen, E. *et al.* (1993) Hepatitis C antibodies in dialysis patients and patients with leukaemia. *Journal of Medical Virology* 40, 318–321.

Lackritz, E.M., Satten, G.A., Aberle-Grasse, J. *et al.* (1995) Estimated risk of transmission of the human immunodeficiency virus by screened blood in the United States. *New England Journal of Medicine* 333, 1721–1725.

Lam, J.P.H., McOmish, F., Burns, S.M. *et al.* (1993) Infrequent vertical transmission of hepatitis C virus. *Journal of Infectious Disease* 167, 572–577.

Lanphear, B.P. (1994) Trends and patterns in the transmission of bloodborne pathogens to health care workers. *Epidemiological Reviews* 16, 437–450.

Lanphear, B.P., Linnemann, C.C., Jr. and Cannon, C.G. (1993) Decline of clinical hepatitis B in workers at a general hospital: relation to increasing vaccine-induced immunity. *Clinical Infectious Disease* 16, 10–14.

Lanphear, B.P., Linnnemann, C.C., Cannon, C.G. *et al.* (1994) Hepatitis C virus infection in healthcare workers: Risk of exposure and infection. *Infection Control and Hospital Epidemiology* 15, 745–750.

Lauer, J.L., Van Drunen, N.A. and Washburn, J.W. (1979) Transmission of hepatitis B virus in clinical laboratory areas. *Journal of Infectious Disease* 140, 513.

Lee, H., Swanson, P., Shorty, V.S. *et al.* (1989) High rate of HTLV-II infection in seropositive IV drug abusers from New Orleans. *Science* 244, 471–475.

Lee, H., Weiss, S., Brown, L. *et al.* (1990) Patterns of HIV-1 and HTLV-I/II in intravenous drug abusers from the middle Atlantic and central regions of the USA. *Journal of Infectious Disease* 162, 347–352.

Lee, H.H., Swanson, P., Rosenblatt, J.D. *et al.* (1991) Relative prevalence and risk factors of HTLV-I and HTLV-II infection in US blood donors. *Lancet* 337, 1435–1439.

Leibowitz, S., Greenwald, L., Cohen, I. *et al.* (1949) Serum hepatitis in a blood bank worker. *Journal of the American Medical Association* 140, 1331–1333.

Lettau, L.A., Alfred, H.J. and Glew R.H. (1986) Nosocomial transmission of delta hepatitis. *Annals of Internal Medicine* 104, 631–635.

Levy, B.S., Harris, J.C. and Smith, J.L. (1977) Hepatitis B in ward and clinical laboratory employees of a general hospital. *American Journal of Epidemiology* 106, 330–335.

Lewis, T.L., Alter, H.J. and Chalmers, T.C. (1973) A comparison of the frequency of hepatitis B antigen and antibody in hospital and non-hospital personnel. *New England Journal of Medicine* 289, 647.

Lin, H.H., Kao, J.H. and Hsu, H.Y. (1994) Possible role of high-titer maternal viremia in perinatal transmission of hepatitis C virus. *Journal of Infectious Disease* 169, 638–641.

Linnemann, C.C., Jr., Hegg, M.E. and Ramundo, N. (1977) Screening hospital patients for hepatitis B surface antigen. *American Journal of Clinical Pathology* 67, 257–259.

Linnen, J., Wages, J., Zhen-Yong, Z-K. *et al.* (1996) Molecular cloning and disease association of hepatitis G virus: a transfusion-transmissible agent. *Science* 271, 505–508.

Lissen, E., Alter, H.J., Abad, M.A. *et al.* (1993) Hepatitis C virus infection among sexually promiscuous groups and the heterosexual partners of hepatitis C virus infected index cases. *European Journal of Clinical Microbiology and Infectious Disease* 12, 827–833.

Locasciulli, A., Bacigalupo, A. and Vanlint, M.T. (1991) Hepatitis C virus infection in patients undergoing allogenic bone marrow transplantation. *Transplantation* 52, 315–318.

Locasciulli, A., Cavalleto, D. and Pontisso, P. (1993) Hepatitis C virus serum markers and liver disease in children with leukaemia during and after chemotherapy. *Blood* 82, 2564–2567.

London, W.T. (1981) Primary hepatocellular carcinoma – etiology, pathogenesis, and prevention. *Human Pathology* 12, 1085–1097.

London, W.T., Di Figlia, M. and Sutnick, A. (1969) An epidemic of hepatitis in a chronic hemodialysis unit: Australia antigen and differences in host response. *New England Journal of Medicine* 281, 571–578.

Louie, M., Low, D.E., Feinman, S.V. *et al.* (1992) Prevalence of blood-borne infective agents among people admitted to a Canadian hospital. *Canadian Medical Association Journal* 146, 1331–1334.

Lubick, H.A. and Schaeffer, L.D. (1986) Occupational risk of dental personnel survey. *Journal of the American Dental Association* 113, 10–12.

Lurman, A. (1855) Eine icterus epidemic. *Berliner Klinische Wochenschrift* 22, 20–23.

MacCallum, F.O. (1945) Transmission of arsenotherapy jaundice by blood: failure with faeces and nasopharyngeal washings. *Lancet* i, 342.

Makris, M., Preston, F.E. and Triger, D.R. (1990) Hepatitis C antibody and chronic liver disease in haemophilia. *Lancet* 335, 1117–1119.

Mann, J.M., Francis, H., Quinn, T.C. *et al.* (1986) HIV seroprevalence among hospital workers in Kinshasa, Zaire. *Journal of the American Medical Association* 256, 3099–3102.

Marcus, R. (1988) The cooperative needlestick surveillance group; CDC's health-care workers surveillance project: an update. Abstract 9015. *4th International Conference on AIDS*, Stockholm, Sweden.

Mast, S.T. and Gerberding, J.L. (1991) Factors predicting infectivity following needlestick exposure to HIV: an in vitro model. *Clinical Research* 39, 58A.

Maynard, J.E. (1978) Viral hepatitis as an occupational hazard in the health care professional In: Vyas, G.N., Cohen S.N. and Schmid, R. (eds) *Viral Hepatitis: A Contemporary Assessment of Etiology, Epidemiology, Pathogenesis and Prevention.* Franklin Institute Press, Philadelphia. p. 321.

Mitsui, T., Iwano, K., Masuko, K. *et al.* (1991) Hepatitis C virus infection in medical personnel after needlestick accident. *Hepatology* 16, 1109–1114.

Monath, T.P. (1990) Colorado tick fever. In: Mandell, G.L., Douglas, R.G. and Bennett, J.E. (eds) *Principles and Practice of Infectious Diseases* 3rd edn. Churchill Livingstone, New York. pp. 1233–1240.

Morgan, H.W. and Williamson, D.A.J. (1943) Jaundice following administration of human blood products. *British Medical Journal* 1, 750.

Morris, I.M., Cattle, D.S. and Smits, B.J. (1975) Endoscopy and transmission of hepatitis B. *Lancet* ii, 1152.

Nahmias, A.J., Weiss, J., Yao, X. *et al.* (1986) Evidence for human infection with an HTLV III/LAV-like virus in central Africa, 1959. (letter) *Lancet* i, 1279–1280.

Nalpas, B., Driss, F., Pol, S. *et al.* (1991) Association between HCV and HBV infection in hepatocellular carcinoma and alcoholic liver disease. *Journal of Hepatology* 12, 70–74.

Niu, M.T., Coleman, P.J. and Alter, M.J. (1993) Multicenter study of hepatitis C virus infection in chronic hemodialysis patients and hemodialysis center staff members. *American Journal of Kidney Disease* 22, 568–573.

Niu, M.T., Alter, M.J., Kristensen, C. *et al.* (1992) Outbreak of hemodialysis-associated non-A, non-B hepatitis and correlation with antibody to hepatitis C virus. *American Journal of Kidney Disease* 4, 345–352.

Noble, R.C., Kane, M.A., Reeves, S.A. *et al.* (1984) Post-transfusion hepatitis A in a neonatal intensive care unit. *Journal of the American Medical Association* 252, 2711–2715.

Ohbayashi, A., Okochi, K. and Mayumi, M. (1972) Familial clustering of asymptomatic carriers of Australia antigen in patients with primary liver disease and primary liver cancer. *Gastroenterology* 62, 618.

Ohto, H., Terazawa, S. and Sasaki, N. (1994) Transmission of hepatitis C virus from mothers to infants. *New England Journal of Medicine* 330, 744–750.

Okada, K., Kainiyama, I. and Inometa, M. (1976) E antigen and anti-e in the serum of asymptomatic carrier mothers as indicators of positive and negative transmission of hepatitis B virus in their infants. *New England Journal of Medicine* 294, 746.

Osmond, D.H., Charlebois, E., Sheppard, H.W. *et al.* (1993) Comparison of risk factors for hepatitis C and hepatitis B virus infection in homosexual men. *Journal of Infectious Disease* 167, 66–70.

Panlilio, A.L., Shapiro, C.N., Schable, C.A. *et al.* (1995) Serosurvey of human immunodeficiency virus, hepatitis B virus, and hepatitis C virus infection among hospital-based surgeons. *Journal of the American College of Surgeons* 180, 16–24.

Papaevanglion, D., Trichopoulos, D. and Kemagtinon, A. (1975) Prevalence of hepatitis B antigen and antibody in prostitutes. *British Medical Journal* 2, 256.

Parritt, R.J. and Olsen, R.E. (1947) Two simultaneous cases of leprosy developing in tattoos. *American Journal of Pathology* 23, 805–817.

Pattison, C.P., Boyer, K.M. and Maynard, J.E. (1974) Epidemic hepatitis in a clinical laboratory: possible association with computer card handling. *Journal of the American Medical Association* 230, 854–857.

Paulsen, A.G., Kvinesdal, B., Aarby, P. *et al.* (1989) Prevalence and mortality from human immunodeficiency virus type 2 in Bissau, West Africa. *Lancet* i, 827–831.

Pike, R.M. (1976) Laboratory-associated infections: summary and analysis of 3,921 cases. *Health Laboratory Science* 13, 105–114.

Piot, P., Plummer, F.A., Mhalu, F.S. *et al.* (1988) AIDS: an international perspective. *Science* 239, 573–579.

Poiesz, B.J., Ruscetti, F.W., Gazdar, A.F. *et al.* (1980) Detection and isolation of type-C retrovirus particles from fresh and cultured lymphocytes of patients with cutaneous T-cell lymphoma. *Proceedings of the National Academy of Science USA* 77, 7415–7419.

Polish, L.B., Tong, M.J., Co, R.L. *et al.* (1993) Risk factors for hepatitis C virus infection among health care personnel in a community hospital. *American Journal of Infection Control* 21, 196–200.

Propert, S.A. (1938) Hepatitis after prophylactic serum. *British Medical Journal* 2, 677.

Puro, V., Petrosilio, N., Ippolito, G. *et al.* (1995) Hepatitis C virus infection in healthcare workers. *Infection Control and Hospital Epidemiology* 16, 324–326.

Rizzetto, M. (1983) The delta agent. *Hepatology* 3, 729.

Robert-Guroff, M., Weiss, S.H., Giron, J.A. *et al.* (1986) Prevalence of antibodies to HTLV-I and -II in intravenous drug abusers from an AIDS endemic region. *Journal of the American Medical Association* 255, 3133–3137.

Saag, M.S., Crain, M.J., Decker, W.D. *et al.* (1991) High-level viremia in adults and children infected with human immunodeficiency virus, relation to disease stage and CD4+ lymphocyte levels. *Journal of Infectious Disease* 164, 72–80.

Saeed, A.A., Ahmed, M.A. and Abdulmohsin, A.R. (1991) Hepatitis C virus infection in Egyptian volunteer blood donors in Riyadh. *Lancet* 338, 459–460.

Schneider, L. and Geha, R. (1994) Outbreak of hepatitis C associated with intravenous immunoglobulin administration-United States, October 1993–June 1994. *Morbidity and Mortality Weekly Reports* 43, 505–509.

Schneider, W.J. (1979) Hepatitis B: an occupational hazard of health care facilities. *Journal of Occupation Medicine* 21, 807–810.

Schweitzer, I.L., Dunn, A.E.F. and Peters, R.L. (1973) Viral hepatitis in neonates and infants. *American Journal of Medicine* 55, 762–771.

Seeff, L.B., Wright, E.C. and Zimmerman, H.J. (1978) Type b hepatitis after needlestick exposure: prevention with hepatitis b immunoglobulin; final report of the Veterans Administration Cooperative Study. *Annals of Internal Medicine* 88, 285–293.

Seeff, L.B., Buskell-Bales, Z. and Wright, E.C. (1992) Long-term mortality after transfusion-associated non-A, non-B hepatitis. *New England Journal of Medicine* 327, 1906.

Sexton, D.J., Gallis, H.A., McRae, J.R. *et al.* (1975) Possible needle-associated Rocky Mountain spotted fever. *New England Journal of Medicine* 292, 645.

Shanson, D.C., Evans, R. and Lai, L. (1985) Incidence and risk of transmission of HTLV-III infection to a staff at a London hospital, 1982–1985. *Journal of Hospital Infection* Suppl. C:15–22.

Shikata, T., Karasawa, T., Abe, K. *et al.* (1977) Hepatitis B c antigen and infectivity of hepatitis B virus. *Journal of Infectious Disease* 136, 571–576.

Simmonds, P., McOmish, F., McCullough, P. *et al.* (1992) Contamination of immunoassay controls with hepatitis C virus. *Lancet* 338, 1539–1542.

Simonetti, R.G., Camma, C., Fiorello, F. *et al.* (1992) Hepatitis C virus infection as a risk factor for hepatocellular carcinoma in patients with cirrhosis. *Annals of Internal Medicine* 116, 97–102.

Simons, J.N., Leary, T.P., Dawson, G.J. *et al.* (1995) Isolation of novel virus-like sequences associated with human hepatitis. *National Medicine* 1, 564–569.

Sirchia, G., Bellobuono, A. and Giovanetti, A. (1989) Antibodies to hepatitis C virus in Italian blood donors. *Lancet* ii, 797.

Smith, D.H., Johnson, B.K., Isaacson, M. *et al.* (1982) Marburg virus disease in Kenya. *Lancet* i, 816–820.

Smith, R.P., Evans, A.T., Popovsky, M. *et al.* (1986) Transfusion-acquired babesiosis and failure of antibiotic treatment. *Journal of the American Medical Association* 256, 2726–2727.

Snydman, D.R., Bryan, J.A. and Macon, E.J. (1976) Hemodialysis-associated hepatitis: report of an epidemic with further evidence on mechanisms of transmission. *American Journal of Epidemiology* 104, 563–570.

Stokes, J.H., Beerman, H. and Ingraham, N.R. (1945) *Modern Clinical Syphilology, Diagnosis, Treatment, Case study*, 3rd ed. The W.B. Saunders Co., Philadelphia.

Szmuness, W. (1978) Hepatocellular carcinoma and the hepatitis B virus: Evidence for a causal association. *Proceedings in Medical Virology* 24, 40.

Szmuness, W., Neurath, A.R. and Stevens, C.E. (1981) Prevalence of hepatitis B 'e' antigen and its antibody in various HBsAg carrier populations. *American Journal of Epidemiology* 113, 113–121.

Tabor, E., Purcell, R.H. and Gerety, R.J. (1983) Primate models and titered inocula for the study of human hepatitis A, hepatitis B, and non-A, non-B hepatitis. *Journal of Medical Primatology* 12, 305–318.

Tanaka, K., Hirohata, T., Koga, S. *et al.* (1991) Hepatitis C and hepatitis B in the etiology of hepatocellular carcinoma in the Japanese population. *Cancer Research* 51, 2842–2847.

Tokars, J.I., Marcus, R. and Culver, D.H. (1993) Surveillance of HIV infection and zidovudine use among health care workers after occupational exposure to HIV-infected blood. *Annals of Internal Medicine* 118, 913–919.

van der Poel, C.L. (1994) Hepatitis C virus: epidemiology, transmission and prevention. *Current Studies in Hematology and Blood Transfusion* 61, 137–163.

van der Poel, C.L., Reesink, H.W., Mauser-Bunschoten, E.P. *et al.* (1991) Prevalence of anti-HCV antibodies confirmed by recombinant immunoblot in different population subsets in the Netherlands. *Vox Sanguinis* 61, 30–36.

Watanabe, J., Minegishi, K. and Mitsumori, T. (1990) Prevalence of anti-HCV antibody in blood donors in the Tokyo area. *Vox Sanguinis* 59, 86–88.

Weiss, S.H., Saxinger, W.C., Rechtman, D. *et al.* (1985) HTLV-III infection among health care workers: association with needlestick injuries. *Journal of the American Medical Association* 254, 2089–2093.

Weiss, S.H., Goedert, J.J., Gartner, S. *et al.* (1988) Risk of human immunodeficiency virus (HIV-1) infection among laboratory workers. *Science* 239, 68–71.

Wejstal, R., Widell, A., Mansson, A.S. *et al.* (1992) Mother to infant transmission of hepatitis C virus. *Annals of Internal Medicine* 117, 887.

Williams, S.V., Huff, J.C. and Feinglass, E.J. (1974) Epidemic viral hepatitis type B in hospital personnel. *American Journal of Medicine* 57, 904–911.

Wong-Staal, F. and Gallo, R.C. (1985) Human T-lymphotropic viruses. *Nature* 317, 395–403.

WHO (1995) The current global situation of the HIV/AIDS pandemic as of 15 December 1995. *World Health Organization Weekly Epidemiolological Record*. December 15, 1995.

Yu, M.C., Tong, M.J., Coursaget, P. *et al.* (1990) Prevalence of hepatitis B and C viral markers in black and white patients with hepatocellular carcinoma in the United States. *Journal of the National Cancer Institute* 82, 1038–1041.

CHAPTER 4
Blood exposure data in Europe

D. Abiteboul

INTRODUCTION

In Europe (Europe of the Twelve plus Austria, Finland, Norway, Sweden and Switzerland) 6.5 million workers (or 6.8 including students) are potentially exposed to the hazards of blood contact (Van Damme *et al.*, 1995).

A number of pathogens can be transmitted in the event of percutaneous (needlestick and sharps injuries) or mucocutaneous exposures to blood (Collins and Kennedy, 1987; Jeffries, Chapter 1 this volume; Collins, Chapter 2 this volume). However, hepatitis B (HBV) and C (HCV) viruses and the human immunodeficiency virus (HIV) may be considered to be the three major blood-borne pathogens because of their prevalence in the population receiving care, the frequency of chronic carrier status in infected patients and the seriousness of the infection. Before studying the incidence and the causes of blood exposures (BE) the characteristics and risk factors of accidents that have led to infection by these viruses will be reviewed, based on European data.

TRANSMISSION RISK FACTORS

Hepatitis B Virus (HBV)

A very large number of incidence and prevalence surveys have shown that the frequency of BE and of needlestick injuries is a major factor in transmission. Studies conducted in the 1970s and 1980s indicate risk levels 2–10 times higher for health-care workers than for the population at large (Van Damme *et al.*, 1995). It is of note that for a non-immunized person the risk of acquiring HBV infection following percutaneous exposure can be as high as 30–40% when the patient is HBe antigen-positive.

The availability of a vaccine since 1982 has helped to reduce substantially the frequency of occupationally-acquired hepatitis B (Abiteboul *et al.*, 1990; Van Damme *et al.*, 1995). The risk of HBV infection remains, however, because, on average, only 40–50% of the population concerned in Europe are vaccinated, even though it is recommended and in some cases compulsory. Few countries have exhaustive reporting schemes that record the exact number of cases of occupationally-acquired hepatitis B. In their model Van Damme *et al.* (1995) set

Occupational Blood-borne Infections (eds C.H. Collins and D.A. Kennedy)   59

the number of new cases of infection among European health-care professionals each year at approximately 10,000, in the hypothesis of a 40% vaccination coverage.

Hepatitis C Virus (HCV)

Unlike HBV, the seroprevalence of HCV among health-care staff is generally comparable to that of blood donors, as demonstrated in Germany (Jochen, 1992), Belgium (De Brouwer and Lecomte, 1994), Italy (Campello *et al.*, 1992; Puro *et al.*, 1995), the UK (Zuckerman *et al.*, 1994) and France (Djeriri *et al.*, 1996). When the seroprevalence is higher, the same risk factors are found as in the general population, while occupational factors have little significance (Petrosillo *et al.*, 1995). These data reveal low transmissibility in the event of occupational exposure. However, ten well-documented cases of seroconversion have been recorded (Table 4.1).

The risk of seroconversion after percutaneous exposure to HCV has been studied by seven European teams. Transmission rates of between 0 and 1.9% were reported (Table 4.2). By pooling the data from nine European, American and Japanese studies, Puro *et al.* (1996) calculated a mean seroconversion rate of 2.1% (95% CI 1.2–3.4%), higher than the figures mentioned earlier because of the Japanese data included (rate of 3–6%). The higher rate in Japan may be explained by the different infectivity of the HCV strains concerned, and probably also by the differences in the tests used and by the small number of subjects studied. Further studies are needed on larger subject groups in order to learn more about the risk factors of occupational HCV transmission, although this virus is clearly less easily occupationally transmitted than HBV.

Human Immunodeficiency Virus (HIV)

Twenty-eight documented cases of seroconversion after a specific occupational exposure to a known HIV-infected source had been reported among European health-care workers by June 1996 (Tables 4.3 and 4.4). There were also 57 cases of possible occupationally-acquired infection. This inventory was drawn up on the basis of data collected by Heptonstall *et al.* (1995) and by Fitch *et al.* (1995). Since these two reviews, 15 additional cases have been described in Germany (Jarke, 1996) and one in an Italian surgeon (Ippolito *et al.*, 1996).

All the proven cases have certain characteristics in common: the majority concern nursing staff. Needlestick injuries are the most common: 24, compared with two cuts and two mucocutaneous projections. Of 18 needlestick injuries where details of the type of needle are given, 16 involved hollow needles, and 14 of these involved venepuncture. In most cases, the source patient had AIDS.

The summary by Heptonstall *et al.* (1995) of the results of prospective surveillances of staff exposed to infected blood makes it possible to estimate the risk of HIV seroconversion at 0.32% (95%, CI 0.18–0.45) following percutaneous exposure (PCE). Only one case of seroconversion after mucocutaneous exposure (MCE) was observed in these cohorts, and the risk would appear to be of the order of 0.03% (95%, CI 0.066–0.18).

Table 4.1. Reports of documented hepatitis C conversions.

Author, year reported	Country	Occupation	Source patient	Exposure details	HCV antibody after exposure		Clinical outcome
					−ve	First +ve	
Vaglia et al., 1990	Italy	Surgeon	HIV+ drug user	Needlestick during surgery	0	98 days	Acute hepatitis (25 days) → chronic
Cariani et al., 1991	Italy	Nurse	HCV RNA+ Dialysed	Needlestick	0	16 weeks	Acute hepatitis (13 weeks)
Marranconi et al., 1992	Italy	Surgeon	Post-transfusion chronic hepatitis	Needlestick	0	14 weeks	Acute hepatitis (5 weeks) → chronic
Sartori et al., 1993	Italy	Nurse	Dialysed	Mucocutaneous exposure	0	4 months	ALT ↗
Perez-Trallero et al., 1994	Spain	Physician	HCV RNA+ HIV+ Drug user	Needlestick (with biopsy forceps)	0	6 months	Asymptomatic
Puro et al. 1995	Italy	Nurse	?	Needlestick (IM)	0	6 months	Asymptomatic
Puro et al. 1995	Italy	Surgeon	HIV+	Needlestick (blood drawing)	0	3 months	Acute hepatitis (4 weeks) → chronic
Puro et al. 1995	Italy	Surgeon	HIV+	Needlestick (blood drawing)	0	6 months	Asymptomatic
Puro et al. 1995	Italy	Nurse	?	Needlestick (blood drawing)	0	3 months	ALT ↗
Domart et al. 1996a	France	Nurse	Dialysed	Needlestick	<4 weeks	6 weeks	ALT → chronic

Table 4.2. Incidence studies of HCV seroconversion rates among health-care workers (HCWs) after a percutaneous exposure to HCV-positive blood (follow-up: at least 5 months).

Country	HCWs	Seroconversion	%	References
Italy	30	0		Stellini *et al.*, 1993
Italy	61	0		Petrosillo *et al.*, 1994
Italy	331	4	1.2	Puro *et al.*, 1996*
Spain	69	0		Hernandez *et al.*, 1992
Spain	53	1	1.9	Perez-Trallero *et al.*, 1994
UK	24	0		Zuckerman *et al.*, 1994
France	48	0		Domart *et al.* 1996b

*Only hollow-bore needlestick injuries.

Table 4.3. Reported occupationally-acquired HIV infections in health-care workers and AIDS cases by country in Europe, June 1996.

Country	Cumulative AIDS cases*	Documented	Possible	Total
France	41,948	10	27	37
Spain	39,670	4	–	4
Italy	33,701	4	2	6
Germany	15,093	3	16	19
UK	12,437	4	8	12
Switzerland	5,639	1	–	1
Holland	3,341	–	2	2
Belgium	2,124	2	1	3
Denmark	1,872	–	1	1
Totals	155,825	28	57	85

*December 1995.

Table 4.4. Occupationally-acquired HIV infections by occupation in Europe – June 1996.

Occupation	Documented	Possible	Total
Nurse (including student nurse)	21	18	39
Doctor (including medical student)	–	10	10
Laboratory worker	1*	6	7
Surgeon	1	5	6
Health-aide/nurse-aide	–	3	3
Dentist/dental worker	–	2	2
Housekeeper/maintenance worker	–	2	2
Surgical attendance staff	–	1	1
Other/unspecified HCW staff	5	10	15
Total	28	57	85

*Phlebotomist.

These rates, however, are merely the mean rates for different types of exposure with source patients at different stages of the disease. According to a case–control study conducted by the Centers for Disease Control and Prevention (CDC, 1995) in collaboration with France and the United Kingdom, the factors which significantly increase the risk of HIV transmission in the event of percutaneous exposure are: depth of wound, source patient in the terminal phase of AIDS, a device visibly contaminated with blood, a procedure involving a needle placed directly in the patient's vein or artery. The study also shows a decrease in risk of almost 79% (43–94%) in health-care workers who took Zidovudine following exposure.

Frequency and Circumstances of Blood Exposure

A very large number of studies have been carried out in Europe in the last 10 years to investigate the incidence and causes of BE. Some, like those of the Italian Study Group on Occupational Risk of HIV Infection (SIROH) in Italy or the Blood Exposures Study Group (GERES) in France, have set up voluntary hospital networks where active surveillance is carried out. Comparative analysis of the results of these surveys is difficult because of the different methodologies: only observational studies make rigorous assessment possible, but they are difficult to implement, and therefore rare. Furthermore, denominator figures are either not available or vary considerably (number of employees, hours worked, type of device used, etc.). Also, the activities taken into account tend to vary.

In the studies based on self-reported cases (Abiteboul et al., 1992; Ippolito et al., 1993, 1994; Iten et al., 1995; Iwatsubo et al., 1996), most BEs involve nursing staff (61–67%), followed by doctors on a level with nurse aids and housekeepers, both with 14–17% reporting. Needlestick injuries are the most frequent type of injury; they account for 58–75% of exposures, and half of these occur at the patient's bedside. Blood-drawing procedures are involved in 30–35% of cases. Most injuries are sustained after use, especially with non-protected devices before disposal. Recapping is at the origin of 5–17% of BEs, although this practice has shown a tendency to decline in the more recent studies. Mucocutaneous exposures (MCEs), which are reported mainly in operating rooms and laboratories, account for 1 1–33% of BEs.

This type of study is open to substantial biases because of under-reporting (see also Collins and Kennedy, 1987). In certain of these works the incidences are presented by the authors, or may be calculated (studies 1–3, Table 4.5): PCEs vary with the study and the occupational category from 0.01–0.1 per 100 HCW years. These rates are 10–100 times smaller than those obtained by questionnaire or interview surveys of exposure recall in the last days' or months' work (studies 4–7, Table 4.5) or by monitoring by the survey team (study 8). Under-reporting is particularly common among doctors, especially surgeons, as shown by the vast difference between the numbers of cases reported to the official bodies and those calculated from the questionnaire surveys. Luthi and Dubois-Arber (1995) reported that only 3.4% of BEs of doctors were reported, compared with 39.7% for nurses. The reasons most

Table 4.5. Annual incidence of percutaneous and mucocutaneous exposures (PCE and MCE) as presented by the authors or calculated from their figures.

Reference	Type of study	Location	Population	Incidence (per 100 HCW years)			
				Nursing staff	Medical staff	Associated staff*	Laboratory staff
1	Self-reporting to Occupational Health Service	50 hospitals (France)	60,500 HCWs	PCE = 4.2 MCE = 0.6	PCE = 1.1 MCE = 0.2	PCE = 1.0 MCE = 0.1	PCE = 1.6 MCE = 0.7
2	Self-reporting to Occupational Health Service	40 hospitals (France)	47,000 HCWS	PCE = 7.5 MCE = 0.8	PCE = 3.1 MCE = 0.5	PCE = 2.0 MCE = 0.2	PCE = 3.0 MCE = 0.9
3	Self-reporting to declaration system	1 hospital (Switzerland)	561 nurses	PCE = 9.0			
4	Self-administered question; PCE during previous 4 months	1 hospital (Switzerland)	561 nurses	PCE = 109			
5	Interview: PCE during previous year	98 hospitals (Italy)	20,005 HCWs	PCE = 35.3	Surgeons: PCE = 54.9 Physicians PCE = 26.5	PCE = 18.7	PCE = 14.7
6	Self-administered question: PCE during previous 4 weeks	7 hospitals (Switzerland)	3116 HCWs	PCE = 49	PCE = 428**	PCE = 11	
7	Self-administered question: PCE during employment on unit (Denmark)	1 hospital (Denmark)	135 HCWs	PCE = 11 MCE = 3	PCE = 13-51*** MCE = 17-21***	PCE = 9 MCE = 11	
8	Active surveillance of PCE by trained investigators	17 hospitals	518 nurses	PCE = 32			

References:
1. Abiteboul et al., 1991, 1992. 2. Iwatsubo et al., 1994, 1996. 3. Saghafi et al., 1992. 4. Saghafi et al., 1992.
5. Albertoni et al., 1986, 1992. 6. Luthi and Dubois-Arber, 1995. 7. Nelsing et al., 1993, 1996a.
8. Bouvet et al., 1990, 1991.
*Associated staff: nurse aids, housekeepers, domestics. **Only surgeons and anaesthetists. ***Depending on function (junior or senior).

frequently given for this failure to report BE are: exposure considered to be without risk, takes up too much time, reporting procedures too complicated or not known (Nelsing *et al.*, 1993; Luthi and Dubois-Arber, 1995; Burke *et al.*, 1996). Conversely, in interview or self-administered questionnaire surveys, employees sometimes tend to overestimate the risk.

Although it is difficult to compare figures, one gets the impression that, on the whole, the frequency of blood exposures is changing little, or even tending to increase, even though health-care workers have been trained for more than 10 years now to comply with Universal Precautions, and in spite of the availability of safety devices and protective equipment. Figures of the order of those summarized in Table 4.5 were cited by Collins and Kennedy (1987) in a review of similar studies carried out between 1975 and 1986.

Most of the studies mentioned above reveal insufficient compliance with Universal Precautions: more than half the BEs could theoretically have been avoided by better compliance, which may explain why there has been no decrease in BE. Decreases in BEs where Universal Precautions had been followed were reported, however, in several works, such as those of Kristensen *et al.* (1992), Saghafi *et al.* (1992) and Nelsing *et al.* (1997a). One possible explanation may be that the impact of better compliance with precautions is masked in many studies by better compliance with reporting.

Be that as it may, compliance with Universal Precautions alone is insufficient. In order to further reduce BEs it is essential to identify the most dangerous procedures and devices, with a view to safer design of work and devices in the different specialities.

This was the object of a study carried out from 1990 to 1992 by the SIROH (Ippolito *et al.*, 1994) in 33 voluntary hospitals in Italy. In order to determine which medical devices caused needlestick injuries, each hospital was asked to communicate the total number of needlestick injuries reported in each device category, and the total number of needles in each category used in the course of the year. This information would be used to calculate device-specific injury rates. Intravenous catheter stylets were found to have the highest rate (15.7/100,000 devices used), followed by winged needles (10.1/100,000), vacuum-tube phlebotomy needles (4.3/100,000) and disposable syringes (3.8/100,000).

This type of survey is useful in encouraging product manufacturers to design safer devices. Self-sheathing needle devices are now available on the market and have already been adopted by some hospitals. There has been little research in Europe on their impact on injury frequency. Baseline device-specific rates are useful for comparison with the rates obtained subsequent to the introduction of new safety devices.

Device category is not the only criterion that must be taken into consideration, however, since each device may be used for more than one type of procedure, and some procedures may involve greater risk than others. Winged needles intravenous (IV) sets, for example, are used, connected to vacuum tubes for difficult venous access, drawing blood for culture, or for infusion, with different injury mechanisms in each case. In certain countries, such as the United States, needles are used for accessing IV lines, which explains the high injury rate (36.7/100,000) with this type of needle reported in the study by Jagger *et al.* (1988), but not by Ippolito *et al.* (1994), since Italy, like other European countries, uses needleless access. Accident rates thus vary considerably with the uses to which devices are put.

What is more, these studies refer to hollow needles, which are certainly the most dangerous, but not the only devices that cause injuries; they are also caused, for example, by devices specific to each specialty (surgery, catheterization, etc.).

RISK FACTORS FOR THE MAIN SECTORS OF ACTIVITY

Operating Room

Operating room risks have been the subject of numerous questionnaire surveys and a number of observational studies, mainly in the United States (Short and Bell, 1993). In Europe the most substantial study was carried out by Antona *et al.* (1994) and Johanet *et al.* (1995) in 12 French hospitals, covering 3,554 surgical procedures. At least BE occurred in 11.7% of procedures, 4.2% involved at least one PCE, and 8.7% one MCE. Comparison with other studies is difficult in view of the wide variations in the rates observed: 6–50% for overall rates depending on the study, and 1.7–15% for percutaneous injuries (Short and Bell, 1993). In one British study on 427 urological operations, more than 30% of procedures resulted in blood contamination. These variations are probably due to a number of factors: different methodologies, the specialties observed, surgical practices that differ from country to country. Nevertheless, the greatest exposure risk is always found in lengthy operations with bleeding; among the specialities with highest exposure are vascular surgery, gynaecological surgery and orthopaedic surgery (Antona *et al.*, 1994; Johanet *et al.*, 1995; Williams *et al.*, 1996; Nelsing *et al.*, 1997a,b).

Antona et *al.* (1994) and Johanet *et al.* (1995) reported that the operating surgeon was the most exposed with two PCEs and 5.6 MCEs per 100 person-procedures. Less exposed were interns (1.4 PCEs and 3 MCEs) and attending nurses (1.1 PCEs and 1.6 MCEs). If we consider that a surgeon carries out between 300 and 500 procedures per year on average, he will be a victim of 6–10 PCEs per year, which is the same order of frequency as that reported by surgeons in questionnaire surveys (Porteous, 1990a; Luthi and Dubois-Arber, 1995; Nelsing *et al.*, 1996a,b).

More than 75% of injuries occur during suturing, often on the non-dominant index finger due to the use of fingers to hold tissue while suturing or to blind suturing. Most of these injuries occur during wound suturing (Antona *et al.*, 1994; Johanet *et al.*, 1995; Consten *et al.*, 1995; Nelsing *et al.*, 1997a). The 'no touch' technique reduces the risk of PCE during this phase (Corlett *et al.*, 1993).

Blood contamination of the hands is significantly reduced by double gloving. Only a minority of surgeons were double gloved, however: 38% at the time of the studies of Antona *et al.* (1994) and Johanet *et al.* (1995); only 25% of

Belgian surgeons reported wearing them with HIV-infected patients (Szpalski *et al.*, 1992). In the study by Luthi and Dubois-Arber (1995) 19% of surgeons systematically wore double gloves.

Half of all MCEs involved the face in the Antona *et al.* (1994) study; 25–52% of procedures were observed to involve projections to the face, rising to up to 84% during total hip replacement (Brearley and Buist, 1989; Porteous, 1990b).

While there is hope of reducing the frequency of PCEs in the operating room thanks to progress with laparoscopic procedures, the MCE risk will continue to require vigilance. McNicholas *et al.* (1989) highlighted the frequency of blood contacts (32.6%) during urological endoscopic surgery.

All these data are commensurate with those reported in the main American studies (Short and Bell, 1993).

Because of the frequency with which they are exposed to blood, surgeons are amongst the high-risk staff for hepatitis B, as demonstrated by the prevalence studies carried out before vaccination (Van Damme *et al.*, 1995). Their vaccination coverage remains insufficient, however, with between 53 and 69% of surgeons vaccinated according to the different studies.

The risk is lower where HIV is concerned, because solid needles are used: to date only one documented case of a sharps injury (none of needlestick injuries) has been published (Ippolito *et al.*, 1996). However, when HIV prevalence is high among patients, the risk ceases to be negligible (Houweling and Coutinho, 1991; Consten *et al.*, 1995). For surgeons who travel to Africa to operate, for example, Consten *et al.* (1995) considered that the risk of seroconversion was 15 times greater than for a colleague working in Europe (i.e. one seroconversion per 300 surgeons per year). It is of note that four of five cases of possibly occupationally-acquired HIV infection among European surgeons were contracted in Africa (Heptonstall *et al.*, 1995).

The risk of hepatitis C is less well documented; three cases of seroconversion have nevertheless been described among surgeons (cf. Table 4.1).

Acute-care Medical Wards

After surgical procedures, patient care in medical wards is the next most frequent cause of BE, mainly affecting doctors and nurses. The studies where under-reporting is not too substantial (studies 4–8, Table 4.5) show PCE incidences of between 26 and 51 per 100 person years in medical staff (interns, residents and physicians) and between 32 and 109 in nursing staff. The frequencies vary in particular with the number of exposure-prone gestures performed by each individual. A Danish study by Nelsing *et al.* (1993) found that interns and residents experienced five times as many PCEs as did nurses since it is they who perform all the invasive procedures. In other countries, such as France, nurses are at higher risk since they perform many of these tasks. Although the incidences are lower than for surgeons, the contamination risk is probably higher because of the daily performance of invasive procedures using hollow needles, as demonstrated by the predominance of nurses among cases of HIV and HCV seroconversion.

In order to investigate the risk linked to each invasive procedure, the GERES carried out a prospective study of needlestick injuries among nurses (Bouvet *et al.*, 1991): 17 French hospitals participated with seven intensive care units and ten medical wards. Nurses were asked to report every injury systematically to trained investigators. At the same time, cross-sectional studies were carried out in order to collect data on routine invasive procedures: nurses recorded each invasive act they performed using a needle.

Eighty-one per cent of injuries occurred when performing invasive acts. In order to estimate the risk linked to each procedure, rates of invasive acts including needlestick injuries were compared with rates of invasive acts performed routinely by nurses. The results suggested that some invasive procedures might involve higher risk than others: i.e. 100,000 acts on implantable access (IA) systems induced 230 needlestick injuries, when the same number of invasive tests (fingerstick procedures) induced only four. Thus, actions on IA systems involved 57 times more risk than invasive tests. From lowest to highest risk the order of graduation was: invasive tests, hypodermic injections, arterial blood drawing, venous sampling, infusions, sampling for blood culture, and acts on IA systems. A similar survey, carried out in 1992, showed a decrease of injuries in relation to blood culture and infusions, but not significant enough to change the graduation of risk (Abiteboul *et al.*, 1993).

Although the methodology is different, certain of these results may be compared with those of Ippolito *et al.* (1994) presented above, showing rates of a similar order and a similar graduation of risk.

In the different surveys shown in Table 4.5, the main mechanisms of injury among nurses related to disposal (50–60%); 21–35% of PCEs are related to non-protected used devices, frequently after injections and venous blood drawing. The container is involved in 10–15% of PCEs (injury by needle protruding from overfilled container or while introducing devices, particularly winged needles).

During the procedure, PCE occurred while withdrawing the needle from the patient, especially when inserting infusion (with the stylet), or attempting to penetrate a plastic stopper (blood culture collection). Rebounding of needle when withdrawn from an IA system should be stressed as a specific mechanism (Bouvet *et al.*, 1991; Abiteboul *et al.*, 1993). Although difficult to quantify, contributing circumstances such as workload, interruptions or emergencies are often mentioned by nurses (Abiteboul *et al.*, 1992; Luthi and Dubois-Arber, 1995).

With regard to doctors, the risk is not so well documented. Nelsing *et al.* (1997a,b) performed a nation-wide survey in order to assess the risk factors of BE among hospital employed doctors. The incidence of PCEs and MCEs reported by 6,005 Danish physicians varied with the medical specialties: from the highest in cardiology (2.9 PCEs per person-risk year) and pulmonary medicine (6.8 MCEs per person-risk year) to the lowest in infectious medicine (0.4 PCEs and 0.5 MCEs per person-risk year). The majority of PCEs were caused by hollow needles, the mechanisms of injury being much the same as those observed for nurses. Nevertheless, certain medical procedures are particularly at risk: arterial and venous catheterization, especially in emergency. Unsafe working practices were involved in up to 50% of PCEs. For example, recapping injuries were more frequent than in the case of nurses (up to 39% for arterial blood sampling).

In contrast to surgeons, junior doctors (interns and residents) working in medical wards were at higher risk than senior doctors (1.7 PCEs and 3.9 MCEs per person-risk year versus 0.9 PCEs and 1.9 MCEs per person-risk year) because they perform a greater number of invasive procedures. Inexperience also seemed to have an impact on BE sustained by the youngest graduates. Two other surveys carried out among medical students confirm that trainees are at higher risk – about 50% of students had a significant BE during the third or fourth year of medical school (Gompertz, 1990; Tarantola et al., 1996).

The hazards of BE for housekeepers and associated staff, although often neglected, are also significant as shown in Table 4.5. The easily preventable injuries sustained by these people are due to used needles and other sharps left on trays, in linen or improperly discarded.

Dialysis Practice

Dialysis has long been known as a high-risk activity for transmission of blood-borne infections to both patients and staff. Before vaccination, BEs in dialysis units were responsible for a high number of cases of hepatitis B. At present, health-care workers are exposed above all to HCV, and to a lesser degree to HIV. The prevalence of HCV among patients undergoing dialysis varies from country to country, but it is invariably higher than in the general population: 23% in 50 dialysis units in France (Poignet et al., 1995); 39% in nine dialysis centres in Italy (Petrosillo et al., 1995). Three cases of HCV seroconversion recorded in Table 4.1 had their origin in dialysis.

The BE risk factors specific to dialysis have been studied in Europe by Poignet et al. (1995) and Petrosillo et al. (1995). These authors report that needlestick injuries are the most frequent injuries: 3–3.9 per 10,000 dialyses, caused mainly by large-bore needles used for establishing vascular access or for blood drawing. The risk is greatest when removing the needle from the arterio-venous fistula.

Mucocutaneous exposures, according to both authors, are less frequent (0.8–1.7 per 10,000 dialyses), but they nevertheless account for 20–30% of BEs in the dialysis environment. The most frequent causes are: cannulating the arteriovenous fistula and disconnecting or break in the blood circuit.

Laboratory Settings

Laboratories are permanent blood exposure sites: numerous cases of hepatitis B were recorded in the 1970s and 1980s, and blood transmission of a wide variety of infections were reported (Collins, 1993; Jagger and Bentley, Chapter 12 this volume). Few of the studies mentioned in Table 4.5 took any interest in labora-tories; those which did found relatively low PCE frequencies but more MCEs reported than in other sectors. Pelletier et al. (1996) studied 136 BEs in clinical laboratories. Unlike in health-care wards, 66% of injuries occurred during the procedure rather than at the disposal stage. Sharps injuries and cuts account for 46% of BEs, because of the wide variety of sharp objects used: broken tubes, Pasteur pipettes or scalpels, especially during extemporaneous examinations and autopsies conducted in anatomical pathology units.

Needlestick injuries are less frequent in laboratory staff, except for phlebotomists. As distinct from United States practice, blood sampling in Europe is rarely done by laboratory staff in Europe, but rather by nurses, which explains why the 17 cases of HIV seroconversion reported by Heptonstall *et al.* (1995) among laboratory staff in the Western world include only one case in Europe (a phlebotomist). As for blood projections and aerosols, they are largely underestimated (see Kennedy and Collins, Chapter 11 this volume).

The increasing use of automation may make it possible to reduce these risks by reducing handling, but it could also generate new risks in the event of poor design or during maintenance (Truchaud *et al.*, 1996).

Other Settings

Since most of the studies were carried out in hospitals, we lack data on staff who, although less exposed, are potentially at risk (ambulance staff, prison staff, firemen, funeral service practitioners, waste disposal staff: see Collins and Kennedy, Chapter 14 this volume). In France, Lot and Abiteboul (1994) reported two HIV seroconversions outside the health-care professions: one concerned a waste disposal truck driver who sustained a needlestick injury when handling a container opened by accident; the other concerns a refuse collector who was injured by a used needle that was left in amongst household waste.

CONCLUSIONS

Blood exposures must continue to be monitored and surveyed. In spite of the prevention efforts of the last 10 years, they are still a major occupational risk, as shown by the studies presented. In Europe these studies mainly concern percutaneous exposures of doctors and nurses, who are the most exposed populations. But data are lacking for preventive action in favour of other occupational categories (dentists, employees operating outside the hospital environment, for example in homecare) or of certain very specific activities (catheterization, endoscopy, etc.). Furthermore, it is essential to evaluate the impact of the measures taken, in particular the introduction of new safety devices, on BE incidence.

REFERENCES

Abiteboul, D., Gouaille, B. and Proteau, J. (1990) Prevention de l'hepatite B a l'Assistance Publique-Hôpitaux de Paris. Bilan de 7 ans de vaccination par les médecins du travail. *Archives des Maladies Professionnelles* 51, 405–412.

Abiteboul, D., Antona, D., Azoulay, S. *et al.* (1992) Surveillance of occupational exposure to blood in Assistance Publique-Hôpital de Paris (AP-HP) in 1990–1991. In: *Proceedings of 1st International Congress on Occupational Health for Heath Care Workers, Freiburg, September 1992.*

Abiteboul, D., Antona, D., Descamps, J.M. *et al.* (1993) Procedures a risque d'exposition au sang pour le personnel infirmier: surveillance et evolution de 1990 a 1992 dans 10 hôpitaux. *Bulletin Epidemiologique Hebdomadaire* 43, 195–196.

Albertoni, F., Ippolito, I.G., Petrosillo, N. *et al.* (1992) Needlestick injury in hospital personnel: A multicenter survey from central Italy. *Infection Control and Hospital Epidemiology* 13, 540–544.

Antona, D., Johanet, H., Abiteboul, D. *et al.* (1994) Accidental blood exposures during surgery: evaluating the risk. Conference on Prevention of Transmission of Pathogens in Surgery and Obstetrics; Atlanta, 1994. *Infection Control and Hospital Epidemiology* 15, 347.

Bouvet, E., Abiteboul, D., Fourrier, A. *et al.* (1991) A one year prospective ongoing survey on blood exposures among 518 nurses in 17 hospitals in France. In: *Proceedings of VIIth International Conference on AIDS. Florence, June 1991.* Abstract 4173.

Brearley, S. and Buist, L.J. (1989) Blood splashes: an underestimated hazard to surgeons. *British Medical Journal* 299, 1315.

Burke, S., Jankowski, R. and Madan, I. (1996) Contamination incidents among doctors and midwives: reasons for non-reporting and knowledge of risks. In: *Proceedings of an International Colloquium on Blood-borne Diseases.* Paris, June 1995. Health Services Section of the International Social Security Association, Hamburg, pp. 343–345.

Campello, C., Majori, S., Poli, A. *et al.* (1992) Prevalence of HCV antibodies in health care workers from northern Italy. *Infection* 20, 224–226.

Cariani, E., Zonaro, A., Primi, D. *et al.* (1991) Detection of HCV RNA and antibodies to HCV after needlestick injury. *Lancet* 337, 850.

CDC (1995) Case–control study of HIV seroconversion in health-care workers after percutaneous exposure to HIV-infected blood. *Morbidity and Mortality Weekly Reports* 44, 929–933.

Collins, C. H. (1993) *Laboratory-Acquired Infections,* 3rd edn. Butterworth-Heinemann, Oxford.

Collins, C.H. and Kennedy, D.A. (1987) Microbiological hazards of occupational needlestick and 'sharps' injuries. *Journal of Applied Bacteriology* 62, 385–402.

Consten, E., Van Lanschot, J., Henny, P. *et al.* (1995) A prospective study on the risk of exposure to HIV during surgery in Zambia. *AIDS* 9, 585–588.

Corlett, M.P., England, D.W., Kidner, N.R. *et al.* (1993) Reduction in incidence of glove perforation during laparotomy wound closure by 'no touch' technique. *Annals of the Royal College of Surgeons* 75, 330–332.

De Brouwer, C. and Lecomte, A. (1994) HCV antibodies in clinical health-care workers. *Lancet* 344, 962.

Djeriri, K., Fontana, L., Laurichesse, H. *et al.* (1996) Seroprevalence des marqueurs des hepatites virales A, B et C parmi le personnel hospitalier du centre hospitalo-universitaire de Clermont-Ferrand. *Presse Medicale* 25, 145–150.

Domart, M., Abiteboul, D. and GERES (1996a) HCV occupational risk: retrospective study on Non A–Non B and C hepatitis reported as occupational diseases by occupational physicians of AP-HP from 1988 to 1992. In: *Proceedings of an Inter-*

national Colloquium on Blood-borne Diseases. Paris, June 1995. Health Services Section of the International Social Security Association, Hamburg, pp. 298–301.

Domart, M.. Hamidi, K., Antona, D. *et al.* (1996b) HCV occupational risk among healthcare workers after an accidental exposure. In: *Proceedings of an International Colloquium on Blood-borne Infections.* Paris, June 1995. Health Services Section of the International Social Security Association, Hamburg, pp. 292–295.

Fitch, K.M., Perez Alvarez, L., De Andres Medina, R. *et al.* (1995) Occupational transmission of HIV in health care workers. *European Journal of Public Health* 5, 175–186.

Gompertz, S. (1990) Needlestick injuries in medical students. *Journal of the Society of Occupational Medicine* 40, 19–20.

Heptonstall, J., Porter, K. and Gill, O.N. (1995) Occupational transmission of HIV. Summary of published reports. Unpublished Report of the Public Health Laboratory Service. London.

Hernandez, M.E., Bruguera, M., Puyuelo, T. *et al.* (1992) Risk of needle-stick injuries in the transmission of hepatitis C virus in hospital personnel. *Journal of Hepatology* 16, 56–58.

Houweling, H. and Coutinho, R.A. (1991) Risk of HIV infection among Dutch expatriates in sub-Saharan Africa. *International Journal of STD & AIDS* 2, 252–257.

Ippolito, G., Puro, V., De Carli, G. *et al.* (1993) The risk of occupational HIV infection in health care workers. *Archives of Internal Medicine* 153, 1451–1458.

Ippolito, G., De Carli, G., Puro, V. *et al.* (1994) Device-specific risk of needlestick injury in Italian health care workers. *Journal of the American Medical Association* 272, 607–610.

Ippolito, G. and the Italian Study Group on Occupational Risk of HIV Infection [SIROH] (1996) Scalpel injury and HIV infection in a surgeon. *Lancet* 347, 1042.

Iten, A., Maziero, A., Jost, J. *et al.* (1995) Surveillance des expositions professionnelles a du sang ou des liquides biologiques: la situation en Suisse au 31 decembre 1994. *Bulletin Office Federal de la Santo Publique* 24, 4–7.

Iwatsubo, Y., Azoulay, S., Bonnet, N. *et al.* (1996). *Accidents Exposant au Sang chez le Personnel de IAP-HP: Analyse des Accidents Survenus en 1994 et Bilan de 5 Ans de Suivi.* Rapport, Assisance Publique, Hôpital de Paris, 42 pp.

Jagger, J., Hunt, E.H., Brand-Einaggar, J. *et al.* (1988) Rates of needle-stick injury caused by various devices in a university hospital. *New England Journal of Medicine* 319, 284–288.

Jarke, J. (1996) Berufsbedingte HIV-infektion bei medizinischem personnal. 19 faligeschichten aus deutschland. *Robert Koch Institute Info* 1, 12–17.

Jochen, A.B.B. (1992) Occupationally acquired hepatitis C virus infection. *Lancet* 339, 304.

Johanet, H., Antona, D., Bouvet, E. *et al.* (1995) Risques d'exposition accidentelle au sang au bloc operatoire. Resultats d'une étude prospective multicentrique. *Annales de Chirurgie*, 49, 5, 403–410.

Kristensen, M.S., Wernberg, N.M. and Anker-Moller, E. (1992) Health care workers' risk of contact with body fluids in a hospital: the effect of complying with the universal precautions. *Infection Control and Hospital Epidemiology* 13, 719–724.

Lot, F. and Abiteboul, D. (1994) Infections professionnelles par le VIH en France. Le point au 31 decembre 1993. *Bulletin Epidemiologique Hebdomadaire* 25, 111–113.

Luthi, J.C. and Dubois-Arber, F. (1995) Evaluation de la strategie de prevention du sida en Suisse, 1993–1995. *Personnel hospitalier: Étude Suisse sur les Expositions au VIH et aux Hepatites chez le Personnel Hospitalier.* IUMSP, Lausanne, 67 pp.

McNicholas, T.A., Jones, D.J. and Sibley, N.A. (1989) AIDS: the contamination risk in urological surgery. *British Journal of Urology* 63, 565–568.

Marranconi, F., Mecenero, V., Pellizer, G.P. *et al.* (1992) HCV infection after accidental needlestick injury in health-care workers. *Infection* 20, 111.

Nelsing, S., Nielsen, T.L. and Nielsen, J.-O. (1993) Occupational blood exposure among health care workers: 1. Frequency and reporting. *Scandinavian Journal of Infectious Diseases* 25, 193–198.

Nelsing, S., Nielsen, T.L. and Nielsen, J.O. (1997a) Percutaneous blood exposure among Danish doctors: exposure mechanisms and strategies for prevention. *European Journal of Epidemiology* 13, 387–393.

Nelsing, S., Nielsen, T.L., Bronnum-Hansen, H. *et al.* (1997b) Incidence and risk factors of occupational blood exposure – A nation-wide survey among Danish doctors. *European Journal of Epidemiology* 13, 1–8.

Pelletier, A., Neuville, K., Iwatsubo, Y. *et al.* (1996) Accidental exposures to blood and other biological fluids in the laboratories of AP-HP in 1993. In: *Proceedings of an International Colloquium on Blood-borne Diseases.* Paris, June 1995. Health Services Section of the International Social Security Association. Hamburg, pp. 374–376.

Perez-Trallero, E., Cilia, G. and Saenz, J.R. (1994) Occupational transmission of HCV. *Lancet* 344, 548.

Petrosillo, N., Puro, V., Ippolito, G. *et al.* (1994) Prevalence of hepatitis C antibodies in hospital workers. *Lancet* 344, 339–340.

Petrosillo, N., Puro, V., Jagger, J. *et al.* (1995) The risks of occupational exposure and infection by human immunodeficiency virus, hepatitis B virus and hepatitis C virus in dialysis settings. *American Journal of Infection Control* 23, 278–285.

Poignet, J.L., Litchinko, M.B., Huo, J.F. (1995) Infection par le VHC et le VIH en hemodialyse: facteurs de risques, infections protessionnelles en Ile-de-France au 1er mar, 1995, prevention. *Bulletin Epidemiologique Hebdomadaire* 37, 166–167.

Porteous, M.J. Le, F. (1990a) Hazards of blood splashes. *British Medical Journal* 300, 466.

Porteous, M.J. Le, F. (1990b) Operating practices of and precautions taken by orthopaedic surgeons to avoid infection with HIV and hepatitis B virus during surgery. *British Medical Journal* 301, 167–169.

Puro, V., Petrosillo, N., Ippolito, G. *et al.* (1995) Occupational hepatitis C virus infection in Italian health care workers. *American Journal of Public Health* 85, 1272–1275.

Puro, V., Petrosillo, N., Ippolito, G. *et al.* (1996) Update on occupational HCV infection incidence studies: literature review. In: *Proceedings of an International Colloquium on Blood-borne Diseases,* Paris, June 1995. Health Services Section of the International Social Security Association. Hamburg, pp. 296–297.

Saghafi, L., Raselli, P., Francillon, C. *et al.* (1992) Exposure to blood during various procedures: results of two surveys before and after the implementation of universal precautions. *American Journal of Infection Control* 20, 53–57.

Sartori, M., La Terra, G., Agliettas, M. *et al.* (1993) Transmission of hepatitis C via blood splash into conjunctiva. *Scandinavian Journal of Infectious Disease* 25, 270–271.

Short, L.J. and Bell, D.M. (1993) Risk of occupational infection with blood-borne pathogens in operating and delivery room settings. *American Journal of Infection Control* 21, 343–350.

Stellini, R., Calzini, A.S., Gussago, A. *et al.* (1993) Low prevalence of anti-HCV antibodies in hospital workers. *European Journal of Epidemiology* 9, 674–675.

Szpalski, M., Grumbers, A.F. and Cukier, D. (1992) Contamination professionnelle par HIV et par le virus de l'Hepatite B. Attitude des chirurgiens et anesthesistes belges. Une enquíte d'opinion. *Acta Chirurgica Belgica* 92, 159–163.

Tarantola, A., Casalino, E., Gadjos, V. *et al.* (1996) Frequency and perceived risk of blood exposure among medical students: results of a survey. In: *Proceedings of XIth International Conference on AIDS.* Vancouver, June 1996.

Truchaud, A., Schnipelsky, P., Pardue, H.L. *et al.* (1996) Integrating biological safety into the design of analytical systems in the clinical laboratory. In: *Proceedings of an International Colloquium on Blood-borne Diseases,* Paris, June 1995. Health Services Section of the International Social Security Association, Hamburg, pp. 129–133.

Vaglia, A., Nicolin, R., Puro, V. *et al.* (1990) Needlestick hepatitis C virus seroconversion in a surgeon. *Lancet* 336, 1315–1316.

Van Damme, P., Tormans, G., Van Doorslaer, E. *et al.* (1995) A European risk model for hepatitis B among health care workers. *European Journal of Public Health* 5, 245–252.

Williams, S., Gooch, C. and Cockcroft, A. (1996) A study of blood exposure incidents in surgical practice: their incidence and causation. In: *Proceedings of an International Colloquium on Blood-borne Diseases,* Paris, June 1995. Health Services Section of the International Social Security Association, Hamburg, pp. 354–356.

Zuckerman, J., Ciewley, G., Griffiths, P. *et al.* (1994) Prevalence of hepatitis C antibodies in clinical health care workers. *Lancet* 343, 1618–1620.

CHAPTER 5
Percutaneous blood exposure data: 58 hospitals in the USA

J. Jagger and M. Bentley

INTRODUCTION

The United States has the largest number of recognized cases of occupationally-acquired human immunodeficiency virus (HIV) infections in the world. Although no surveillance system in the US or elsewhere identifies all, or even most, occupationally-infected health-care workers, the number of recognized cases alone is cause for serious concern. A review of international literature up until June 1993 identified 176 documented or probable cases of occupationally-acquired HIV infection worldwide (Ippolito et al., 1993a,b); two-thirds of those were in the United States. One use that can be made of these unfortunately large and growing numbers is to identify transmission patterns, which can lead in turn to focused and effective prevention measures.

The US Center for Disease Control and Prevention (CDC) identified 163 documented or possible cases of occupationally-acquired HIV in the United States until December 1996 (CDC, 1996). Two professional categories, nurses and clinical laboratory workers – primarily phlebotomists – ranked first among HIV-infected health-care workers, accounting for 24% each of the 123 reported cases. Physicians ranked next, accounting for 12% of cases. The high number of cases among nurses is consistent with the fact that they comprise the largest group of health-care workers in the US, numbering 2.2 million (source: American Nurses Association). Similarly, the number of HIV-infected physicians is consistent with the smaller number of physicians – approximately 670,000 in the US workforce (source: American Medical Association). The number of infected phlebotomists, however, is disproportionate to their numbers in the workforce: less than 100,000 phlebotomists are employed in the US according to the National Phlebotomy Association and the American Society of Phlebotomy Technicians, less than 1/20th the number of nurses.

A notable difference between nurses and phlebotomists is that the latter consistently perform blood-drawing procedures, while nurses perform a wider variety of procedures. When a phlebotomist sustains a needlestick injury the device causing injury is most likely to be a blood-filled needle. In contrast, a nurse may be injured by a needle used for an intramuscular injection or

Occupational Blood-borne Infections (eds C.H. Collins and D.A. Kennedy)

intravenous infusion, which would not be blood-filled, and less often by a needle used for blood drawing.

Needles used for different purposes appear to carry different risks for transmitting HIV (Berry, 1993). The same may be true for other blood-borne pathogens, such as hepatitis B and hepatitis C, but there are insufficient surveillance data to confirm this. In a report by the CDC on the exposure circumstances of workers with occupationally-acquired HIV infection, those who were exposed by needlestick had all been stuck by hollow bore, blood-filled needles. Furthermore, phlebotomy was the procedure most frequently associated with HIV exposures (Metler *et al.*, 1992). Similar conclusions have been drawn from data reported in other countries (Ippolito *et al.* 1993a,b).

This evidence suggests that blood drawing presents a high risk of exposure to blood-borne pathogens and that the risk profiles of different professional groups are in part linked to the frequency with which they perform blood drawing procedures. These are compelling reasons for focusing prevention efforts on the devices, procedures, and professional groups involved in blood drawing.

The data presented here describe the patterns of percutaneous injuries associated with blood-drawing procedures in a network of hospitals in the US.

METHODS

Fifty-eight hospitals which voluntarily participated in three data-sharing networks contributed one year of data for this report. All the hospitals use the standardized Exposure Prevention Information Network (EPINet) system for tracking percutaneous injuries in their institutions (Jagger *et al.*, 1994). Network A includes nine hospitals located in six states in the eastern half of the US; they report their data to the University of Virginia. Network B consists of 50 hospitals in South Carolina that report their data to the Palmetto Hospital Trust Needlestick Prevention Demonstration Project in Columbia, South Carolina. Network C includes 11 Sisters of Providence hospitals in the Pacific Northwest that report their data to Johnson & Higgins of Washington, Inc., in Seattle. The hospitals represent a cross-section of institutions in diverse geographic locations; of the hospitals included in this report, 26 had an average daily census of less than 100 beds, 16 had from 100 to 299 occupied beds, and 16 had 300 or more occupied beds. Seventeen were teaching hospitals.

Data included all percutaneous injuries reported by health-care workers to the employee health departments or similar designated authorities in their institutions. Data collection began in September 1992. Each participating facility provided one year of data on disk to investigators; of the 70 hospitals in the three networks, 58 with complete data at the time of this report were included.

RESULTS

There was an overall total of 3,829 percutaneous injuries in the merged database, and a cumulative total of 11,978 occupied beds in the 58 hospitals. Needlestick incidents were selected in which the device associated with the injury was: (i) a syringe used for drawing venous or arterial blood; (ii) a winged steel needle (butterfly) used for blood drawing; or (iii) a vacuum tube phlebotomy needle. Four hundred and seventy-one cases met these criteria – 12.3% of all injuries from the 58 hospitals. Table 5.1 shows the job categories of workers reporting needlestick injuries from blood drawing needles. Nurses and phlebotomists together accounted for two-thirds of cases. The remaining cases were reported primarily by respiratory therapists (a job category that does not exist in many countries outside the US), who often draw blood for arterial blood gas analysis, and by clinical laboratory technicians and physicians (mainly medical residents).

Table 5.2 shows the location of these incidents. Most injuries occurred in patient rooms; the next most frequent locations were emergency departments and intensive or critical care units. Less frequent locations were out-patient clinics, operating rooms, and clinical laboratories.

Figures 5.1–5.4 compare the mechanisms of needlestick injuries for four major needle devices used for blood drawing. This breakdown of injuries shows a profile of when the injuries occurred during the use/disposal cycle; each device has a unique profile, reflecting the design characteristics of the device and the handling requirements for performing specific procedures. For instance, recapping injuries are more frequently associated with syringes used for drawing arterial blood (Fig. 5.1) than for similar syringes used for drawing venous blood (Fig. 5.2). This may be due to the requirement for removing needles from arterial blood gas syringes before delivering the filled syringes to the laboratory. On the other hand, injuries that occur when withdrawing a needle from the stopper of a tube are more frequent with syringes used for drawing venous

Table 5.1. Job classification of health workers reporting needlestick injuries associated with blood drawing (58 hospitals, one year).

Job	Number	Percentage
Nurse	157	33.3
Phlebotomist	150	31.8
Respiratory therapist	43	9.1
Clinical laboratory worker	40	8.5
Physician*	39	8.3
Attendant (non-surgical)	14	3.0
Other	28	5.9
Total	471	100

*30/39 reported needlestick injuries were to residents.

Table 5.2. Place of occurrence of needlestick injuries associated with blood drawing (58 hospitals, one year).

Job	Number	Percentage
Patient room	251	53.3
Emergency department	66	14.0
Intensive/critical care unit	39	8.3
Out-patient department	24	5.1
Clinical laboratory	21	4.5
Operating theatre	13	2.8
Venepuncture	12	2.5
Treatment room	9	1.9
Other	36	7.6
Total	471	100

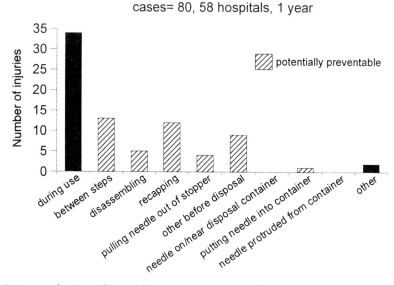

Fig. 5.1. Mechanism of injury from syringes used for drawing arterial blood. (Courtesy of the International Health Care Worker Safety Center, University of Virginia.)

blood because the blood is often injected into tubes before delivery to the laboratory.

The risks associated with winged steel needles (butterflies) are largely related to problems in transporting the devices safely to the disposal containers and difficulties in pushing the needles through the openings of containers (Fig. 5.3). The coiled tubing attached to the needles makes the devices awkward to handle.

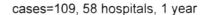

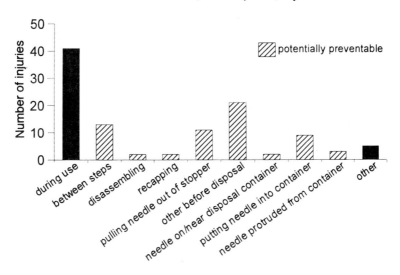

Fig. 5.2.   Mechanism of injury from syringes used for drawing venous blood. (Courtesy of the International Health Care Worker Safety Center, University of Virginia.)

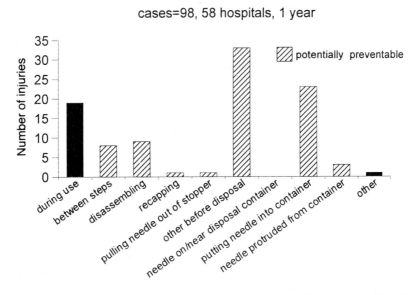

Fig. 5.3.   Mechanism of injury from winged steel needles used for drawing blood.

cases=184, 58 hospitals, 1 year

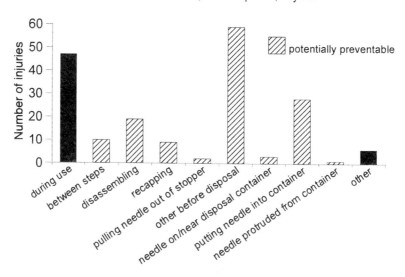

Fig. 5.4. Mechanism of injury from vacuum tube phlebotomy needles.

There were 184 injuries from vacuum tube phlebotomy needles (Fig. 5.4). The circumstances of the injuries were similar to those for winged steel needles: that is, they were caused by problems in transporting needles to disposal containers and in introducing needles into the containers.

Hollow-bore needles were not the only devices associated with injuries incurred during blood drawing procedures. Lancets used for fingersticks and heelsticks, glass capillary tubes, and glass vacuum tubes containing blood samples also caused injuries. Most numerous among these were 108 injuries caused by fingerstick and heelstick lancets that lacked an automatic retracting safety feature. Fifty-four per cent of the injuries occurred when disengaging a disposable lancet from a reusable holder or when handling the used lancet prior to disposal. These injuries reflect the difficulty of handling such small devices which, if no protective shield is provided, require the fingers to remain close to the exposed point as the device is disassembled and discarded.

Eighteen injuries were caused by glass capillary tubes used to contain small-volume blood samples following fingerstick or heelsticks. Injuries occurred during all phases of handling. Force must be applied when pushing the tubes into putty to close off one end, and again the tubes are subject to significant force when they are centrifuged for haematocrit determination. Under such stresses the fragile glass easily fractures. The devices are blood-filled and have the potential to produce sizable lacerations with significant inoculation of blood.

Nine injuries were caused by broken glass vacuum tubes. Although there were relatively few such injuries when compared with the 184 from phlebotomy needles used to draw blood into the tubes, vacuum tube injuries are particularly serious because of the large amounts of blood that can be introduced into the lacerations. Breakage often occurred at the top of the tubes, when the tight-fitting stoppers were removed either to introduce or withdraw blood samples.

PREVENTABLE INJURIES

The EPINet system is designed to identify the proportion of injuries that are potentially preventable through improvements in the design of sharp medical devices. Injuries are considered preventable if an alternative already exists that can eliminate the sharp device or unsafe feature. For instance, all injuries from needles used to inject fluid into or withdraw fluid from intravenous access ports are preventable, because needleless equipment can be used. All injuries caused by the breakage of glass devices for which non-breakable alternatives are available can also be prevented.

Some, but not all, needles can be eliminated. Many devices have needles that are used to pierce skin or tissue; these are necessary needles. Needlestick injuries that occur after use or between uses of a necessary needle, however, are potentially preventable, when a safety feature that shields the hand from the needle can be put into place. The percentage of needlestick injuries that occur during the performance of the procedure, when the needle must be exposed for use, is not included in the preventable fraction for that device.

The concept of the 'preventable fraction' is intended to project a target that may be achieved by the implementation of feasible measures. It is not intended to imply that there is a limit to the potential reductions that can be achieved. It is possible that some safety devices on the market may already exceed the estimated preventable fraction in a given device category. It is also possible that in the future additional prevention strategies, such as procedure changes or new technology, may increase the preventable fraction or even eliminate injuries in a specific device category altogether.

Figures 5.1–5.4 highlight the preventable fraction of needlestick injuries for each device. They were: (i) 58% for syringes used for drawing venous blood; (ii) 55% for syringes used for drawing arterial blood; (iii) 80% for winged steel needles (butterflies); and (iv) 71% for vacuum tube phlebotomy needles. Across all four devices, 67% (316/471) of injuries fell into the potentially preventable fraction.

Seventy-four per cent of injuries from fingerstick or heelstick lancets fell into the preventable fraction. Nearly all injuries, however, caused by such lancets are preventable, because there are safety lancets presently on the market with a built-in spring action that automatically withdraws the sharp point into a shielded position immediately after being discharged. The sharp lancet retracts so quickly that it is highly improbable that a health-care worker could sustain a puncture injury after performing a fingerstick or heelstick procedure. The percentage of preventable injuries, therefore, is likely to be significantly greater than 74%. The safest retracting lancets are those that do not permit a contaminated lancet to be inadvertently discharged a second time.

An additional advantage of the safety lancets is that they reduce the potential for patient-to-patient cross-contamination, because when the used lancet is disposed of, all parts of the device that have come into contact with the patient are automatically discarded with the lancet. The problem of cross contamination from spring-loaded lancets was linked to an outbreak of hepatitis B and was subsequently the subject of a 1990 FDA Safety Alert (Food and Drugs Administration, 1990; Douvin et al., 1990).

All injuries (100%) from glass capillary tubes used for haematocrit determination are preventable, because it is not necessary to use these devices – unbreakable plastic capillary tubes are available, as well as a system for haematocrit determination that does not require the use of capillary tubes at all. These injuries could be eliminated tomorrow with the appropriate selection of products.

It is difficult to estimate the proportion of preventable injuries from broken vacuum tubes. Plastic vacuum tubes are available as an alternative to glass; however, vacuum tubes are used for many different applications and the proportion of glass tubes that could be eliminated by the plastic ones is not certain. There is also a redesigned vacuum tube stopper for glass vacuum tubes that is intended to reduce stresses on the top edge of the tube and thereby lower the risk of breakage if the stopper is removed.

As a final point, the handling requirements of blood drawing equipment must be taken into consideration when implementing prevention programmes or evaluating safer devices. After withdrawing a phlebotomy needle from a patient, the health-care worker must apply pressure to stem bleeding at the puncture site. This leaves only one hand free to handle the exposed needle. The health-care worker must either dispose of the needle with one hand or leave it on a nearby surface until he or she is free to move away from the patient. This situation points to the need for an appropriate disposal container within arm's reach of the patient. It also emphasizes the advantage of having a safety blood drawing device that requires only one hand for activation. If two hands are needed to activate a safety feature after the needle has been removed from the patient, the health-care worker is in the same situation as with a conventional device, and cannot activate the needle protection feature until both hands are free.

Figure 5.5 illustrates six hazardous practices that should be avoided when blood is drawn.

We have compiled a selection of devices that are currently on the market and which are designed to prevent percutaneous injuries associated with blood-drawing procedures. These devices are illustrated in Fig. 5.6. Inclusion in this selection does not imply an endorsement of any specific product. Nor is it inclusive (see also Bouvet, Chapter 7 this volume). Safety devices are likely to vary in their effectiveness in preventing injuries. Institutions that purchase them must evaluate them thoroughly for: (i) health-care worker safety; (ii) patient safety; and (iii) their impact on the reliability of laboratory tests.

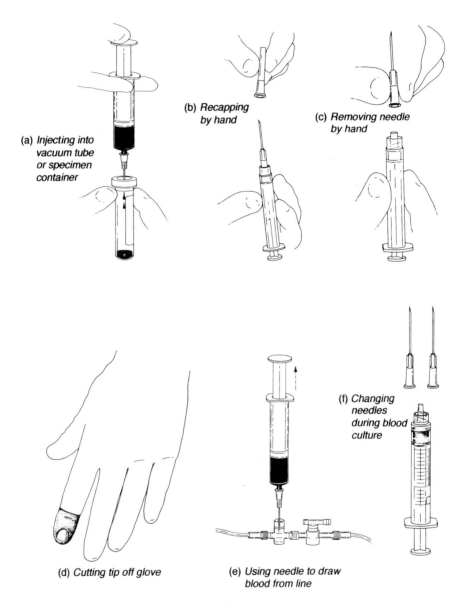

(a) *Injecting into vacuum tube or specimen container*

(b) *Recapping by hand*

(c) *Removing needle by hand*

(d) *Cutting tip off glove*

(e) *Using needle to draw blood from line*

(f) *Changing needles during blood culture*

Fig. 5.5. Six hazardous practices to avoid.

(a)

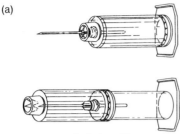

protected position

VACUTAINER® Brand Safety-Lok™
■ Becton Dickinson
*Single use vacuum tube/needle holder
with protective sliding sleeve that pushes
forward after use and locks in place*

(b)

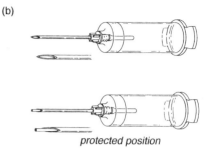

protected position

Punctur-Guard™ ■ Bio-Plexus
*After final tube of blood is drawn, blunt
internal needle is activated by forward
pressure of vacuum tube. Needle point is
blunted before it is removed from patient*

(c)

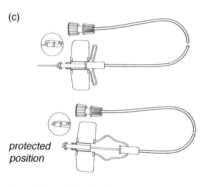

*protected
position*

Angel Wing™ Safety Needle
■ Sherwood Medical
*Stainless steel barrier tip is advanced forward
to end of needle, locking over point as needle
is withdrawn from patient; one-handed activation*

(d)

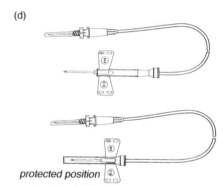

protected position

VACUTAINER® Brand Safety-Loc™
Winged Needle ■ Becton Dickinson
*After removal from patient, safety shield is
advanced forward and locks in place
beyond needle tip*

(e)

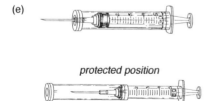

protected position

B-D Safety-Lock™ ■ Becton Dickinson
*Needle guard has protective sliding sleeve
that pushes forward after use and locks in
place. Note: the 10cc syringe with the shield
locked in place can accept a 3cc to 10cc
vacuum tube, allowing injection of blood
into tube with shielded needle*

(f)

In use

After use

Glucolet 2™ Retracting Lancet ■ Miles,
Inc./Diagnostic Division
*Disposable lancet is fitted to reusable
spring-loaded holder. When activated,
lancet instantly protracts and retracts;
retracted lancet is removed from holder
for disposal*

Fig. 5.6. Safety products related to blood drawing.

(g)

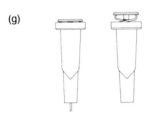

In use   After use

MICROTAINER™ Brand Safety Flow
Lancet ■ Becton Dickinson
*Self-contained lancet is manually activated
and automatically retracts when activating
lever is released*

(h)

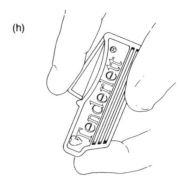

Tenderlett® Automated Skin Incision Device
■ International Technidyne Corporation
*When device is triggered, surgical steel blade
swiftly protracts and then automatically
retracts. Design precludes inadvertent reuse*

(i)

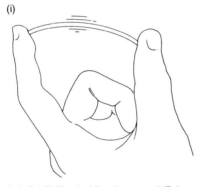

SafeCrit™ Plastic Microhematocrit Tube
■ Norfolk Scientific
*Capillary tube made of plastic avoids hazard
of glass breakage*

(j)

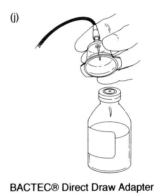

BACTEC® Direct Draw Adapter
■ Becton Dickinson
*Designed for blood culture procedures.
Vacuum vial and covered needle safety
adapter allow blood to be drawn directly
into culture medium, avoiding need to
inject into specimen container*

(k)

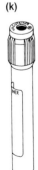

HEMOGARD VACUTAINER® Brand
Vacuum Tube Stopper ■ Becton Dickinson
*Rigid stopper grips outside of tube; intended
to reduce risk of tube breakage and blood
splash when removing stoppers*

Fig. 5.6.   *Continued.*

(l)

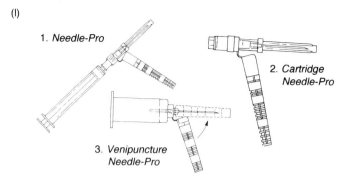

1. *Needle-Pro*

2. *Cartridge Needle-Pro*

3. *Venipuncture Needle-Pro*

Needle-Pro™ Needle Protection Devices ■SIMS: Smiths Industries Medical Systems
Hinged sheath engages over needle; used needle is pressed into Needle-Pro device using one hand. Comes in 3 configurations. (1) Needle-Pro: Basic needle protection device; can be used for arterial blood drawing. (2) Cartridge Needle-Pro: Combines hypodermic needle cartridge with Needle-Pro sheath. (3) Venipuncture Needle-Pro: Disposable blood collection tube holder and integral needle protection device

(m)

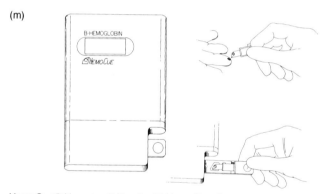

HemoCue® Hematocrit Reader ■ HemoCue, Inc.
Uses flat plastic cuvette to contain blood sample for hematocrit determination. Cuvette is inserted directly into reader, avoiding need for centrifugation

Fig. 5.6. *Continued.*

ACKNOWLEDGEMENTS

International Health Care Worker Safety Center, University of Virginia

Network Coordinators: Janine Jagger, Beth Blackwell, Melanie Balon Florida Hospital (Orlando, FL), Dianne Ross, Carol Griffin; Martha Jefferson Hospital (Charlottesville, VA), Pam Jones, Edwina Juillet; North Broward Hospital (Ft. Lauderdale, FL), Marc Gomez, Janet Narushko; Saint Joseph Hospital (Omaha, NE), Ann Lorenzen; Saint Vincent Hospital (Indianapolis, IN), Dianne Spiller; Saint Vincent Hospital (Erie, PA), Mary Jo Dolecki, Diane Dougan; Shands Hospital (Gainesville, FL), Deborah Boeff, Suzanne Hench;

University Hospitals of Cleveland (Cleveland, OH), Pamela Parker; University of Virginia Hospitals (Charlottesville, VA), Betty Joe Coyner, Vickie Pugh

Sisters of Providence, Johnson & Higgins of Washington, Inc., Seattle

Network Coordinators: Eileen Bradshaw, Janet Swapp
Providence General Hospital and Medical Center (Everett, WA), Catherine Murray; Providence Hospital (Anchorae, AK), Veronica Allarmas; Providence Hospital (Centralia, WA), Joan McKenzie, RN, MN; Providence Hospital (Medford, WA), Ruth DeVee; Providence Hospital (Toppenish, WA), Leslie Simmons, RN; Providence Medical Center (Portland, OR), Joanne Henkel, RN; Providence Medical Center (Seattle, WA), Catherine Elliott-Rostykus, RN, Betsy Blessing-Hubbard, RN, Barb Schubert; Providence Milwaukie Hospital (Milwaukie, OR), Debbie Marshall, RN; Providence Seaside Hospital (Seaside, OR), Leonard Naidoff, RN; St. Elizabeth Medical Center (Yakima, WA), Judy Hanratty, RN; St. Joseph Medical Center (Burbank, CA), Cathy McDonald; St. Peter Hospital (Olympia, WA), Bev Masini, RN, Barbara Soule, RN, CIC; St. Vincent Hospital and Medical Center (Porland, OR), Becky Fuller, RN, COHN

Palmetto Hospital Trust Needlestick Prevention Demonstration Project, South Carolina

Network Coordinators: Marshall Fowler, Susan Austin, Kim Carter
Abbeville County Hospital (Abbeville, SC), Nancy Brock, RN; Allen Bennet Memorial Hospital (Greer, SC), LouAnne Weber, RN-CSMSN; Anderson Memorial Hospital (Anderson, SC), Rhonda Chalfant; Baptist Medical Center Columbia (Columbia, SC), Gwen Floyd; Baptist Medical Center Easley (Easley, SC), Lois McCready, RN; Barnwell County Hospital (Barnwell, SC), Allene Townes; Beaufort Memorial Hospital (Beaufort, SC), Helena Gregg, RN; Bruce Hospital System (Florence, SC), Laurie Horton, RN; The Byerly Hospital (Hartsville, SC), Susan Nash, RN; Cannon Memorial Hospital (Pickens, SC), Linda Masters; Charleston Memorial Hospital (Charleston, SC), Robin Smith; Chester County Hospital (Chester, SC), Marian Bagley, RN; Clarendon Memorial Hospital (Manning, SC), Lynne Bowen, RN; Conway Hospital (Conway, SC), Lenora Thompson, RH; Edgefield County Hospital (Edgefield, SC), Pat Robinson, RN; Elliott White Springs Memorial Hospital (Lancaster, SC), Julie Bowers, RN; Fairfield Memorial Hospital (Winnsboro, SC), Debra Gudenas, RN; Florence General Hospital (Florence, SC), Debby Rapp, RN; Georgetown Memorial Hospital (Georgetown, SC), Emma Miller, RN; Greenville Memorial Medical Center (Greenville, SC), LouAnne Weber, RN; Greenwood Methodist Home (Greenwood, SC), Juanita J. Butler; Hillcrest Hospital (Simpsonville, SC), LouAnne Weber; Hilton Head Hospital (Hilton Head Island, SC), jane Binns, RN, CDE; Kershaw County Memorial Hospital (Camden, SC), Margaret Perry; Laurens County Hospital (Clinton, SC), Linda Casey, RN; Lexington Medical Center and Drug Abuse Program (Lexington, SC), Cathy Blanks, RN; The Loman Home (White Rock, SC), Sheila Bantz, RN; Loris Community Hospital (Loris, SC), Linda Mills, RN; Lower Florence County Hospital (Lake City, SC), Martha Lyerly, RN; Marion Memorial Hospital (Marion, SC),

Margaret Perritt, RN; Marshall I. Pickens Hospital (Greenville, SC), LouAnne Weber, RN; Mary Black Memorial Hospital (Spartansburg, SC), Eleanor Mabry; McLeod Regional Medical Center (Florence, SC), Vicky Zelenka, RN, CIC, Celia Atkinson, RN; Medical University of South Carolina (Charleston, SC), Nancy Sifford, RN; Mullins Hospital (Mullins, SC), Wendy Meares, RN; Newberry County Memorial Hospital (Newberry, SC), Suzie Arthur, RN; North Greenville Hospital (Grenvile, SC), LouAnne Weber, RN; Oconee Memorial Hospital (Seneca, SC), Rickey Herring; The Regional Medical Center of Orangeburg and Calhoun Counties (Orangeburg, SC), Paula Bailey; Richland Memorial Hospital (Columbia, SC), Doris Wadford, RN; Roger Huntington Nursing Center (Greenville, SC), LouAnne Weber, RN; Roger C. Peace Hospital (Grrenville, SC), LouAnne Weber, RN; Saint Francis Hospital, Inc. (Greenville, SC), Donna Chastain; Self Memorial Hospital (Greenwood, SC), Susan Milford, RN; Spartansburg Regional Medical Center (Spartansburg, SC), Kathie Guthrie, RN; South Carolina Department of Health and Environmental Control (Columbia, SC), Robert Ball, MD, MPH; Tuomey Regional Medical Center (Sumter, SC), Gloria Bateman, RN; Wallace Thomson Hospital (Union, SC), Susan Latham; Williamsburg County Memorial Hospital (Kingstree, SC), Nancy Floyd, RN; Women's Center of Carolina's Hospital System (Florence, SC), Debby Rapp, RN

REFERENCES

Berry, A.J. (1993) Are some types of needles more likely to transmit HIV to health care workers? *American Journal of Infection Control* 21, 216–218.

CDC (1996) *HIV/AIDS Surveillance Report.* 8, 21.

Douvin, C., Simon, D. and Zinelabidine, H. (1990) An outbreak of hepatitis B in an endocrinology unit traced to a capillary blood sampling device. *New England Journal of Medicine* 4, 57–58.

Food and Drug Administration, Center for Devices and Radiological Health (1990) *FDA Safety Alert: Hepatitis B Transmission via Spring-loaded Lancet Devices.* Washington, FDA.

Ippolito, G., Puro, V. and De Carli, G. (1993a) Infezione professionale da HIV in operatori sanitari: descrizione dei casi con sieroconversione documentata segnalati al 30 giugno 1993. *Giornale Italiano della AIDS* 4, 63–75.

Ippolito, G., Puro, V., De Carli, G. (1993b) Infezione professionale da HIV in operatori sanitari: 2. descrizione dei casi senza sieroconversione documentata segnalati al 30 giugno 1993. *Giornale Italiano della AIDS* 4, 186–193.

Jagger, J., Cohen, M. and Blackwell, B. (1994) EPINet: a tool for surveillance of blood exposures in health care settings. In: Charney, W. (ed) *Essentials of Modern Hospital Safety*, vol. 3. CRC Press, Ann Arbor, pp. 223–239.

Metler, R., Ciesielski, C., Marcus, R. and Ward, J. (1992) Exposure circumstances of workers with occupationally acquired HIV infection. Presented at Frontline Healthcare Workers: a National Conference on Prevention of Device-mediated Bloodborne Infections, Washington, DC.

CHAPTER 6
Detection of surface and air-borne blood contamination

D.A. Kennedy

It is inevitable that during work involving liquids, including work with blood, surface contamination will occur, for example, as a result of deliberate contact with the liquid, and accidental escape caused by leakage from closures, breakage of containers, spillage, splashing, spraying and droplet production. In addition, there may be air-borne contamination arising from aerosol production. These topics are reviewed below.

There have been three basic approaches in the investigation of environmental surface and airborne blood contamination. These are:

1. Methods based on the detection of haemoglobin.
2. Detection of hepatitis B surface antigen (HBsAg).
3. Detection of tracer materials that have been deliberately added to blood.

DETECTION OF HAEMOGLOBIN

Tests Used in Forensic Science: the Kastle-Mayer Test

Some tests for blood on surfaces are based on the fact that haemoglobin and a number of its derivatives have a catalytic activity which is analogous to that of the enzyme peroxidase. Catalytic tests for blood are reviewed by Gaensslen (1983). The basic principle is that haemoglobin catalyses the breakdown of a peroxide, usually hydrogen peroxide, yielding oxygen. The oxygen reacts with a colourless chromogen and a coloured product indicates a positive reaction. Solutions of benzidine, *o*-tolidine or *o*-dianisidine were used as chromogens, until fears of carcinogenicity caused a rekindling of interest in the use of reduced phenolphthalein. Traditionally, in forensic science circles this is known as the Kastle-Mayer reagent. A formula for this reagent, which is unstable and not available commercially, and a blood detection method involving its use are given by Culliford (1971).

The method is simple. A circle of filter paper is folded to obtain a quadrant. The point of the quadrant is rubbed on the surface to be tested. Then a drop of each of the reagents is added on to the point of the paper quadrant. Ethyl

Occupational Blood-borne Infections (eds C.H. Collins and D.A. Kennedy)

alcohol is added first, followed by Kastle-Mayer reagent, and finally hydrogen peroxide. If a deep pink colour develops rapidly, the test is positive.

Culliford (1971) stated that the benzidine and Kastle-Mayer tests are so sensitive that 'quantities of blood far too small to see will give a strong positive reaction'. Gaensslen (1983) reviewed various reports of the sensitivity of the Kastle-Mayer tests and it appears that in most hands it can be used to detect blood at a dilution of $1:10^6$. However, the benzidine and Kastle-Mayer tests are not specific and false-positive reactions can be obtained with various chemicals, plant peroxidases and animal tissues and secretions. Holton and Prince (1986) used the Kastle-Mayer test in studies of blood contamination in venepuncture clinics and in laboratories.

Haemoglobin Test-strips

The exact formulation of these is kept secret by manufacturers. Generally, however, they contain a chromogen, such as tetramethyl benzidine, and an organic peroxide. The formulation is usually such that a dark green or blue reaction indicates a positive test. Hemastix (Bayer Diagnostics) have been used to detect blood on environmental surfaces, e.g. in haemodialysis centres (Bond *et al.*, 1977), and on objects in schools and homes (Peterson *et al.*, 1976). Cole *et al.* (1992) evaluated four different urine testing test strips that were designed to detect blood, among other constituents, and found that they were all capable of detecting blood in urine at a concentration of 10 red blood cells μl^{-1}.

Beaumont (1987) used Hemastix to detect blood contamination of surfaces in an autopsy suite. A swab moistened with physiological saline was rubbed over the surface of interest. A saline eluate derived from the swab was then tested with the test strips. For convenience, this is called the swab-rinse method. They found that Hemastix was capable of detecting sheep blood diluted $1:10^5$ in saline and dried on a stainless steel surface.

Table 6.1 shows the sensitivity of different blood detection methods. While the sensitivity is expressed in different ways, nevertheless it can be seen that in all cases very small amounts of blood can be detected by them all. False positive reactions may be caused by oxidizing agents such as hydrogen peroxide and sodium hypochlorite (Beaumont, 1987). Myoglobin, which may be liberated

Table 6.1. Sensitivity of some blood detection tests.

Test	Sensitivity	Reference
Kastle-Mayer	Blood diluted $1:10^6$	Gaensslen (1983)
Hemastix[1]	Blood diluted $1:10^5$	Beaumont (1987)
Chemstrip 9[2]	About 5 red blood cells per high-power field	Daum *et al.* (1988)
Multistix 10SG[3]		
Combur Test[4]	$\Big\}$ 10 rbc μl^{-1}	Cole *et al.* (1992)
nephroPHAN[5]		
Combi screen[6]		

Suppliers: 1 and 3: Bayer Diagnostics; 2 and 4: Boehringer Mannheim Diagnostics; 5: Chemopol UK; 6: Cambridge Selfcare Diagnostics.

during tissues from surgery, is unlikely to give a false-positive reaction with Hemastix (Heinsohn and Jewett, 1993).

DETECTION OF HEPATITIS B SURFACE ANTIGEN (HBsAg)

Transmission of blood-borne hepatitis B virus (HBV) is a significant occupational risk for health-care workers and many studies have been done on environmental surface contamination with HBV, as indicated by the presence of hepatitis B surface antigen (HBsAg), derived from some of the blood samples that had been processed. In effect, such studies are indirect studies of blood contamination in which HBsAg is used as a tracer substance. As such, generally they would tend to underestimate the extent of blood contamination because not every blood sample that had been handled would have contained HBsAg. The best estimates of environmental surface contamination would have been obtained where blood came from a population with a high incidence of hepatitis B. For example, Piazza *et al.* (1987) studied contamination by HBsAg in dental surgeries in different parts of Naples where the prevalence of HBsAg carriers may lie between 4% and 13% of the population. In general these studies used a swab-rinse method to sample environmental surfaces coupled with Ausria (Abbott Laboratories) radioimmunoassay for HBsAg (Bond *et al.*, 1977). This assay has been claimed to be able to detect as little as 1 ng of HBsAg. Modern commercial assays for HBsAg have a high level of specificity and are very sensitive, with most being able to detect HBsAg at a concentration of 0.125 IU ml^{-1} (Palmer *et al.*, 1995).

TRACER METHODS

Three types of tracer materials have been used to study contamination in workplaces. These are radioactive materials, of which Technetium 99^{m} has been most commonly used; suspensions of microorganisms; and fluorescent agents such as sodium fluorescein (Lewis and Wardle, 1978) and Uvitex NFW (Ciba Geigy, 1977), a liquid optical brightener (Kennedy *et al.*, 1988). Very small quantities of sodium fluorescein can be detected using photographic techniques (Taylor, 1990). At a dilution of $1:10^{6}$ in non-fluorescent distilled water, Uvitex NFW is still visibly fluorescent when viewed under a chromatography lamp.

Microorganisms

It is important that organisms used in tracer studies are sufficiently robust to survive workplace manipulations and environmental conditions. Suspensions of *Serratia marcescens* were used until pathogenicity of the organisms was demonstrated. Holton and Prince (1986) used a suspension of *Micrococcus luteus* to study contamination associated with venepuncture. The spores of *Bacillus subtilis*, var. *globigii*, NCTC 10073 (BG) are very robust and have been widely used in aerobiological and surface contamination studies. They are generally considered to be harmless (Darlow *et al.*, 1969). Colonies of BG produce a characteristic orange pigment that simplifies identification in the presence of background flora. T3 bacteriophage, a particle with a size of the same order as that of the hepatitis B virus, has been used as a tracer in studies by the Medical Research Council workers of blood contamination in a haemodialysis unit (MRC, 1975).

The Double-tracer Method

While haemoglobin and HBsAg detection methods are sensitive, a significant drawback is that blood contamination which cannot be detected by the naked eye will be identified only if blood is present on the surface which is swabbed. Generally, it is not practicable, or may be impossible, to swab the whole of the environmental surfaces that may possibly have been contaminated with blood. As a consequence, unless the blood contamination is extensive, it may be missed simply because it was not picked up from the areas that were swabbed.

In an attempt to circumvent this problem, and to give a more reliable estimate of the extent of blood contamination undetectable by the naked eye, the double-tracer method was devised. The method was used originally to study the extent of contamination with serum samples of a clinical biochemistry laboratory (Kennedy *et al.*, 1988). Horse serum was used as the base material, to which was added BG spores and Uvitex NFW at concentrations of 10^8 cfu ml^{-1} and 0.1% respectively. Uvitex NFW is a liquid distyryl biphenyl-derivative optical brightener of very low toxicity (Ciba-Geigy, 1977). It is an intensely fluorescent agent which, when excited at 350 nm, emits light at 438 nm. This subvisible emission was visualized and photographed, at a long time-exposure using an ultraviolet light source, e.g. a chromatography lamp. The incorporation of 0.1% v/v Univex NFW into nutrient agar or horse serum did not inhibit colony formation from BG spores (Kennedy, 1988). It was reasoned that the use of the fluorescent tracer would enable large areas to be screened for evidence of contamination. However, since previous experience had demonstrated that there was often background fluorescence on the surfaces of clinical laboratories, e.g. as a result of the commonplace spillage of reagents, it could not be certain that a fluorescent area was the result of contamination with horse serum containing the double tracers. It was therefore necessary to incorporate a confirmatory test into the procedure. Recovery of BG spores, as evidenced by the growth of colonies with the characteristic pigment, would provide the confirmatory test.

Horse serum containing the tracers was carefully pipetted into a number of screw-capped tubes of the type used in the laboratory for postal specimens. After ensuring that caps of the tube were tight and that there was no external fluorescence, indicating contamination, the tubes were labelled appropriately, and corresponding request forms were prepared to ensure that the samples were analysed by a wide range of laboratory instrumentation. The samples containing tracers and their request forms were placed in a tray in the laboratory's sample reception area to ensure that they would be handled as any other samples.

Box 6.1. Surface blood contamination in clinical laboratories.

Lauer *et al.* (1979)
Outer surface of sample container
Gloves and bare hands
Ballpoint pens and felt markers
Reagent container
Surface of cell counter
Blood film spreader

Holton and Prince (1986)
Outside of specimen tubes
Gloves worn for venepuncture
Needle holder used for venepuncture
Working surfaces

Kennedy *et al.* (1988)
Outside of specimen tubes
Disposable gloves
Bench surface
Floor
Analytical equipment
Inside of discard bucket
Computer keyboard

Later in the day, after work had stopped, a UV lamp was used to examine the areas where the double-tracer sera had been handled. Wherever fluorescence was detected, the area was swabbed using a bacteriological swab. A non-fluorescent area was swabbed as a control. Swabs were plated-out on nutrient agar. The plates were incubated aerobically at 37°C for 16 hours, then kept for 10 hours at ambient temperature to develop the BG orange pigment before examination. The double-tracer serum gave a good yield of BG colonies and the control sample yielded no BG colonies. Other samples yielded BG colonies, thus indicating contamination caused by the handling and processing of the double-tracer sera. The third section of Box 6.1 outlines the results.

Some preliminary work demonstrated that 0.1% sodium fluorescein when incorporated into nutrient agar did not significantly inhibit BG colony formation, although growth was retarded (Kennedy, 1988). Sodium fluorescein together with BG spores are considered to be a candidate double-tracer combination, but more developmental work is needed.

STUDIES ON BLOOD CONTAMINATION OF SURFACES

A number of studies have indicated the ubiquity of surface contamination in workplaces where blood is handled or is shed. The results of some of these are outlined below.

Clinical Laboratories
The results of three studies are outlined in Box 6.1.

Dental Surgeries
The results of a study carried out by Piazza *et al.* (1987) are outlined in Box 6.2.

Autopsy Suite
The results of a study by Beaumont (1987) are outlined in Box 6.3.

Haemodialysis Units
Box 6.4 outlines the findings of a study by Bond *et al.* (1977).

Box 6.2. Surface blood contamination in dental surgeries.

- Work benches
- Dental units for instruments
- Dental chair headrest
- Dental chair arms
- Miscellaneous instruments (after use, before disinfection)

Box 6.3. Surface blood contamination in an autopsy suite.

- Cold and hot water faucet handles
- Oscillating bone saw handles
- Disinfectant bottle
- Refrigerator door handle
- Formalin dispensing valve
- Desk top
- Cabinet knobs
- Telephone receiver
- Drawer handles
- Door knob

Box 6.4. Surface blood contamination in haemodialysis units.

- Interior of connecting tube on positive pressure monitor
- IV drip pole on dialysis machine
- Control knobs on dialysis machine
- Bed rail
- Wall in hall-way
- Walls of dialysis sites
- Tops of centrifuge buckets used for haematocrit determination
- Erythrocyte sedimentation rate tube racks
- Handgrip of telephone

Skin Contamination by Blood Splashing

Visible blood splashing was found to be a common event during routine haemodialysis (MRC, 1975). Thus there would be an opportunity for direct skin contamination by impaction and indirect contamination as a consequence of skin contact with blood on surfaces.

Bull *et al.* (1991) demonstrated that during autopsies there was a risk of blood splashes contaminating skin around the eyes of pathologists. The study was performed by counting with the aid of a hand lens blood splashes on the internal and external surfaces of safety spectacles. Most blood was seen on the front external surfaces but in some cases the sides and interior surfaces were contaminated with blood.

A number of studies, e.g. those by Brearley and Buist (1989), Porteous (1990), and Gerberding *et al.* (1990), have shown that surgeons are similarly at risk of blood splashes. Taylor (1990), used sodium fluorescein as a tracer and showed that the face of a surgeon was at risk of contamination by blood during transurethral resection. While there was a random distribution of splashes, there was a recurring pattern of the orbit of the eye that looks through the endoscope and the tip of the nose. These studies indicate the benefits afforded by wearing safety spectacles with side-shields, or better still a full-face visor, wherever there is a risk of blood splashing.

AIR-BORNE BLOOD CONTAMINATION: INTRODUCTION

This title means contamination of air with an aerosol derived from blood. Because there has been some confusion in the past over the use of the word aerosol, it is necessary to offer some explanation.

Aerosols are composed of fine particles with aerodynamic diameter in the region of 10 μm or less which are suspended in a gas. They have a very low mass and consequently settle very slowly in still air. They have almost no motion independent of the gas in which they are suspended and consequently can easily diffuse or move with local air currents some distance away from the site of production. On the other hand, droplets (which are sometimes erroneously called aerosols) are much larger particles that have a definite, sometimes visible, trajectory away from the site of production. This means that, after being transiently air-borne, droplets come to rest close to the site of production giving rise to surface contamination. Blood 'splatter' droplets are reported to be approximately 50 μm in size (Heinsohn and Jewett, 1993).

If aerosol particles are inhaled they can lodge in different parts of the respiratory system, depending upon the aerodynamic diameter (see Collins, 1993). Particles less than 5 μm in diameter pose the biggest risk because they can become lodged in the vulnerable alveolar regions of the lungs where gas exchange takes place. If a suspension of microorganisms is aerosolized, the liquid phase of the aerosol particle very rapidly evaporates leaving whatever organisms, e.g. pathogenic viruses, bacteria or fungal spores, suspended in air, thereby posing a risk of infection if inhaled. This topic is discussed in more detail elsewhere (Kennedy, 1995; Collins *et al.*, 1996).

Production of Blood Aerosol

In order to produce aerosol (or droplets) energy must be introduced into the starting material to overcome the cohesive forces of the material. The amount of energy that is required is proportional to the new surface that is formed. For example, if 1 ml of water is converted into 5 µm particles, more than 1.5×10^{10} are produced and the new surface area is approximately $1.2 \times 10^4\,cm^2$. Thus, a relatively small amount of a suspension of microorganisms has the potential to be converted into a large cloud of infectious aerosol.

Hemeon (1963) stated that aerosols are formed by two events in sequence:

1. *Pulvation*, which is a mechanical or pneumatic process that projects fine particles at high velocity from the coagulated state into the air of the immediate vicinity as individual particles; and
2. *Secondary air currents*, which transport the localized particles away from the site of formation.

Pulvation mechanisms for blood include the formation of fine ligaments or thin sheets from the parent liquid, which then break up into smaller particles. Common situations in which these mechanisms are to be found include: vibration of a fine needle, e.g. when withdrawn from the septum cap of a vial (Fig. 6.1); when expelling fluid blood from a pipette or syringe (Fig. 6.2); bursting of bubbles; rupture of films of blood, e.g. as are sometimes found at the mouths of blood sample containers after removal of the closure following mixing; and centrifugation of blood, especially if there is a breakage of the container or leakage.

Blood plasma and serum can be aerosolized, perhaps more readily than can whole blood. The ease at which blood can be aerosolized is related to its

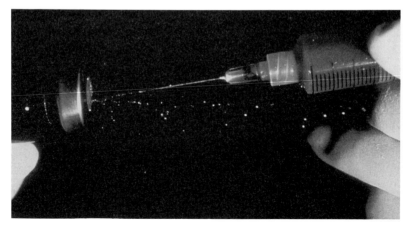

Fig. 6.1.   Aerosols produced when a needle is withdrawn from a septum-capped vial.

Fig. 6.2. Aerosols produced when pipette contents are blown out.

Table 6.2. Reports of whole blood aerosol.

Microhaematocrit centrifuge with leaking tube, microtonometer, removal of blood
  wetted plug and screw caps (Rutter and Evans, 1972)
Simulated venepuncture*, spraying from a nebulizer, bursting of bubbles (MRC, 1975)
Simulated venepuncture* (Petersen, 1980)
Use of a blood film spinner (Harper, 1981)
Bursting of bubbles – simulated removal of wetted closure (Kennedy 1988)
Simulated surgery – use of bone saw, drills and electrocautery (Heinsohn *et al.*, 1991)

*i.e. insertion of and withdrawal of hypodermic needle from blood-filled rubber tubing.

viscosity. It can be speculated that the ease at which blood can be aerosolized is related to its viscosity. Blood viscosity is a function of the number of red cells (rbcs) in suspension (Dintenfass, 1980). It would follow that in anaemia with a low rbc, blood viscosity would be lower than normal, and hence relatively easier to aerosolize. Where there is a high rbc count, as in polycythaemia, blood will be more viscous, and hence more difficult to aerosolize. Table 6.2 outlines reports of whole blood aerosol that was found during experiments using air sampling equipment. There is clearly need for more experimental work in this area.

It has been shown that dried-out spillages can by foot movement become airborne as respirable particles (Jones *et al.*, 1966) and that microorganisms can be dispersed into the air by sweeping dry floors (Lidwell, 1967). It seems reasonable to assume that aerosols could be produced from dried-out blood spillages by such methods.

Air-borne Blood Contamination in Health-care Premises

Following the conjecture by Almeida *et al.* (1971), that air-borne droplets may have been implicated in a cluster of hepatitis B cases in a haemodialysis unit, there was started a debate on whether a true aerosol can be produced from whole blood. Today, despite the publication of experimental results to the contrary (Table 6.2) the debate continues. But is there evidence that respirable blood aerosols are produced in health-care workplaces?

BG spore tracer studies in haemodialysis units using slit samplers, multi-stage impingers and settle plates (MRC, 1975) demonstrated that most blood aerosol contamination was associated with the use of needles as a patient was connected to and disconnected from the dialyser. A study was carried out in a dental surgery during which time HBsAg-positive patients were treated (Petersen *et al.*, 1979). One air sampler was located behind the patient's head on the headrest of the dental chair and another was clipped to the chest pocket of the dentist's coat, i.e. in the breathing zone. All air samples gave negative results for HBsAg and blood (using Hemastix). Negative results were also obtained by Bond *et al.* (1977), when they carried out air sampling experiments in two haemodialysis units. However, a recent study (Heinsohn and Jewett, 1993) in which cascade impactors were mounted in the breathing zones of primary and assistant surgeons has demonstrated that they may be at risk from respirable blood aerosol during a variety of surgical procedures.

SIGNIFICANCE OF ENVIRONMENTAL BLOOD CONTAMINATION

Clearly, accidental contamination of environmental surfaces with blood is a strong possibility and any blood-borne microorganism that can survive in deposited blood presents an infection risk as a consequence of skin contact. The risk would be increased if damaged skin came into contact with contaminated surfaces. Overall, the risk posed by surface contamination is considered to be significant as evidenced by the emphasis on the need for regular decontamination of surfaces, and the use of personal protective equipment, including gloves and protective eyewear, that is to be found in official biosafety guidelines.

Blood can be aerosolized under experimental conditions and blood aerosols have been detected in health-care workplaces. Similarly, any microorganism that can survive the forces involved in aerosolization, drying out of the particles in which they are suspended, and the general environmental conditions may present an infection risk if it can gain entry to the respiratory system.

However, epidemiological studies into the risk of air-borne transmission are not feasible because of the need to control all risk factors, to identify a group of control subjects who are definitely not exposed to blood aerosols and to establish the potential infectivity of patients. Infectivity studies using animal models offer the most definitive approach and until reliable evidence to the contrary is available it would be unwise to presume that transmission of blood-borne pathogens, such as HIV, does not occur (Heinsohn and Jewett, 1993).

REFERENCES

Almeida, J.D., Chisholm, G.D, Kulatilake, A.E. *et al.* (1971) Possible airborne spread of serum-hepatitis virus within a haemodialysis unit. *Lancet* ii, 849–850.

Beaumont, L.R. (1987) The detection of blood on non-porous environmental surfaces: an approach for assessing factors contributing to the risk of occupational exposure to blood in the autopsy suite. *Infection Control* 8, 424–426.

Bond, W.W., Petersen, N.J. and Favero, M.S. (1977) Viral hepatitis B: Aspects of environmental control. *Health Laboratory Science* 14, 235–252.

Brearley, S. and Buist, L.J. (1989) Blood splashes: an underestimated hazard to surgeons. *British Medical Journal* 299, 1315.

Bull, A.D., Channer, J., Cross, R.D. *et al.* (1991) Should eye protection be worn when performing necropsies? *Journal of Clinical Pathology* 44, 782.

Ciba-Geigy (1977) Uvitex NFW. Technical Circular 6150E. Manchester, Ciba-Geigy Dyestuffs and Chemicals.

Cole, C., Bennitt, W. and Halloran, S.P. (1992) Urine reagent strips, MDD Evaluation Report number MDD/92/22. Medical Devices Directorate, London.

Collins, C.H. (1993) *Laboratory Acquired Infections*, 3rd edn. Butterworth-Heienmann, Oxford.

Collins, C.H., Aw, T.C. and Grange, J.M. (1996) *Microbial Diseases of Occupations, Sports and Recreations*. Butterworth-Heinemann, Oxford.

Culliford, B.J. (1971) The examination and typing of bloodstains in the crime laboratory. Stock Number 2700-0083, US Government Printing Office, Washington, DC.

Darlow, H.M., Simmons, D.J.C. and Roe, F.J.C. (1969) Hazards from environmental skin painting of carcinogens. *Archives of Environmental Health* 18, 883–893.

Daum, G.S., Krolikowski, F.J., Reuter, K.L. *et al.* (1988) Dipstick evaluation of hematuria in abdominal trauma. *American Journal of Clinical Pathology* 89, 538–542.

Dintenfass, L. (1980) Molecular rheology of human blood: its role in health and disease (today and tomorrow?). In: Astarita, G., Marrucci, G. and Nicholais, L. (eds) *Rheology*, Volume 3. Plenum Press, New York, pp. 467–480.

Gaensslen, R.E. (1983) *Sourcebook in Forensic Serology, Immunology and Biochemistry*. National Institute of Justice, Washington.

Gerberding, J.L., Littell, C., Tarkington, A. *et al.* (1990) Risk of exposure of surgical personnel to patients blood during surgery at San Francisco General Hospital, *New England Journal of Medicine* 322, 1788–1793.

Harper, C. J. (1981) Contamination of the environment by special purpose centrifuges used in clinical laboratories. *Journal of Clinical Pathology* 34, 1114–1223.

Heinsohn, P. and Jewett, D.L. (1993) Exposure to blood-containing aerosols in the operating room: A preliminary study. *American Industrial Hygiene Association Journal* 54, 446–453.

Heinsohn, P., Jewett, D.L., Balzer, L. *et al.* (1991) Aerosols created by some surgical power tools: particle size distribution and qualitative hemoglobin content. *Applied Occupational Environmental Hygiene* 6, 773–776.

Hemeon, W.C.L. (1963) *Plant and Process Ventilation*, 2nd edn. The Industrial Press, New York.

Holton, J. and Prince, M. (1986) Blood contamination during venepuncture and laboratory manipulations of specimen tubes, *Journal of Hospital Infection* 8, 178–183.

Jones, I.S., Pond, S.F. and Stevens, D.C. (1966) Resuspension factors for plutonium. Paper No. 8. *International Symposium on Radiological Protection of the Worker by the Design and Control of his Environment*. April 1966, Bournemouth.

Kennedy, D.A. (1988) Studies in laboratory-acquired infection with particular reference to the role of equipment. PhD Thesis, University of London.

Kennedy, D.A. (1995) Blood samples and reagents: hazards and risks. In: Lewis, S.M. and Koepke, J.A (eds) *Haematology Laboratory Management and Practice*. Butterworth-Heinemann, Oxford. Chapter 10.

Kennedy, D.A., Stevens, J.F. and Horn, A.N. (1988) Clinical laboratory environmental contamination: use of a fluorescence/bacterial tracer. *Journal of Clinical Pathology* 41, 1229–1232.

Lauer, J.L., Van Drunen, N.A., Washburn, J.W. *et al.* (1979) Transmission of hepatitis B virus in clinical laboratory areas. *Journal of Infectious Diseases* 140, 513–516.

Lewis, S.M. and Wardle, J.M. (1978) An analysis of blood specimen container leakage. *Journal of Clinical Pathology* 31, 888–892.

Lidwell, O.M. (1967) Take-off of bacteria and viruses. In: Gregory, P.H. and Monteith, J.C. (eds) *Airborne Microbes*. Cambridge University Press, Cambridge.

MRC (Medical Research Council) (1975) Experimental studies on environmental contamination with infected blood during haemodialysis. *Journal of Hygiene, Cambridge* 74, 133–148.

Palmer, D., Perry, K.R., Mortimer, P.P. *et al.* (1995) Fifteen HBsAg screening assays, MDA evaluation report number MDA/95/52. Medical Devices Agency, London.

Petersen, N.J. (1980) An assessment of the airborne route in hepatitis B transmission. *Annals of the New York Academy of Sciences* 353, 157–166.

Petersen, N.J., Barrett, D.H., Bond, W.W. *et al.* (1976). Hepatitis B surface antigen in saliva, impetiginous lesions and the environment in two remote Alaskan villages. *Applied and Environmental Microbiology* 32, 572–574.

Petersen, N.J., Bond, W.W. and Favero, M.S. (1979) Air sampling for hepatitis B surface antigen in a dental operatory. *Journal of the American Dental Association* 99, 465–467.

Piazza, M., Guadagnino, V., Picciotto, L. *et al.* (1987) Contamination by hepatitis B surface antigen in dental surgeries. *British Medical Journal* 295, 473–474.

Porteous, M.J.L. (1990) Operating practices of and precautions taken by orthopaedic surgeons to avoid infection with HIV and hepatitis B virus during surgery. *British Medical Journal* 301, 167–169.

Rutter, D.A. and Evans, G.G.T. (1972) Aerosol hazards from clinical laboratory apparatus, *British Medical Journal* i, 594–596.

Taylor, J.D. (1990) AIDS and hepatitis B and C: contamination risk at transurethral resection. A study using sodium fluorescein as a marker. *Medical Journal of Australia* 153, 257–260.

CHAPTER 7
Phlebotomy

E. Bouvet

INTRODUCTION

Health-care workers (HCWs) often perform phlebotomy to collect blood from patients for laboratory tests. This is especially true for those working in a hospital setting, whether in wards or in laboratories. The personnel who carry out this task vary from country to country. Blood samples are usually drawn from in-hospital patients by nurses in France and Switzerland. Other health-care workers may be called upon to perform such procedures: physicians, for instance, can sample blood in addition to their many other tasks. In Denmark, venous blood sampling is performed mainly by interns and residents in hospitals (Nelsing *et al.*, 1993).

In North America, some laboratory technicians have specialized as phlebotomists. The members of this new profession are solely responsible for sampling patients' blood (see Jagger and Bentley, Chapter 5 this volume).

Accidental blood exposure (ABE) during phlebotomy is a major concern which requires constant attention and renewed prevention efforts. Indeed, phlebotomy presents a particularly high risk for accidental injuries and accidental blood exposure. Furthermore, occupationally-acquired human immune deficiency virus (HIV) infection surveillance data collected throughout the world, and especially in those countries which have published their data, show that accidental blood exposure during phlebotomy is the first cause of occupationally-acquired HIV infection. Phlebotomy is therefore a high-risk procedure in terms of accidental blood exposure and a high-risk one as well in terms of viral transmission in case of ABE.

RISKS: WHAT TYPE OF ACCIDENT?

Occupationally-acquired Infections and Phlebotomy

Many different viruses may be transmitted to health-care workers in case of cutaneous or mucocutaneous exposure during phlebotomy: hepatitis B virus (HBV) is the first known transmissible agent and infection may be prevented by vaccination. Hepatitis C virus (HCV) is probably the most frequently transmitted agent in the 1990s since the transmission rate following ABE is approximately 3% (10 times higher than in the case of HIV) and the number of people with HCV in the general population is much higher than is the case for either HIV or HBV. Furthermore, in most cases, HCV-infected patients are completely symptom-free and usually unaware that they harbour the virus. In the study which he carried out in collaboration with the French Public Health Network (Réseau National de Santé Publique (RNSP)), Desenclos *et al.* (1996) showed that 75% of subjects with HCV viraemia are unaware of their sero-status and that nearly half of them have no known risk factor.

Among the 79 well-documented HIV infections acquired occupationally in industrialized countries and described by Abiteboul (Chapter 4, this volume), 53 were described in detail in the report by Heptonstall *et al.* (1995) (see Fitch *et al.*, 1995). At least 28 of these cases (53%) involved phlebotomy. Nine of the ten documented French cases (Lot and Abiteboul, 1995) – all of which occurred in nurses – resulted from an injury during or after phlebotomy. In eight cases phlebotomy aimed at sampling blood using a vacuum tube. Blood sampling had been carried out to perform blood cultures in the remaining case. By the end of 1993, 39 cases of occupationally-acquired HIV infection had been documented in the United States, 15 of which had occurred in phlebotomists. These health-care workers are the most represented professional category in occupationally-acquired HIV infection in the United States. These data clearly show that accidents involving phlebotomy are among those with the highest risk of occupationally-acquired viral infection. This is especially true for HIV (Centers for Disease Control and Infection: CDC, 1995).

Epidemiology of Accidental Blood Exposure during Phlebotomy

Prospective or retrospective investigations on accidental blood exposure, which are usually carried out in a collaborative network of health-care structures throughout the world (USA, Switzerland, Denmark, Italy, France, etc.), enable us to document the respective role played by various procedures in accidental injuries and blood exposures (Jagger *et al.*, 1988; Bouvet *et al.*, 1991; Abiteboul *et al.*, 1992, 1993; Albertoni *et al.*, 1992; Ippolito *et al.*, 1994; Iten *et al.*, l995; Luthi and Dubois-Archer,1995).

Jagger (1994) described the characteristics of percutaneous injuries associated with blood sampling in a national panel composed of 58 hospitals in the United States (see Jagger and Bentley, Chapter 5 this volume). These data include all percutaneous injuries in health-care workers over a one-year period starting in September of 1992. The participating hospitals varied in size (26 of them had less than 100 beds, 16 of them had between 100 and 300 beds and 26 of them had over 300 beds).

Three-thousand eight-hundred and twenty-nine percutaneous injuries occurred during the study period. Four hundred and seventy-one were associated with phlebotomy (12.3% of all percutaneous accidents in these participating hospitals): venous or arterial blood sampling using a needle mounted on a syringe, vacuum tube sampling, or sampling using a winged needle. Approximately two out of three cases occurred in nurses or phlebotomists. Other accidents were reported in 'respiratory therapists' – a category of health-care worker which does not exist in many countries – and in laboratory technicians and biologists. Accidents occur most often at the patient's bedside (53.3%), followed by emergency rooms and intensive care units. The mechanism of the accident is quite specific for each type of procedure. Accidents which are associated with the use of needles mounted on syringes for blood sampling, for instance, often occur as the health-care worker removes the needle from the rubber stopper of the tube after injecting the blood into the sample tube. In cases where vacuum systems or winged needles are used, most percutaneous exposures occur as the needle is being carried to the puncture-resistant container or as the blood-soiled equipment is being inserted into the container. In winged needles, the tubing which is connected to the needle kinks easily and hinders risk-free handling of the device after use. Approximately 60–80% of accidental blood exposures occur after the procedure has been completed. They are potentially avoidable with strict observance of universal safety precautions or the systematic use of safety devices. The remaining 20–40% of accidents occur during the blood-drawing procedure itself. These exposures are much more dependent on the conditions in which the sampling is being performed, the quality of the patient's vein supply, of his or her movement, and are therefore less easily avoided.

The study conducted by Halduven et al. (1995) at Santa Clara Valley Medical Center in San Jose aimed at analysing the 881 accidental percutaneous exposures which occurred over an 8-year period in that hospital, during which a prevention programme was implemented. Percutaneous accidents associated with phlebotomy accounted for 3–7% of the total number of percutaneous accidents.

Since 1994, the Prevention of Nosocomial Infection Committee (CCLIN Paris Nord) has been conducting a prospective study in the northern part of France as well as Paris (unpublished data). Amongst the 2381 accidental blood exposures reported in 1995, 634 occurred during blood sampling, phlebotomy being associated with the accident in over 40% of cases (20% using a vacuum system, 10% using needles mounted on syringes and the remaining 10% occurring during blood cultures). Ninety-three per cent of the accidental blood exposures were percutaneous. These exposures therefore entailed percutaneous injuries using hollow needles soiled with blood. Such exposures carry a high risk of viral transmission since this risk is correlated with the quantity of blood which is inoculated to the victim during the accidental injury. The volume of blood inoculated during a percutaneous exposure using a hollow needle has been estimated using in vitro models (Mast et al., 1993; Bennett and Howard, 1994). The volume of inoculum varies with the needle's gauge and the depth of the injury. It has been estimated at between 0.3 and 0.5 µml of blood in the case of a 5-mm deep injury using a 22-gauge needle (0.71 mm in diameter). In the model developed by Mast et al., an 18-gauge needle holds 3 µml of blood and the volume of the inoculum fell to 1.5 µml – in other words it was reduced by 50% when the needle had to pierce a single pair of gloves and to 1.1 µml when it had to puncture two pairs of gloves. The volume of the inoculum transferred during an injury using a suture needle is much lower and has been estimated at

0.2 μml of blood. This volume is reduced by over 50% after it punctures a single glove barrier. Phlebotomy needles are therefore responsible for transmitting the highest quantity of blood, thereby carrying the highest risk of occupationally-acquired viral infection (Chamberland *et al.*, 1995).

Injuries

Injuries which occur after phlebotomy involve a hollow needle which has been used to draw venous blood. The hollow needle varies in gauge and is filled with blood. Injuries usually involve the ball of the thumb of the minor hand. The risk of occupationally-acquired infection by some viruses varies according to the depth of injury, the needle gauge, the quantity of injected blood, the injury and needle site, and the amount of infectious particles present in the injected blood.

Injuries occur during the procedure as the health-care worker removes the needle from the patient's vein, as the patient makes a sudden or unexpected move, as both hands join (one holding the needle and the other to apply pressure to the vein), as health-care workers recap their needles or discard the needle into a puncture-resistant container, or as the needle is left lying unprotected (on a tray, in a bed, on the floor, etc.). The risk for injury increases with the time span between the removal of the needle from the patient's vein and the time at which it is disposed of using a puncture-resistant container.

Blood culture presents a specific problem (Bouvet *et al.*, 1991). This procedure sometimes uses a device which carries two distinct risks of injury. The needle which is used to puncture the patient's vein carries the risk commonly associated with phlebotomy and is connected to a rubber tubing. The other, distal, end of this tubing features a second needle which is inserted into two injection caps on the blood culture bottles. This distal needle often slips from the health-care worker's hand as he or she tries to insert it into the bottle's cap and injures the HCW's minor hand which is holding the bottles. This type of injury carries the maximum risks of occupationally-acquired viral infection: large-bore needle containing blood; usually deep injury (Gerberding, 1995). This is why holders are now available in order to avoid these accidents by protecting the hand which is holding the blood culture bottles (see below).

Mucocutaneous Exposure to Blood

Accidental contact with patients' blood during sampling was common when simple needles and open tubes were still used to collect blood. In such cases, the chances of coming in contact with blood were high: the health-care worker's hands which were holding the system were often exposed. In comparison the risk of accidental blood exposure is greatly reduced if vacuum tube systems are used. There remains, however, a risk of accidental exposure when many tubes have to be filled and inserted through the multiple sample adapter. Indeed, such multiple sample adapters sometimes leak at the base of the needle which is inserted into the tubes.

Blood projection may occur if the system unexpectedly and suddenly comes apart. This may occur if a vacuum system's needle and multiple sampler separate, if a winged needle becomes disconnected from its tubing or if the distal end of the tubing is mishandled. Such occurrences are sometimes due to flaws in the devices themselves which can burst, thereby projecting massive amounts of blood on the health-care worker's bare skin areas located near the patient, especially the HCW's face, hands, eye or mouth mucosa (Puro *et al.*, 1990; Saghafi *et al.*, 1992; Jagger, 1994).

PREVENTION

General Aspects of Prevention

The prevention of phlebotomy-linked injuries is a priority. Indeed, such injuries are among the most hazardous ABEs occurring in the health-care setting. Prevention objectives should be as follows:

- *Immediately following blood sample collection*: HCWs should be able to discard or shield the blood-stained needle as quickly as possible after it has been withdrawn from the patient's vein. This can effectively be done by supplying HCWs with an adequate supply of puncture-resistant needle containers and/or safety equipment. The danger of injury persists as long as the needle has not been disposed of or adequately covered. If possible, the procedure required to safely discard or cover the soiled needle should require the use of a single hand only, since HCWs must avoid bringing their two hands together if they are holding a soiled needle in one of them (Nelsing *et al.*, 1996a).
- *During blood sample collection*: risks linked with sudden or unexpected moves and difficulties arising from needle placement into the patient's vein must be avoided whenever possible. Such occurrences are liable to cause HCWs to injure themselves and can be effectively prevented by ensuring that they are properly placed, work in adequately designed areas, and use adequate and high-quality equipment. The patient must be comfortable and properly prepared, if possible, and HCWs performing the phlebotomy must not be needlessly disturbed.
- *Avoiding contact with the patient's blood*: leaks at the junction between the multiple sample adapter and the needle, the needle and a syringe or the junction between winged needles and the tubing.

Achieving these goals requires a well-defined and wilful policy for the enhancement of risk reduction in the health-care setting. Such a policy demands that the following preconditions be met (Martin *et al.*, 1992):

1. A standardized, active surveillance programme for accidental blood exposure including an analysis of such ABEs should they occur and information feedback to HCWs.
2. A well thought-out policy for blood sampling and related safety equipment; implementation should be preceded by evaluation.

3.   The elaboration by nurses or phlebotomists who carry out phlebotomy on a daily basis of a well-defined procedure taking the local conditions into account so that phlebotomy may be carried out with the least risk of injury.

4.   A training programme for health-care personnel: training is required for each health-care worker who performs phlebotomy. Safety training should be available to all students before they have to perform invasive procedures and should be regularly reinforced in practising HCWs (Tarantola *et al.*, 1996; Nelsing 1997a,b). This continuous training enables these workers to update their knowledge on new safety devices, to analyse conditions and causes of avoidable injuries with accidental blood exposure which may have occurred in the health-care setting.

ORGANIZATION – TRAINING – INFORMATION

Organization is a fundamental aspect of safety in performing invasive procedures in the health-care setting. Accident-causing conditions should be foreseen if possible and each step of the procedure should be planned so as to reduce the risk of injury to the utmost. Blood sampling should be performed only if certain prerequisites have been met: the patient should be properly seated and comfortable, thereby reducing the risk of sudden or unexpected moves. The health-care worker should also be comfortably seated before beginning the procedure and should be within reach of a tray or trolley containing the devices required for phlebotomy: needle; multiple sample adapter; sampling tubes; cotton swabs; antiseptics; rubber tourniquet; puncture-resistant container large enough to receive the blood-soiled needle as soon as the sampling has been completed. This last point is an essential one. Many accidents occur after procedure completion because the blood-soiled equipment has not been properly and immediately discarded. A puncture-resistant container must be available to health-care workers at the blood sampling site. This requires a stable surface on which to position the container, such as a roll-in trolley if phlebotomy is to be carried out at the patient's bedside. It will avoid going back and forth between the patient and the device preparation site while phlebotomy is ongoing. Phlebotomy should be continuous, well thought-out, and should neither be interrupted nor hasty. Work should be organized in such a way that HCWs are not disturbed during phlebotomy by telephone calls, questions or conversations. This rule must be observed and enforced in the health-care, as well as the laboratory setting. Health-care supervisors and hospital managers are responsible for ensuring that the organization of health-care procedures meet maximum security standards, that HCWs are properly informed of prevention methods and receive continuous training and that proper safety equipment be made available to them.

Universal Precautions are the very minimum measures that should be observed in all cases where phlebotomy is required. These precautions must be applied in each and every patient, since their blood is to be considered infected a priori. The main principles of Universal Precautions are:

- never recapping needles;
- wearing gloves during phlebotomy and sampling;
- immediately disposing of soiled devices in adequate puncture-proof containers;
- washing hands immediately in case of accidental cutaneous exposure to blood;
- never disconnecting a needle from syringes or hand-held sampling systems;
- immediately disinfecting of blood-soiled sampling devices and surfaces using a recently-diluted 10% hypochlorite solution or another efficient disinfectant;
- discarding disposable waste in leak-proof containers featuring a warning sign.

Venous blood sampling procedures must be performed only in strict observance of precautions destined to protect persons responsible for carrying and handling sample tubes used for blood analyses. Implementation of Universal Precautions appears to have contributed to a decrease in injuries in hospitals. (Beekmann *et al.*, 1994). Accidents which could have been avoided by the strict observance of these Universal Precautions are referred to as 'preventable accidents'. Most injuries which occur after the sampling procedure has been completed are preventable. The preventable rate is lower when Universal Precautions are observed. This rate can therefore be used as an indicator of ABE prevention strategies in the health-care setting (Kristensen *et al.*, 1992). Non-preventable accidents are those which occur while the sampling needle and device are in use, when they are bared to perform sampling and immediately prior to disposal, since there is always some time lag between the displacement of the needle from the vein and the disposal of the blood-soiled needle. This is where safety devices have not yet been discarded. The availability of new safety devices featuring needle covers changes the notion of preventable accidents as the number of potentially preventable ABEs increases. The main obstacle to their widespread use is their high cost.

SAFETY DEVICES AND SAFE PRACTICES

Over the past two decades, many blood-sampling devices have been improved, e.g. the Safety Lock system shown in Fig. 7.1 (see also Jagger and Bentley, Chapter 5 this volume). Some of these changes greatly improved health-care workers' safety, although this may not have been the designers' principal intent. Conversely, others have been designed more recently to improve HCW safety but sometimes complicate the procedure. In industrialized countries, certain blood sampling device characteristics are considered necessary and their use is now unquestioned, although devices do not always meet these standards. For

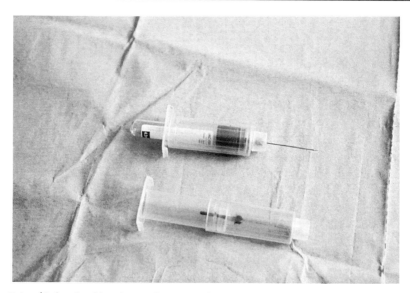

Fig. 7.1. 'Safety Lock' syringe. (Courtesy of M.Y. Cousson, Institut National de Recherche et Sécurité, Paris.)

instance, HCWs must always carry out phlebotomy using a vacuum tube by puncturing rubber stoppers with a reusable or non-reusable multiple sample adapter. This multiple sample adapter prevents blood contact or injury as tubes are filled with the blood samples (see Jagger and Bentley, Chapter 5 this volume). In France blood sampling with a needle mounted on a syringe is not regarded as good practice. Rubber stopper puncture should never be performed without protection. Direct sampling through a dismounted needle and blood drawing into open sample tubes should be proscribed as this technique entails an important risk of cutaneous blood exposure and injury as the needle is positioned.

Phlebotomy may be carried out using either a needle mounted on a multiple sample adapter which allows blood sample tube positioning without risk of cutaneous blood contact, or a small-gauge, easily manoeuvrable winged needle connected to a tubing. This needle is mounted on a multiple sample adapter which covers the distal needle used to puncture the blood sample tubes' stopper (Fig. 7.2). The first of these techniques is preferred by some HCWs. The short distance between the needle and the tubes, however, allows neither the patient's limb nor the phlebotomist to move and entails a risk of needle displacement from the patient's vein.

Gloves

Gloving is required when performing phlebotomy. Stotka *et al.* (1991) showed in a hospital prospective study that the majority of exposures incurred by medical staff was caused by lack of glove use. This protection may be useful in two ways: firstly, it prevents the HCW's hand from coming in contact with the patient's blood; secondly, although it does not prevent injury and should injury

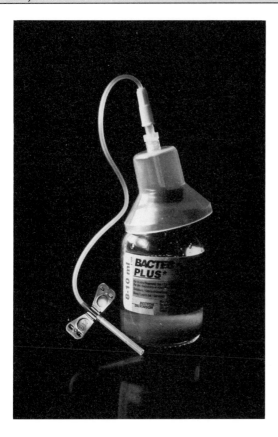

Fig. 7.2. Device for safe blood culture: BACTEC™ (Becton Dickinson). (Courtesy of M.Y. Cousson, Institut National de Recherche et Sécurité, Paris.)

occur, the passage of the hollow needle through a single glove barrier reduces by 35% to 50% the quantity of the inoculum of the patient's blood which will be inserted under the HCW's skin. Considering that the volume of this inoculum is an important parameter in the risk of viral transmission, gloving is an essential aspect of the prevention of viral transmission in case of injury during phlebotomy. This last assertion, however, has not been substantiated by retrospective case–control studies (see also Morgan, Chapter 8 this volume).

Puncture-resistant Sharps Containers

Phlebotomy should be carried out only if the health-care worker is within reach of a puncture-resistant container so that the soiled needle may be safely discarded immediately after use. Containers must be suitable for their daily use by HCWs. A container which is used during phlebotomy must be large enough to receive the needle and its multiple sample adapter, the needle and the syringe or the winged needle, its tubing and its multiple sample adapter. As already stated, maximum safety levels are met only when there is minimal handling of the phlebotomy device. This means that the needle should not be dismounted from the syringe or the multiple sample adapter. If the HCW chooses to discard the entire device, instead of reusing the multiple sample adapter, the container used must be larger. Such containers will be filled more quickly and will have to be changed more often. Over-filled containers, in turn, become a possible source of injury as it exposes health-care personnel to soiled devices protruding from its opening. Container stability must also come into consideration. Furthermore, containers must never be filled to the top. The container must be filled only to a certain level which is visible to the HCW. The use of transparent containers enables HCWs to easily evaluate their saturation. Criteria for container selection are defined by Abiteboul *et al.* (1997) (see also Collins and Kennedy, Chapter 18 this volume).

CONCLUSION

Phlebotomy is one of the procedures with the highest risk of accidental blood exposure for health-care workers and laboratory phlebotomists. In case injury does occur, the risk of viral transmission is high. Venous phlebotomy is the main source of injury in documented occupationally-acquired HIV infection. Most of the accidents which occur during phlebotomy are avoidable and can be prevented if universal safety precautions are observed. Most of the injuries occur after the procedure has been completed, with a needle which should have been immediately disposed of in a nearby puncture-resistant container. Prevention plays an essential role. It is based on the observance of certain simple safety rules and the use of certain safety devices: phlebotomy needle for vacuum blood-sampling; multiple sample adapter (non-reusable if possible); puncture-resistant needle container and safety sampling devices which cover the needle. The implementation of these safety precautions must be part of a genuine prevention strategy in the health-care setting or the laboratory and must include health-care worker evaluation, training and information as well as a surveillance and evaluation system in case of occupational injury.

REFERENCES

Abiteboul, D., Antona, D., Azoulay S. *et al.* (1992) Surveillance of occupational exposure to blood in Assistance Publique – Hospitals of Paris (AP-HP) in 1990–1991. In: *Proceedings of 1st International Congress on Occupational Health for Health Care Workers.* Freibourg, September 1992.

Abiteboul, D., Antona, D., Descamps, J.M. *et al.* (1993) Procedures à risque d'exposition au sang pour le personnel infirmier: surveillance et devolution de 1990 a 1992 dans 10 hôpitaux. *Bulletin Epidemiologique Hebdomadaire* 43, 195–196.

Abiteboul, D., Bouvet, E., Decamps J.M. *et al.* (1997) Comment choisir un conteneur pour objects piquant/tranchant. *Hygienes,* in press.

Albertoni, F., Ippolito, G., Petrosillo, N. *et al.* (1992) Needlestick injury in hospital personnel: A multicenter survey from central Italy. *Infection Control and Hospital Epidemiology* 13, 540–544.

Beekmann, S.E., Vlabov, D., Koziol, D.E. *et al.* (1994) Temporal association between implementation of universal precautions and a sustained, progressive decrease in percutaneous exposures to blood. *Clinical Infectious Diseases* 18, 562–569.

Bennett, N.T. and Howard, R.J. (1994) Quantity of blood inoculated in a needlestick injury from suture needles. *Journal of the American College of Surgery* 178, 107.

Bouvet, E., Abiteboul, D., Fourrier, A. *et al.* (1991) A one year prospective ongoing survey on blood exposures among 518 nurses in 17 hospitals in France. In: *Proceedings of VIIth International Conference on AIDS. Florence, June 1991.* Abstract WD 4173.

CDC (1995) Case–control study of HIV seroconversion in health-care workers after percutaneous exposure to HIV-infected blood. *Morbidity and Mortality Weekly Reports* 44, 929–933.

Chamberland, M.E., Ciesielski, C.A., Howard, R.J. *et al.* (1995) Occupational risk of infection with human immunodeficiency virus. *Surgical Clinics of North America* 75, 1057–1070.

Desenclos, J.C., Dubois, F., Couterier, E. *et al.* (1996) Estimation du nombre de sujets infectés par le HCV en France, 1994–1995. *Bulletin Epidémiologique Hebdomadaire* 5, 22–23.

Fitch, K.M., Perez Alvarez, L., De Andres Medina, R. *et al.* (1995) Occupational transmission of HIV in health care workers. *European Journal of Public Health* 5, 175–186.

Gerberding, J.L. (1995) Management of occupational exposures to blood-borne viruses. *New England Journal of Medicine* 332, 444–451.

Halduven, D.J., Phillips, E.S., Clermons, K.V. *et al.* (1995) Percutaneous injury analysis; consistent categorization, effective reduction methods, and future strategies. *Infection Control and Hospital Epidemiology* 16, 582–589.

Heptonstall, J., Porter, K. and Gill, O.N. (1995) Occupational transmission of HIV. Summary of published reports. Unpublished Report of the PHLS, London.

Ippolito, G., De Carli, G., Puro, V. *et al.* (1994) Device-specific risk of needlestick injury in Italian health care workers. *Journal of the American Medical Association* 272, 607–610.

Iten, A., Maziero, A., Jost, J. *et al.* (1995) Surveillance des expositions professionnelles du sang ou des liquides biologiques: la situation en Suisse au 31 decembre 1994. *Bulletin Office Federal de la Santé Publique* 24, 4–7.

Jagger, J. (1994) Risky procedures, risky devices, risky job. *Advances in Exposure Prevention,* 1, 1,4–9

Jagger, J, Hunt, E.H., Brand-Elnaggar, J. *et al.* (1988) Rates of needle-stick injury caused by various devices in a university hospital. *New England Journal of Medicine* 319, 284–288.

Kristensen, M.S., Weinberg, N.M. and Anker-Moller, E. (1992) Healthcare workers' risk of contact with body fluids in a hospital: the effect of complying with the universal precautions. *Infection Control and Hospital Epidemiology* 13, 719–724.

Lot, F. and Abiteboul, D. (1995) Infections professionnelles par le VIH en France. Le point au 31 decembre 1993. *Bulletin Epidemiologique Hebdomadaire* 44, 111–113.

Luthi, J.C and Dubois-Arber, F. (1995) Evaluation de la stratégie de prevention du SIDA en suisse, 1993–1995. Personnel hospitalier: *Etude Suisse sur les Expositions au VIH et aux Hepatites chez le Personnel Hospitalier.* IUMSP, Lausanne, 67 pp.

Martin, L.S., Hudson, C.A. and Strine, P.W. (1992) Continued need for strategies to prevent needlestick injuries and occupational exposures to bloodborne pathogens. *Scandinavian Journal of Working Environment and Health* 18, Suppl. 2, 94–96.

Mast, ST, Woolwine, J.D. and Gerberding, J.L. (1993) Efficacy of gloves in reducing blood volumes transferred during simulated needlesticks injury. *Journal of Infectious Disease* 168, 1589–1592.

Nelsing, S., Nielsen, T.L. and Nielsen, J.O. (1993) Occupational blood exposure among health care workers: 1. Frequency and reporting. *Scandinavian Journal of Infectious Diseases* 25, 193–198.

Nelsing, S., Nielsen, T.L. and Nielsen, J.O. (1997a) Percutaneous blood exposure among Danish doctors: exposure mechanisms and strategies for prevention. *European Journal of Epidemiology*, in press.

Nelsing, S., Nielsen, T.L., Bronnum-Hansen, H. *et al.* (1997b) Incidence and risks factors of occupational blood exposure – A nation-wide survey among Danish doctors. *European Journal of Epidemiology*, in press.

Puro, V., Ranchino, M. and Profili, F. (1990) Occupational exposures to blood and risk of HIV transmission in a general hospital. *European Journal of Epidemiology* 6, 67–70.

Saghafi, L., Raselli, P., Francillon, C. *et al.* (1992) Exposure to blood during various procedures: Results of two surveys before and after the implementation of universal precautions. *American Journal of Infection Control* 20, 53–57.

Stotka, J.L., Wong, E.S., Williams, D.S. *et al.* (1991) An analysis of blood and body fluid exposures sustained by house officers, medical students, and nursing personnel on acute-care general medical wards: a prospective study. *Infection Control and Hospital Epidemiology* 12, 583–590.

Tarantola, A., Casalino, E., Gadjos, V. *et al.* (1996) Frequency and perceived risk of blood exposure among medical students. Results of a survey study. XI International Conference on AIDS, Vancouver.

CHAPTER 8
The medical profession

D.R. Morgan

Surgeons and pathologists have always risked infection if a scalpel slipped or infectious blood splashed onto open cuts or mucous membranes, when infection, septicaemia and death could result. By the 1940s, with the availability of antibiotics and vaccines, the consequences of bacterial infections could be controlled to a great extent – but the risk from viruses remained.

With the recognition that 'serum hepatitis' was caused by a virus and with the development of techniques to identify markers of hepatitis B virus (HBV) in the 1970s it became evident that needlesticks and sharps injuries continued to present occupational hazards (Dienstag and Ryan, 1982). Prior to the development of the first hepatitis B vaccine in 1982, up to 8000 US health-care workers contracted hepatitis B occupationally each year, resulting in 200 deaths annually, and leaving many hundreds more at risk from cirrhosis and liver cancer. Today, hepatitis B infection continues to be an occupational hazard worldwide, although certainly in developed countries it is increasingly being controlled by immunizing personnel at risk. Moreover, a newly identified hepatitis virus, hepatitis C virus (HCV), has been found to be widespread and capable of being transmitted by blood-to-blood contact during clinical procedures, particularly following percutaneous injury with contaminated sharps (British Medical Association: BMA, 1996). This agent could replace HBV as an important health risk for doctors, their colleagues, and potentially for patients. In addition, the human immunodeficiency virus (HIV) has spread worldwide during the past decade and more than 220 health-care workers are reported to have been infected with HIV as a consequence of clinical or laboratory work (Heptonstall et al., 1995). The occupational risks of HIV infection will continue to be important as the disease spreads within communities, particularly in developing countries where health-care resources and laboratory facilities are poor. Any blood sample or any patient may therefore present a risk of infection from some form of blood-borne pathogen. Patients will also be at risk of infection if their surgeon or gynaecologist carries an infection which could be passed on during an invasive procedure; HBV is known to have been transmitted during invasive procedures from 34 infected health workers to at least 350 patients in the US and elsewhere since the early 1970s (Bell et al., 1993).

It is likely that health-care workers who underwent training in the 'pre-AIDS era' will have received little education or training about occupational risks from cross-infection in today's health-care environment. Exposure to blood and blood-containing body fluids have been extremely commonplace and health-care workers may have developed cavalier attitudes towards such exposures. Medical students and all new generations of medical staff will continue to risk infection from blood-borne viruses, if they do not acknowledge the presence of the risk and take steps to reduce it. Knowledge about occupational risks alone may not be enough to influence worker behaviour and modification may depend on a complex set of interactions. The way that individuals assess their personal risk within the workplace and compare this with the risks in society in general may also be important.

REPORTING OF EXPOSURES AND SHARPS INJURIES BY MEDICAL STAFF

It is difficult to ascertain which staff are most at risk of sharps injuries as these injuries are under-reported across the complete spectrum of health-care staff (Collins and Kennedy, 1987). There is evidence, however, that doctors and medical students consistently under-report sharps injuries, despite regular contact with sharps (McCormick and Maki, 1981; Mangione *et al.*, 1991; Chiarello, 1993). Whether this is due to lack of knowledge of the correct reporting procedures, unavailability of an occupational health department, lack of concern about sharps injuries, pressure of work or a fear that the occurrence of an accidental sharps injury may reflect on the skill of the individual, etc. is unclear. Doctors may often treat themselves, and other staff may consider the wound too trivial or are not aware of any requirement to report or any mechanism for reporting. Astbury and Baxter (1990) found that although 18% of injuries had been reported on an accident form only 5% had been notified to personnel in the occupational health department.

THE CAUSES OF NEEDLESTICK AND SHARPS INJURY

Jagger *et al.* (1988) showed that sharps injuries could be categorized into three types: those occurring (i) before use; (ii) during and after use but before disposal; and (iii) during or after disposal. The study found that only 17% of incidents occurred during use of a medical procedure, e.g. when the user withdrew a needle from a patient, but the majority occurred when staff were preparing the items for disposal (70%). Members of a surgical team were shown to risk injury which could expose either the operator, assistant or patient to a blood-borne infection. Operator exposures mainly involved injury with a solid suture needle or from eye contamination with blood splashes (see also Jagger and Bentley, Chapter 5 this volume).

DOCTORS

Clinical procedures which involve the use of sharps or where regular exposure to patient blood can be expected, e.g. during operations, present a high risk for infection. The type of injury may be important. It is relevant to consider any difference between injuries caused by hollow bore needles, solid needles used for suturing, and other sharps, e.g. whether the injury was deep and penetrating, or superficial, and whether patient blood was likely to be carried into the wound. Techniques used for handling contaminated needles, e.g. re-sheathing by hand or by using a 'device', and the methods used to transport and finally dispose of contaminated sharps are also important considerations.

Tokars *et al.* (1992) reported that most injuries affecting the surgeon were caused by suture needles, with the use of fingers rather than an instrument to hold tissue being important, and 24% of incidents involved an instrument being held by a co-worker. Gerberding *et al.* (1990) reported that sharps exposure occurred in 1.7% of operational procedures observed, and in one study found that there were ten needlesticks with a suture needle, one needlestick with a hollow needle, six hand lacerations (one with scalpel, one with bone and four with sharp instruments). Both of these observational studies concluded that percutaneous injuries occur regularly during surgery and that staff are at risk for intra-operative exposure to blood. Porteous and Lee (1990) sent a postal questionnaire to 1220 UK orthopaedic surgeons; 64% of respondents reported that they had sustained a needlestick injury or received body fluid in their eyes within the previous month.

Injuries sustained during surgical procedures are often inflicted on an assistant and are caused by scalpel blades. Eisenstein and Smith (1992) found that 5% of total reported incidents occurred when a colleague was injured by a co-worker and Gompertz (1990) suggested that injuries could be due to loss of concentration by medical students or by one of the surgical staff during an operation. At least two health-care workers have become infected with HIV following needlesticks caused by colleagues during patient resuscitation procedures (Marcus *et al.*, 1989) and a recent report confirms HIV seroconversion in a surgeon, following a scalpel injury (Ippolito, 1996). In the documented seroconversions of health-care workers for the United Kingdom (1984–1992) percutaneous exposure to patient blood *via* a hollow needle was documented for all four cases (Heptonstall *et al.*, 1993) suggesting that high-risk exposure to freshly drawn HIV-positive blood from a hollow needle represents potentially the greatest risk for infection. Of the 223 documented or 'possibly' occupationally-acquired HIV infections reported worldwide, 35 cases involve doctors, surgeons or medical students. The details of documented cases are shown in Table 8.1.

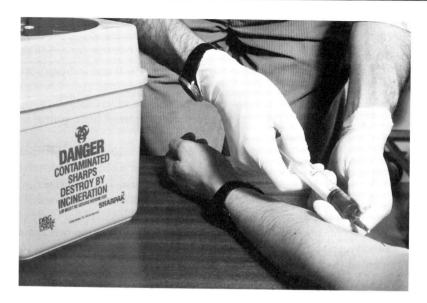

Fig. 8.1. Sharps container at site of venepuncture for immediate disposal of needle and syringe. (Courtesy of the British Medical Association.)

Table 8.3. Re-sheathing techniques reported by 315 medical students who would always or sometimes re-sheathe a needle.

Method	Number	%
'Scooping'	193	61.3
Hold sheath and needle	114	36.2
Some other method, e.g. a 'device'	8	2.5

From Morgan (1994).

the double-ended Vacutainer needle so that the plastic barrel could be re-used, contrary to guidance that vacuum collection systems should be discarded as a single unit and not re-used (BMA, 1990). Re-sheathing is sometimes under-taken so that a contaminated needle can be removed from a syringe before transferring blood into a pathological specimen tube. It has been recommended that where needles must be removed, they should be taken off by hand and placed immediately in a sharps container which should be placed at the point of use. This is probably less hazardous than using forceps or attempting to re-sheathe the needle (BMA, 1990).

Only 42% (816/1950) of students stated that they would report a future needlestick injury; 15% said that they would definitely not report and the remaining 43% that it would depend upon circumstances. Reasons for non-reporting included:

• no one else reports;

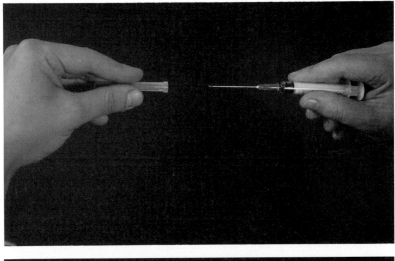

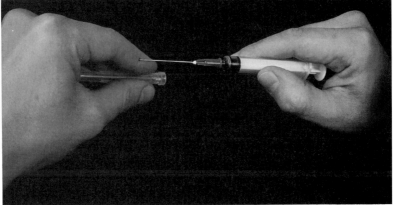

(b)

Fig. 8.2. Up to 40% of all needlestick injuries involve re-sheathing or recapping the needle after use. (a) The health-care worker aims the needle towards the resheathing cap and (b) misses, jabbing the hand holding the cap. (Courtesy of the British Medical Association.)

- did not know the reporting procedure;
- fear that others would think they were making a fuss.

A number of students (355) provided further information about their reasons for not reporting; it is of note that 40.8% (145/355) indicated that they would not know who to report to (Table 8.4). Some students indicated that it is most unlikely that medical students who have experienced adverse reactions about reporting from colleagues, particularly from more senior doctors or the occupational health department staff would be inclined to report injuries subsequently. It is likely that this situation constitutes an important stressor

Table 8.4. Categories selected by 355 students, from the options provided, indicating why they definitely would not or may not report a future sharps injury.

	Number (n = 355)	%
People would think that I was making an unnecessary fuss	53	15
No-one else reports	22	6.2
I do not know to whom I should report	145	40.8
Combinations of reasons	105	29.6
Some other reason	30	8.4

From Morgan (1994).

for students right at the beginning of their medical career. Importantly, non-reporters would be unable to benefit from follow-up counselling and the use of post-exposure prophylactic treatment where there was a risk of hepatitis B infection.

RISK MANAGEMENT

The actual risk to the individual doctor or medical student of contracting any blood-borne infection from a patient will depend upon a range of variables. These factors include:

- seroprevalence of a blood-borne virus or other organism in the patient population;
- known or suspected infectious state of the individual patient;
- risk category of the clinical procedure undertaken;
- degree of exposure to blood at the time of the incident, e.g. the volume of the blood inoculated;
- the availability of any pre-exposure or post-exposure prophylaxis or vaccine.

The prevalence of the three viral agents, HIV, HBV and HCV, differs widely according to the country, the area and the age of a given population. In the United Kingdom, current evidence shows that nationwide there is a very uneven pattern of distribution for all three viruses and that HIV, in particular, continues to be prevalent in major UK cities such as London, Edinburgh and Brighton and therefore, at a basic level, the risk of acquiring infection with HIV may be much higher for a doctor or medical student working in major cities, compared with another colleague working in rural parts of the United Kingdom. This pattern of spread affecting certain major centres will be common to many other countries including the USA, all states of Europe, and also in African countries. However, medical practitioners in areas with an apparently low prevalence of infection should not relax basic standards of precautions, which should be uniform throughout the health-care sector in each country.

The degree of infectivity, for example following needlestick injury, is a major factor and appears to differ according to the agent concerned. Hepatitis B virus is highly infectious and it has been shown that a very small quantity such as 0.04 µl of blood can transfer infection in humans (Napoli and McGowan, 1987; see also Abiteboul, Chapter 4 this volume).

Although the number of case reports is small, HCV transmission after percutaneous exposure to blood in health-care workers has been documented. The workers developed clinical hepatitis with reported incubation periods ranging from 20 days to 13 weeks. Kiyosawa *et al.* (1991) followed up 110 Japanese health-care workers from whom serial serum samples for serology were obtained, after percutaneous exposures to infected patients. Four developed clinical hepatitis and three became anti-HCV positive, representing a transmission rate of 2.7%. The average risk after a needlestick injury involving HCV-infected blood is reported to be in the range 3–10% which is between the risk estimates for infection, following parenteral injury, involving infected patients, with HBV (2–40%) and HIV (0.2–0.5%) (Gerberding, 1995). The potential for transmission of infection to patients must be considered by infected health-care staff.

UNIVERSAL PRECAUTIONS

In 1985 the Centres for Disease Control and Prevention (CDC, 1985) in the US developed the strategy of 'Universal blood and body fluid precautions' to address concerns regarding transmission of HIV in the health-care setting. This was later embodied in US legislation under the *Occupational Safety and Health Regulations* (OSHA, 1991; see Hunt, Chapter 19 this volume). The concept now referred to generally as 'Universal Precautions' emphasizes that all patients should be assumed to be infectious for HIV and other blood-borne pathogens. Although the serology for some patients may be known to a surgical team, for the majority there will be no confirmed carriage of virus and in some cases any test undertaken could be inaccurate, if the patient is in a 'window' following infection and has yet to develop detectable antibodies. In the UK the Department of Health (DH, 1990) and the British Medical Association (BMA, 1990) separately produced guidelines describing primary measures for the prevention of occupational exposure to HIV and blood-borne hepatitis viruses using procedures based on the American Universal Precaution model. Such policies have the benefit of removing the previous two-tier system based on positive identification of infected patients which could be confusing for staff. Universal Precautions have the advantage of providing protection against all blood-borne infections and by removing the need to identify a patient's infectious status, it also serves to improve patient confidentiality.

A range of infection control measures including Universal Precautions have been recommended by the UK Department of Health Advisory Group on AIDS and the Department of Health Advisory Group on Hepatitis to prevent transmission of blood-borne viruses in health-care settings (DH, 1990, 1993).

The Royal College of Surgeons of England has not produced specific guidance on Universal Precautions for its members but advise that individual surgeons should take appropriate protective actions in particular circumstances. The British Orthopaedic Association provides detailed guidelines for protection of the surgical team and, where appropriate, recommends the use of face, eye and mouth protection to provide general protection of the head from splashing with blood or tissue. Detailed guidance is provided for the use of gowns, gloves and footwear (British Orthopaedic Association, 1991). Staff working in areas in which high-risk procedures are undertaken such as operating theatres, intensive care units, mortuaries, casualty and obstetric departments, require the greatest degree of barrier protection. Such levels of protection are also required for staff in contact with patients who have been defined as high risk, such as known HIV- or HBV-positive patients or those attending clinics for sexually transmitted diseases. Universal Precautions, however, have been called into question and may well be superseded by Standard Precautions (see Kibbler, Chapter 20 this volume).

Theatre staff who may come into contact with blood and bodily fluids should wear a disposable waterproof plastic apron under a disposable waterproof gown, face mask, protective footwear and eye wear as well as gloves. Substantial levels of eye contamination may occur during orthopaedic surgery and therefore protective goggles may be necessary and gloves must be changed between patients or when punctured or torn. In recent years, the practice of double gloving has been adopted by some surgical staff. This may help to protect the surgeon from exposure to the patient's blood. A new glove system is now available, however, in which a green glove is worn under a normal, outer, glove. If the outer glove is punctured in the presence of body fluids, a dark green patch quickly develops around the perforation, indicating that the glove should be replaced. Double gloving maintains a barrier between the wearer and the patient in the majority of cases where the outer glove has been punctured, with the inner glove remaining intact in 80% of cases (Matta *et al.*, 1988)

HEPATITIS B IMMUNIZATION

The British Medical Association had been campaigning since 1987 for a national policy for vaccination of health-care staff against hepatitis B to ensure protection for both patients and staff. The publication of the Department of Health guidelines (DH, 1993) required UK National Health Service (NHS) trusts and hospitals to assess the requirement to vaccinate staff according to their risk assessment. All doctors who undertake exposure-prone procedures (EPP), where there is a risk that injury to the worker may result in the exposure of the patient's open tissues to the blood of the worker, are required to have a successful course of vaccination. EPPs include procedures where the worker's gloved hands may be in contact with sharp instruments, needletips and sharp tissues (spicules of bone or teeth) inside a patient's open body cavity, wound or confined anatomical space, where the hands or fingertips may not be completely visible at all times. There is increasing evidence that UK NHS trusts now require all categories of medical practitioner to show evidence of successful vaccination to ensure that both patients and the individual doctor are protected from infection. Guidelines published by the Committee of Vice Chancellors and Principals of UK medical schools recommend that applicants for medical and dental courses should be screened for hepatitis B virus and antibody before entry to medical and dental schools, and be immunized where necessary (Lever, 1994).

Hepatitis B vaccination is reviewed in detail by McCloy (Chapter 16 this volume) and by the British Medical Association (BMA, 1995).

PROCEDURE FOR DEALING WITH SHARPS INJURIES AND BLOOD EXPOSURE

Several methods of reducing the incidence of needlestick injuries are available. These include:

- increased education and training of all hospital personnel who come into contact with used needles and sharps;
- safe disposal of used needles without resheathing;
- use of British Standard (BSI, 1990) robust sharps containers only;
- the appropriate use of needlestick-prevention devices.

A comprehensive review of needlestick-prevention devices has been published and a range of products are available, from needle-less injection systems to needleguards and recapping devices (Charney and Schirmer 1993; also see Bouvet, Chapter 7, Jagger and Bentley, Chapter 12, and Appendix). All such devices may have practical disadvantages as well as safety features and they will require careful trial and assessment, before general use within an organization.

Medical staff must be encouraged to seek help if they experience difficulty with a procedure, e.g. venepuncture, and must report any blood exposures or sharps injuries to the local occupational health unit, after following the recognized first aid procedures. Broken skin that has been in contact with blood or body fluids should be washed as should eyes or mouth if these have come into contact with such fluids. Skin punctures should be encouraged to bleed and be washed thoroughly. After undertaking initial first-aid treatment the injured health-care worker should obtain further advice and report the exposure by telephoning the hospital control of infection officer, the occupational health service or the consultant in communicable disease control. A 24-hour 'hotline' telephone number should be established in every hospital to provide immediate advice to injured staff.

All injured health-care workers should have the opportunity to seek guidance and counselling by experienced staff in the occupational health department or from a specially designated doctor. An injured practitioner may be advised to have blood samples stored for possible future testing. If this is agreed, an immediate specimen will be required to establish a base line followed by others at intervals thereafter; the blood should be stored for six months. Blood should also be collected from the source patient with his/her informed consent and the hepatitis B antigen status and HIV antibody status determined. Where practical, it would be helpful to retest the source patient after six months to reassure the injured worker and for the purposes of epidemiological study.

If the source patient is thought to be hepatitis B e-antigen positive and the injured worker is not fully immune to hepatitis B, then consideration should be given to the administration of hepatitis B immunoglobulin within 48 hours. At the same time a course of hepatitis B vaccination should be commenced. Those who have previously received a first dose of vaccine should complete the course in the normal way but have an additional dose of vaccine, given within 48 hours of the injury (see McCloy, Chapter 16 this volume).

Alpha Interferon

At the present time there is no effective post-exposure prophylaxis for treatment of an injured worker where the source patient is known to be hepatitis C positive. However, at the appropriate stage, an infected worker should be assessed for possible treatment with alpha interferon, which may inhibit virus replication leading to long-term remission and possibly eradication of infection, in about 25% of patients (BMA, 1996).

Treatment may be considered for those presenting with raised serum alanine aminotransferase and a positive polymerase chain reaction (PCR) test for viral RNA; however, as the disease progresses at variable rates the indications for the treatment remain undefined (Dusheiko *et al.*, 1996). Side-effects can be serious, ranging from flu-like illness in the early stages to psychological effects and weight loss later on. More rarely, treatment may lead to bacterial infections, seizures, retinal changes or interstitial lung disease, for example. The treatment is very expensive and may need to be continued for more than one year (Dusheiko *et al.*, 1996). Those infected with HCV genotype 1a/1b, and with a high level of viraemia, have a poorer response to alpha interferon treatment. Factors such as increased age, long-lasting infection, excess body weight, and the presence of severe hepatic fibrosis or cirrhosis are associated with a poor response. Individuals with types 2 and 3 infection are more likely to have a sustained response to treatment than patients with type 1. Few data on outcome are currently available for those infected with other HCV genotypes. Treatment with interferon alpha should be considered *before* cirrhosis develops.

After treatment for six months half of the initial responders will relapse. Up to 20% of patients have a prolonged biochemical response to treatment and do not relapse when interferon is stopped. The standard dose is 3 million units (miu), three times a week, but longer treatments (up to a year) or induction with a higher dose (5–6 miu) may be more effective in some patients. As most of the unwanted side-effects of interferon are dose related, the efficacy of lower dose therapy has been investigated. The research carried out has shown conflicting results but one study showed beneficial effects at doses as low as 0.25 miu three times per week (Varagona *et al.*, 1992). Most recently a multicentred randomized trial concluded that certain patients with chronic hepatitis C are very sensitive to alpha interferon and can be successfully treated with low doses (1.5 miu). After several months of treatment the disappearance of HCV-RNA from the liver in responders is the most reliable predictor of no relapse upon discontinuation of therapy. Dusheiko *et al.* (1996) have called for improved and careful standardization of tests for hepatitis C virus RNA due to their importance in determining whether patients are likely to respond to treatment. It appears that patients with fewer than 10^6 copies of hepatitis C virus RNA per ml are more likely to have a sustained response than others. These authors also suggest criteria for deciding on a successful outcome after one year, which include:

- normal serum alanine aminotransferase;
- negative for hepatitis C virus RNA in serum;
- histological improvement;
- no development of cirrhosis.

Interferon may also be of value for some patients infected with HBV and may suppress viral replication in approximately 40% of individuals with chronic infection acquired in adult life. (See also McCloy, Chapter 16 this volume.)

Zidovudine

Interest in post-exposure HIV chemoprophylaxis for occupationally-exposed health-care workers has developed since zidovudine (AZT) has gained acceptance for treating HIV infection in persons with advanced disease, but appears to have little or no benefit for those with asymptomatic infection (Aboulker and Swart, 1993). Theoretically, when AZT or other nucleoside analogues are administered within a few hours after exposure, they may prevent infection, by preventing HIV replication in the initial target cells. The efficacy of AZT for preventing latent infection in exposed health-care workers is extremely difficult to assess epidemiologically because the low rate of seroconversion requires an extremely large sample and a randomized control trial design. Detecting seroconversion in zidovudine treated individuals may be complicated by delayed seroconversion related to drug-induced inhibition of virus replication. Some experience with AZT for short intervals in healthy health-care workers indicates that reversible haematological toxic effects occur rarely but that less serious symptoms of intolerance such as fatigue, insomnia and headaches are common. Failure to protect against HIV infection following zidovudine prophylaxis in eight health-care workers has been reviewed (Heptonstall *et al.*, 1993) and failure to prevent seroconversion when given within 45 minutes has been reported (Lange *et al.*, 1990). The optimal dosage and duration for zidovudine treatment are not defined fully but most protocols employ conventional dosages used to treat AIDS patients for 2–6 weeks. It is recommended that treatment of pregnant women or persons not practising effective contraception should be discouraged and close monitoring for haematological, hepatic, renal and neurological dysfunction is essential.

A recent case-controlled study involving health-care workers in France, the UK and the USA (CDC, 1995) concluded that the risk of seroconversion depends on certain factors related to the exposure incident and can be modified by post-exposure administration of zidovudine.

Factors:

- deep injury;
- visible blood on the needle or sharp;
- procedure involving a needle placed directly in a vein or artery;
- terminal illness in source patient;
- post-exposure use of zidovudine.

Where zidovudine had been offered, it was generally administered as 1000 mg per day for 3–4 weeks. The results clearly indicate an increased risk of seroconversion associated with larger volumes of blood, i.e. from visible blood or from deeper cuts and direct exposure to blood vessels – as well as significant risk from HIV to patients in the latter stages of AIDS. Although the authors are cautious about interpreting the value of zidovudine, they conclude that the risk of HIV infection among health staff receiving zidovudine was reduced by approximately 79% (CDC, 1995).

In an effort to increase the effectiveness of post-exposure with zidovudine some centres are now recommending using a higher dose combination of zidovudine with another anti-retroviral drug, Lamivudine, and a protease inhibitor such as Indinavir. This combination may substantially reduce circulating blood levels of HIV, possibly preventing infection when HIV enters the body as a result of a sharps injury. Prophylaxis should be initiated as soon as possible, preferably within one hour.

Studies are now proceeding, organized by the CDC in the US, to follow up health-care staff who received this triple therapy.

For other views on Zidovudine see Waldron, Chapter 15 this volume, and McCloy, Chapter 16 this volume.

DISCUSSION

Concern that doctors and health-care workers are not maintaining a high level of accurate knowledge or awareness about blood-borne infections particularly regarding infection control has been raised in recent years (Klimes *et al.*, 1989; McKeown and Williamson, 1992). Repeated surveys have shown that health-care workers have a high risk for sharps injury but also that they may disregard or have a poor appreciation of the risk involved (Buss *et al.*, 1991) but can benefit from educational materials (McKinnon *et al.*, 1992). General practitioners and their practice nurses may receive inadequate training and few surgeries have effective policies for infection control (Hoffman *et al.*, 1988; Foy *et al.*, 1990; Morgan *et al.*, 1990; Sharp *et al.*, 1993). Paediatricians may know little about the prevalence of HIV/HBV infection in the population (Buss *et al.*, 1991), and there may be considerable lack of knowledge about the risks of cross-infection and the correct course of action to be followed in the event of a sharps instrument injury, both by medical students (Choudhury and Cleator, 1992) and doctors in training (O'Neill *et al.*, 1992).

Much of this low level of awareness may stem from the fact that little attention may be given to infection control matters within the undergraduate medical curriculum. Choudhury and Cleator (1992) showed that the instruction given to students at the University of Oxford Medical School consisted of a single lecture given in the first two weeks of the course, before ward work had started and many students taking part in the study could not recall the lecture. Medical students and junior doctors require close supervision with an emphasis upon improving safety before they commence venepuncture and undertake invasive procedures with patients. In this way a 'universal policy' may be implemented based on good tuition, avoiding the adoption of unsafe practices by mimicry.

Extensive studies, particularly in the US by Drs Jagger, Fisher and Gerberding and others have highlighted the need for safer needle devices and the adoption of practices to reduce their risk of needlestick injury have an important part to play. Commercial delays and financial cost of such devices may limit their introduction and therefore other ways of reducing the risk of injury need to be considered.

Both medical students and qualified doctors should be given induction training and refresher training in safe phlebotomy and injection technique, with advice about not re-sheathing contaminated needles. Choudhury and Cleator (1992) were unable to show a relationship between re-sheathing and injury rates but a larger UK study (Morgan, 1994) suggests that individuals who regularly re-sheathe contaminated needles may be at higher risk of injury. It could be that re-sheathing is a 'marker' for other practices or attitudes to practical safety measures, which exposes individuals to injury from other behaviour, although an obvious deficit in hand dexterity alone may not be responsible for individuals having a high risk of needlestick injury (Casanova *et al.*, 1993).

'Safe' manual techniques, e.g. by dropping the sheath onto the needle by gravity or by 'scooping' or use of a re-sheathing device as recommended by Anderson *et al.* (1991) have been criticized as being unsafe or impractical; very few students in a national UK study (Morgan, 1993) reported using a re-sheathing device, but the scoop technique was commonly used.

The number and location of sharps containers in all patient areas should be reviewed and needlestick injuries in each location should be audited and the results communicated to staff on an ongoing basis. Strategies used successfully to reduce needlestick injury in one department should be communicated to managers and staff in other areas who should be encouraged to utilize such concepts. Further research is needed to explore what determines health-care workers' behaviour in relation to occupational hazards. It is known that people tend to underestimate their risk of undesired health hazards despite sufficient knowledge, e.g. cigarette smoking and cancer. We need a better understanding about why health staff and doctors in particular may ignore the recommended guidelines when exposing themselves to potentially infectious materials, whilst simultaneously being conscious of (or even over estimating) the danger (Morgan, 1992).

The style of staff training and support should be tailored according to its objectives. Large meetings provide a useful format for conveying information about blood-borne pathogens and infection control principles but do not encourage discussion of attitudes and anxieties, especially with multidisciplinary groups. Small meetings, particularly of people working together, for example on a ward or in a general practice surgery, should provide a more suitable environment for expressions of uncertainty, for example about technique or practices or previously unspoken concerns about lack of an infection control policy or sterilizing facilities. Peer trainers are potentially the least threatening and most effective in this setting. Posters, such as that in Fig. 8.3, are often effective.

The outcome of staff training and communication needs to be evaluated in terms of the effect on staff knowledge and attitudes but especially on staff behaviour in infection control practices. A hospital 'culture' may exist in which reporting of a sharps injury or blood exposure is not encouraged as this would reflect badly both on the individual, who could be seen to be incompetent or making a fuss, and also reflects on the supervisor. This could become particularly noticeable if assessment of a student after an exposure is badly managed, or if a blood sample is required to be taken from a patient – who may not be agreeable. Reporting of exposures must be encouraged and an effective system of follow-up must be established.

BIN IT

FOLLOW THE SIMPLE RULES BELOW:

- You used it – you bin it
- Do not re-sheath needles
- Discard needle and syringe as one unit
- Dispose of sharps into a safe container, immediately after use
- Do not leave used sharps lying around
- Do not overfill sharps container
- Get immunized against hepatitis B

Fig. 8.3. Model principles for handling sharps – a poster for use in all health-care settings. (Courtesy of the British Medical Association.)

Mandatory testing for HIV and hepatitis viruses is not yet a requirement for medical practice in the United Kingdom. However, any doctors who are found to be carrying HIV, HBV (e-antigen positive) or HCV will be required to notify their employing authority and to seek medical guidance on the implications for them continuing in medical practice. Medical practitioners who routinely undertake exposure-prone procedures (EPPs) may not be allowed to continue with them and will require counselling and guidance on redeployment or retraining, or some other decision regarding their future career.

Education and guidelines will not in themselves bring about changes in health-care worker behaviour, but new methods to impart information such as interactive CD-ROM systems or the use of role playing and problem-solving educational techniques should be examined. Improved clinical safety and infection control will be dependent on the necessary physical environment and the attitudes of senior staff, who must be conducive to change and provide the necessary resources and motivation.

A vaccine for HCV is unlikely to be available in the short term, but a combined vaccine, offering protection against hepatitis A virus HAV and HBV infection, is available in the UK and could be provided to all medical (and dental) students before they come into clinical contact with patients. It could also be offered routinely for all neonates.

REFERENCES

Aboulker, J.P. and Swart, A.M. (1993) Preliminary analysis of the Concorde trial. *Lancet* 341, 889–890.

Anderson, D.C., Blower, A.L., Packer, J.M. *et al.* (1991) Preventing needlestick injuries. *British Medical Journal* 302, 769–770.

Astbury, C. and Baxter, P.J. (1990) Infection risks in hospital staff from blood; hazardous injury rates and acceptance of hepatitis B immunisation. *Journal of the Society of Occupational Medicine* 40, 92–93.

Bell, D.M., Shapiro, C.N. and Gooch, B.F. (1993) Preventing HIV transmission to patients during invasive procedures. *Journal of Public Health Dentistry* 53, 170–173.

BMA (1990) *A Code of Practice for the Safe Use and Disposal of Sharps*. British Medical Association, London.

BMA (1995) *A Code of Practice for the Implementation of the UK Hepatitis B Immunisation Guidelines for the Protection of Patients and Staff*. British Medical Association, London.

BMA (1996) *A Guide to Hepatitis C*. British Medical Association, London.

British Orthopaedic Association (1991) *Guidelines for Prevention of Cross Infection Between Patients and Staff in Orthopaedic Operating Theatres with Special Reference to HIV and the Bloodborne Hepatitis Viruses*. BOA, London.

BSI (1990) *British Standard 7320. Specification for Sharps Containers*. British Standards Institution, London.

Buss, P.W., McCabe, M. and Verrier Jones, E.R. (1991) Attitudes of paediatricians to HIV and hepatitis B virus infection. *Archives of Disease of Children* 66, 961–965.

Casanova, J.E., Barnas, G.P., Gollup, J. *et al.* (1993) Hand dexterity in hospital personnel with multiple needlestick injuries. *Infection Control and Hospital Epidemiology* 14, 473–475.

CDC (1985) Recommendations for preventing transmission of infection with human T-lymphotropic virus type III/LAV in the workplace. *Morbidity and Mortality Weekly Reports* 34, 687–696.

CDC (1995) Case-control study of HIV seroconversion in health care workers after percutaneous exposure to HIV-infected blood – France, United Kingdom and United States, January 1988 – August 1994. *Morbidity and Mortality Weekly Reports* 44, 929–933.

Charney, W. and Schirmer, J. (1993) *Essentials of Modern Hospital Safety,* vol 2. Lewis, Boca Raton.

Chia, H.P., Koh, D. and Jeyaratnam, J. (1993) A study of needlestick injuries among medical undergraduates. *Annals of the Medical Academy of Singapore* 22, 338–341.

Chiarello, L. (1993) Reducing needlestick injuries among health care workers. *AIDS and Clinical Care* 5, 77–79.

Choudhury, R.P. and Cleator, S.J. (1992) An examination of needlestick injury rates, hepatitis B vaccination uptake and instruction on sharps techniques among medical students. *Journal of Hospital Infection* 22, 143–148.

Collins, C.H. and Kennedy, D.A. (1987) Microbiological hazards of occupational needlestick and 'sharps' injuries. *Journal of Applied Bacteriology* 62, 385–402.

Department of Health (DH) (1990) *Guidance for Clinical Health Care Workers: Protection against infection with HIV and hepatitis viruses.* Recommendations of the Expert Advisory Group on AIDS. Department of Health. HMSO, London.

Department of Health (DH) (1993) *Protecting Health Care Workers and Patients From Hepatitis B.* Recommendations of the advisory group on Hepatitis. Department of Health. HMSO, London.

Dienstag, J.L. and Ryan, D.M. (1982) Occupational exposure to hepatitis B virus in hospital personnel: infection or immunisation? *American Journal of Epidemiology* 115, 26–39.

Dusheiko, G.M., Khakoo, S., Soni, P. *et al.* (1996) A rational approach to the management of hepatitis C infection. *British Medical Journal* 312, 357–364.

Eisenstein, H.C. and Smith, D.A. (1992) Epidemiology of reported sharps injuries in a tertiary care hospital. *Journal of Hospital Infection* 20, 271–280.

Foy, C., Gallagher, M., Rhodes, T. *et al.* (1990) HIV and measures to control infection in general practice. *British Medical Journal* 300, 1048–1049.

Gerberding, J.L. (1995) Management of occupational exposures to bloodborne viruses. *New England Journal of Medicine* 332, 444–451.

Gerberding, J.L., Littell, C., Tarkington, A. *et al.* (1990). Risk of exposure or surgical personnel to patients' blood during surgery at San Francisco General Hospital. *New England Journal of Medicine* 322, 1788–1793.

Gompertz, S. (1990) Needlestick injuries in medical students. *Journal of the Society of Occupational Medicine* 40, 19–20.

Heptonstall, J., Gill, O.N., Porter, K. *et al.* (1993) Health care workers and HIV surveillance of occupationally acquired infection in the United Kingdom. *Communicable Disease Report* 3, R147–152.

Heptonstall, J., Porter, K. and Gill, N.O. (1995) *Occupational Transmission of HIV: Summary of Published Reports.* PHLS AIDS Centre (CDSC), London.

Hoffman, P.N., Cooke, E.M., Larkin, D.P. *et al.* (1988) Control of infection in general practice: a survey and recommendations. *British Medical Journal* 297, 34–36.

Ippolito, G. (1996) Scalpel injury and HIV infection in a surgeon. *Lancet* 347, 1042.

Jagger, J., Hunt, E., Brand-Elinaggar, B.A. *et al.* (1988) Rates of needlestick injury caused by various devices in a university hospital. *New England Journal of Medicine* 319, 284–288.

Kiyosawa, K., Sodeyma, T., Tanka, E. *et al.* (1991) Hepatitis C in hospital employees with needlestick injuries. *Annals of Internal Medicine* 115, 367–369.

Klimes, I., Catalan, J., Bond, A. *et al.* (1989) Knowledge and attitudes of health care staff about HIV infection in a health district with low HIV prevalence. *AIDS Care* 1, 313–317.

Lange, J.M.A., Boucher, C.A.B., Hollak, C.E.M. *et al.* (1990) Failure of zidovudine prophylaxis after accidental exposure to HIV. *New England Journal of Medicine* 322, 1375–1377.

Lever, A.M.L. (1994) Hepatitis B and medical student admission. *British Medical Journal* 308, 870–871.

McCormick, R.D. and Maki, D.G. (1981) Epidemiology of needlestick injuries in hospital personnel. *American Journal of Medicine* 70, 928–932.

McKeown, M. and Williamson, D. (1992) Awareness and practice of HIV and hepatitis B guidelines in a psychiatric hospital. *Health Bulletin* 50, 292–295.

McKinnon, M.D., Williams, S., Snashall, D. *et al.* (1992) A study of the detailed circumstances of 'sharps' injuries in health care workers. *Journal of Occupational Medicine* 34, 974–975.

Mangione, C.M., Gerberding, L.J. and Cummings, S.R. (1991) Occupational exposure to HIV: frequency and rates of under reporting of percutaneous and mucocutaneous exposures by medical staff. *American Journal of Medicine* 90, 85–90.

Marcus, R., Kay, K. and Mann, J.M. (1989) Transmission of human immunodeficiency virus (HIV) in health-care settings worldwide. *Bulletin of the World Health Organization* 67, 577–582.

Matta, H., Thompson, A.M. and Rainey, J.B. (1988) Does wearing two pairs of gloves protect operating theatre staff from skin contamination? *British Medical Journal* 297, 597–598.

Morgan, D.R. (1992) Infection control and risk assessment: a review, a pilot survey and recommendations. *International Journal of Risk and Safety in Medicine* 3, 241–252.

Morgan, D.R. (1993) Education and successful prevention programmes. In: Hallauer, J., Kane, M. and McCloy, E. (eds) *Eliminating Hepatitis B as an Occupational Hazard.* Proceedings of an international congress on hepatitis B. Medical Imprint, London, pp. 40–44.

Morgan, D.R. (1994) Infection control in clinical environments with particular reference to the human immunodeficiency virus (HIV) and viral hepatitis. Unpublished PhD thesis, The Open University, UK.

Morgan, D.R., Lamont, R.J., Dawson, J.D. *et al.* (1990) Decontamination of instruments and control of cross infection in general practice. *British Medical Journal* 300, 1379–1380.

Napoli, V. and McGowan, J. (1987) How much blood in a needlestick? *Journal of Infectious Disease* 155, 828.

O'Neill, T., Abbot, A.V. and Radecki, S.E. (1992) Risk of needlesticks and occupational exposures among residents and medical students. *Archives of Internal Medicine* 152, 1451–1456.

OSHA (1991) Occupational Health and Safety Administration. Occupational exposure to blood-borne pathogens. Final rule. *Federal Register* 56, 64004–64182.

Porteous, M.J. and Lee, F. (1990) Operating practices of and precautions taken by orthopaedic surgeons to avoid infection with HIV and hepatitis B virus during surgery. *British Medical Journal* 301, 167–169.

Sharp, C., Maychell, K. and Walton, I. (1993) *Nursing and AIDS: Material Matters; Issues, Information and Teaching Materials on HIV and AIDS for Nurses – a Research Study.* National Foundation for Educational Research, Berkshire.

Tokars, J.I., Bell, D.M., Culver, D.H. *et al.* (1992) Percutaneous injuries during surgical procedures. *Journal of the American Medical Association* 267, 2899–2904.

Varagona, G., Brown, D., Kibbler, H. *et al.* (1992) Response, relapse and re-treatment rates and viraemia in chronic hepatitis C treatment with a2b interferon alpha: A phase III study. *European Journal of Gastroenterology and Hepatology* 4, 701–712.

CHAPTER 9
The dental profession

Crispian M. Scully

INTRODUCTION

Dental staff, like other health-care workers (HCWs), may be exposed to blood-borne viruses carried in blood and oral fluids, and tissues, both in the clinic and laboratory, and there must be a risk of transmission (Lewis *et al.*, 1992). Hepatitis B, hepatitis C, HIV, and the herpesviruses are the main recognized blood-borne pathogens of concern to dental staff in most countries (Scully and Bagg, 1992). Before the almost universal wearing of protective gloves, the risk of transmission of blood-borne viruses across the skin by sharps (needlestick) injuries and through breaks in the epithelium was fairly high. Blood was frequently found on the hands of dental staff after work and after handwashing (Allen and Organ, 1982) and was a potential route for transmission of infection. Studies of hepatitis B virus infections one or two decades ago showed a high rate of infection in dental staff (Scully, 1985).

The possible consequences to dental staff of the transmission of blood-borne viruses can be profound, with considerable morbidity and mortality, as well as the possibility of their transmitting infection to patients, colleagues and family. It has recently been calculated that the annual cumulative risk of infection from the routine dental treatment of patients whose seropositivity is undisclosed is 57 times greater from hepatitis B virus (HBV) than from human immunodeficiency virus (HIV), and that the risk of dying from HBV is 1.7 times greater than the risk of HIV infection, for which mortality is almost certain (Capilouto *et al.*, 1992).

Dental staff and patients now appreciate the need for barrier protection (Bowden *et al.*, 1989) and protective gloves can probably significantly reduce transmission of infection (Noble *et al.*, 1992). Dental instruments are frequently contaminated with blood, saliva and other body fluids, and infections may be transmitted by sharps injuries. Transmission of blood-borne viruses in splatter and aerosols, via the conjunctivae and respiratory tract, is less likely, which is fortunate in view of the increased risk of infection of dental staff by respiratory pathogens (Davies *et al.*, 1994) and the poor efficacy of face masks (Hogan and Samaranayake, 1990).

Dental staff must therefore take exceptional care to avoid sharps injuries and must always protect their hands, eyes and respiratory tract when handling body fluids or tissues, or instruments or surfaces likely to be contaminated by body fluids. They must also take particular care to avoid needlestick injuries to colleagues. The risk of transmission of infection is greater when infected persons are treated, when there are quantities of blood present, when staff have not been immunized, and when infection control breaks down.

Most dental staff experience at least one needlestick (sharps) injury each year (Felix *et al.*, 1994; Gooch *et al.*, 1995). Injuries usually involve burs, local anaesthetic needles, sharp instruments or orthodontic wire (Porter *et al.*, 1990; Siew *et al.*, 1992). Many of these sharps injuries occur outside the mouth (Porter *et al.*, 1990; Porter, 1991; Cleveland *et al.*, 1995). Most dentists (84%) suffer at least one needlestick injury in a 5-year period (Siew *et al.*, 1987). In 1987, US dentists reported about 12 injuries per year but by 1991, this had fallen to about four per year (Siew *et al.*, 1992; Cleveland *et al.*, 1995). The evidence suggests therefore that these injuries may be decreasing. However, dental nurses are the staff most likely to be injured (Porter *et al.*, 1990; Porter, 1991), dental instruments are not consistently sterilized (Scully *et al.*, 1994) and staff are not always compliant with infection control measures, even in a hospital setting (Porter *et al.*, 1995, 1996) despite training (Scully *et al.*, 1992). This is cause for some concern. There is much better compliance, however, with infection control procedures in many countries (Siew *et al.*, 1992) compared with a decade ago. This chapter discusses the known risks of infection of dental staff by blood-borne viruses including hepatitis B and C, HIV, and herpesviruses. The lesser risks, such as from syphilis or various retroviruses other than HIV, are discussed elsewhere (Scully 1995, 1996; Scully *et al.*, 1990b; Scully and Samaranayake, 1992; Siegel, 1996). As yet no data are available on any occupational risk from hepatitis D or from HTLV-I or HTLV-II.

HEPATITIS B

Dental staff have been one of the groups of health-care workers with the highest rate of infection with hepatitis B virus (HBV) and the infection has frequently been transmitted in dental practice (Schiff *et al.*, 1986; Scheutz *et al.*, 1988; Porter *et al.*, 1994; Cottone and Puttaiah, 1996). In the UK, however, and at least some other developed countries over recent years, HBV infection has declined considerably in dental staff, presumably as a consequence of immunization against HBV, and greater care to reduce any possibility of needlestick injury and breakdown in infection control (Polakoff, 1989).

Studies reported about a decade or two ago showed that dental surgeons were one of the groups of HCWs at greatest risk from HBV in USA (West, 1984; Schiff *et al.*, 1986), Australia (Bennet *et al.*, 1985), Germany (Eisenburg *et al.*, 1977) and other countries (Panis *et al.*, 1986; Mori, 1984; Scheutz *et al.*, 1988), although there was only a lowish prevalence in the UK (Jones *et al.*, 1972; Cummings *et al.*, 1986). In the former countries, up to 30% or more of dental surgeons had serological evidence of HBV infection (Feldman and Schiff, 1975), with the risks greatest amongst oral surgeons (Schiff *et al.*, 1986), less in general dental practitioners (Schiff *et al.*, 1986), lower in auxiliary clinical staff (Schiff *et al.*, 1986; Amerena and Andrew, 1987; Savage *et al.*, 1984; James and Sampliner, 1978; Faoagali, 1986), and lowest in clerical and laboratory staff (Savage *et al.*, 1984; Schiff *et al.*, 1986), and in dental students (Goebel and Gitnick, 1979; Hollinger *et al.*, 1977). The risk of HBV infection was also highest in those working in urban hospitals, and the longer the duration of practice (Glenwright *et al.*, 1974; Feldman and Schiff, 1975; Mosley *et al.*, 1975; Goubran *et al.*, 1976; Eisenburg *et al.*, 1977; Nicholas, 1977; Aldershvile *et al.*, 1978; Hardt *et al.*, 1979; Hurlen *et al.*, 1979; Bass *et al.*, 1982; Epstein *et al.*, 1984; Savage *et al.*, 1984; Bennet *et al.*, 1985; Panis *et al.*, 1986; Schiff *et al.*, 1986; Amerena *et al.*, 1987; Siew *et al.*, 1987; Reed and Barrett, 1988; Scheutz *et al.*, 1988).

Since the introduction of vaccines against hepatitis B virus, considerable effort has been put into promoting the immunization of dental staff in many countries. In the UK, early studies showed few dental staff to be immunized but virtually all now are immunized (Scully *et al.*, 1990a, 1993). Since 1990 all dental students have been immunized (Scully and Matthews, 1990), so that all new graduates should now be immune. There are no reports yet of dental staff infected occupationally with hepatitis B variants. Staff in the developing world, however, are often not immunized against hepatitis B, despite the higher prevalence of infection in many of those countries (Scully *et al.*, 1990a).

The success of immunization programmes can be seen in that, by 1987/88, there were no reported cases of acute hepatitis B in UK dental staff (Polakoff, 1989). All UK undergraduate dental students are now offered immunization against HBV, and the UK General Dental Council (GDC) demands that all clinical staff are immune to HBV. However, it is of concern that hospital dentists remain the highest risk group for HBV in USA (and perhaps in some other countries), with a seroprevalence from two different US studies of up to 26–28% (Thomas *et al.*, 1993; Gerberding, 1994); almost one half had not received the HBV vaccine by 1992, despite the requirement stated by the Occupational Safety and Health Administration (OSHA) (1991) that persons with reasonably anticipated contact with blood or other potentially infectious materials be offered the vaccine. Studies reported at the American Dental Association (ADA) national annual meetings have shown a rise to some 70% of dentists being immunized by 1990 (Gruninger *et al.*, 1992) though how representative of the profession as a whole are the delegates is unclear.

Studies of the seroprevalence of HBV markers in hospital dental patients have shown less than 1% to be seropositive and, even in apparently high-risk groups, only 10% are positive, and only 1 or 2% are 'e' antigen (HBeAg) positive (Tullman *et al.*, 1980; Boozer *et al.*, 1986; Matthews *et al.*, 1986; Smith *et al.*, 1987). Saliva from HBV surface antigen (HBsAg)-positive patients may contain HBsAg, HBV-DNA and DNA-polymerase, and might be infective (Broderson, 1974; Heathcote *et al.*, 1974; Macaya *et al.*, 1979; Hurlen, 1980; Ben-Aryeh *et al.*, 1985; Karayiannis *et al.*, 1985; Davison *et al.*, 1987; Jenison *et al.*, 1987; Davis *et al.*, 1989). and there is evidence of transmission by bites (McQuarrie *et al.*, 1974; Cancio-Bello *et al.*, 1982) and within households (Beasley and Hwang, 1983; Leichtner *et al.*, 1981; Powell *et al.*, 1985) and in institutions (Breuer *et al.*, 1985). HBV can also survive on surfaces for at least one week (Bond *et al.*, 1981) and has been identified in dental surgeries (Piazza *et al.*, 1987). However, the level of HBV transmission from such contacts is low, and there are reports of lack of transmission in day-care centres (Shapiro, 1987) and in dental practice (SyWassink and Lutwick, 1983); in the latter study, 19 dental personnel did not contract infection after exposure to HBV-infected oral fluids.

HEPATITIS C

Possibly the lower prevalence of hepatitis C virus (HCV) infection in the general population explains the current low risk of HCV infection to dental staff in most countries (Porter and Scully, 1990; Schiff *et al.*, 1990; Epstein *et al.*, 1992; Porter *et al.*, 1994; Molinari, 1996; Porter and Lodi, 1996), and the rarity of reports of transmission in dental practice (Vaglia *et al.*, 1985). Thus a study from New York City, USA, showed a prevalence in dentists of 1.75% compared with 0.14% in blood donor controls (Klein *et al.*, 1991), but oral surgeons had a particularly high risk. Hospital dentists had a prevalence of HCV up to 6.2% in another US study – higher than that in physicians (Thomas *et al.*, 1993, 1996). Studies elsewhere have not shown this risk and HCV infection in dental staff is at present uncommon in London (Lodi *et al.*, 1996) – as in other health-care workers (Zuckerman *et al.*, 1994), in Wales (Herbert *et al.*, 1992) and Taiwan (Kuo *et al.*, 1993) and no hospital dentist infected with HCV was found in one US study (Gerberding, 1994).

The prevalence of HCV infection in hospital dental patients in USA is about 5% (Shopper *et al.*, 1995). HCV has been demonstrated in saliva (Taka-matsu *et al.*, 1990; Abe and Inchauspe, 1991; Komiyama *et al.*, 1991; Wang *et al.*, 1991, 1992; Liou *et al.*, 1992; Couzigou *et al.*, 1993; Ogasawara *et al.*, 1993; Sugimura *et al.*, 1995; Roy *et al.*, 1996; Young *et al.*, 1993) but not invariably (Takamatsu *et al.*, 1990; Hsu *et al.*, 1991; Fried *et al.*, 1992; Hollinger and Lin, 1992), it can be transmitted experimentally by saliva (Abe *et al.*, 1987), and it has been found on dental equipment (Piazza *et al.*, 1995). HCV may survive for up to one week at room temperature (Fong *et al.*, 1993). It can be transmitted by needlestick injuries (Vaglia *et al.*, 1990). The route of transmission of nearly 50% of cases of HCV is unclear (Alter, 1991) and it has been suggested, though with no evidence that the 'only' risk factor within the previous six months for nearly 10% of patients infected with HCV at one institute, was dental treatment (Mele *et al.*, 1990). Transmission within families (Everhart *et al.*, 1990; Kiyo-sawa *et al.*, 1991; Wang *et al.*, 1992) and in institutions (Chaudhary *et al.*, 1992) is generally uncommon but HCV has been transmitted by a bite (Dusheiko *et al.*, 1990; Figueredo *et al.*, 1994) and by blood splashes on to the conjunctiva (Sartori *et al.*, 1993) and has been transmitted by an unknown route within a hospital medical ward (Allander *et al.*, 1995).

HUMAN IMMUNE DEFICIENCY VIRUS (HIV)

The prevalence of HIV infection among dental patients must vary according to the country, the type of patient seen and other factors. Reported prevalences in dental hospital patients are 4.8% in San Francisco (Dodson *et al.*, 1993), 4% in Italy (Puro *et al.*, 1992), and 9% in Brazil (Almeida *et al.*, 1997). Saliva from persons with HIV infection may contain a variety of agents including HIV and herpes viruses (Groopman *et al.*, 1984; Ho *et al.*, 1985; Levy and Greenspan, 1988; Fox *et al.*, 1989; Atkinson *et al.*, 1990; O'Shea *et al.*, 1990; Goto *et al.*, 1991; Kawashima *et al.*, 1991; Barr *et al.*, 1992; Lucht *et al.*, 1993; Yeung *et al.*, 1993; Kakizawa *et al.*, 1996) and HIV may be found in the mouths of infected neonates and adults (Sawada *et al.*, 1995).

Saliva also contains HIV antibodies and other factors that appear to reduce infectivity (Fultz, 1986; Archibald *et al.*, 1986, 1987; Fox *et al.*, 1988,1989; Atkinson *et al.*, 1990), and the virus appears not to be transmitted by normal social contact within households (Friedland and Klein, 1987). HIV is probably only rarely transmitted by passionate kissing (Piazza, 1989), oral sex (Lyman *et al.*, 1986; Rozenbaum *et al.*, 1988; Spitzer and Weinger, 1989) or bites. Of 16 reports of bites by HIV-infected persons, transmission appears to have occurred in only three cases (Richman and Rickman 1993; Wahn *et al.*, 1986; Anon., 1987; Vidmar *et al.*, 1996) and in one instance HIV was not transmitted but HCV was (Figueredo *et al.*, 1994).

Occupational exposure to saliva of more than 100 HCWs did not result in any seroconversions (Gerberding *et al.*, 1987) and in a study of 1309 dental staff in USA, 94% of whom reported needlestick injuries, and 72% treated high-risk patients, only one male dentist was HIV-seropositive (Klein *et al.*, 1988). Indeed, to date there appear to be only specific reports of three dental staff who may have acquired HIV occupationally in the 15 years of the HIV epidemic (Klein *et al.*, 1988; Lot and Abiteboul, 1992) and a possible six more known to Centers for Disease Control in Atlanta (CDC, 1993b), but there are at least 7000 who have screened seronegative (Ebbesen *et al.*, 1986; Gerberding *et al.*, 1986; Lubick *et al.*, 1986; Flynn *et al.*, 1987; Siew *et al.*, 1987; Doublog *et al.*, 1988; Gruninger *et al.*, 1992).

The Centers for Disease Control (CDC) Needlestick Survey has identified 19 dental staff who suffered percutaneous exposures to HIV-infected material, these being mainly associated with the use of local anaesthetic injection syringes and scalers, but none of these exposures resulted in HIV transmission (Gooch *et al.*, 1995). A recent study from the USA, UK and France followed 710 HCWs who had needlestick injuries involving HIV. Only 31 seroconverted, giving a risk of about 0.3%, and most seroconversions were in persons receiving injuries involving larger quantities of blood, often from terminally ill AIDS patients, and the HCWs did not receive zidovudine prophylaxis (CDC, 1995). Most needlestick injuries in dental staff involve small quantities of blood, are not deep and the needles have very small bore (Gooch *et al.*, 1995).

The number of dental staff infected occupationally with HIV is thus extremely low (Morris and Turgut, 1990; Scully and Porter, 1991), and there are no proven occupational transmissions, although as discussed above, two male dentists and one female assistant appear to have contracted the infection occupationally (Porter and Scully, 1994). It is of course possible that some cases of HIV of unknown route of transmission may have been transmitted in a dental environment (Scully and Mortimer, 1994).

HUMAN LYMPHOTROPIC VIRUS TYPES I AND II (HTLV-I AND HTLV-II)

There are no data on the possible risk to dental staff from these viruses.

BLOOD-BORNE HERPESVIRUSES

Most, if not all, herpesviruses can be found in saliva, particularly in immunocompromised persons (Scully, 1995, 1996). They can pose an occupational risk for dental staff.

Herpes simplex virus (HSV) is commonly found in saliva (Hatherly *et al.*, 1980; Spruance, 1984; Kameyama *et al.*, 1985, 1989), and since many dental staff are non-immune (over 40% of dental students in USA) (Brooks, 1981) there would be a risk were they not to use barriers. Indeed many infections have been recorded in the past in dental staff (Corey and Spear, 1986; Scully, 1989, 1995, 1996).

Recent serological studies, however, show that HSV is no longer a significant risk (Herbert *et al.*, 1995), presumably because protective gloves are now routinely worn. Most dental staff are immune to varicella-zoster virus (VZV) and thus infection is now particularly unlikely.

Cytomegalovirus (CMV) can be found in saliva. There are many examples of outbreaks of infection in day-care and other facilities where young children are in close proximity, and seroepidemiological studies of dental staff treating patients of a general hospital in USA show them to be one of the highest risk groups at enrollment in the study, but with few subsequent seroconversions over a 6 month period (Gerberding, 1994). By contrast, seroepidemiology of dental staff in UK dental hospitals suggests little risk (Jones *et al.*, 1972; Herbert *et al.*, 1995) and outbreaks of transmission to dental staff have not been recorded (Epstein and Scully, 1993).

Epstein–Barr virus (EBV) is also found in saliva (Niederman *et al.*, 1976; Alsip *et al.*, 1988), and is transmitted in saliva, typically during kissing. There is a higher prevalence of EBV serum antibodies in clinical dental students and staff compared with preclinical students (Herbert *et al.*, 1995) but this does not prove that transmission to dental staff is a significant occupational risk.

Human herpesviruses 6,7 and 8 are also found in saliva, but any occupational risk is unclear, and the only available serological data suggest no special risk from HHV-6 (Herbert *et al.*, 1995).

The use of gloves and protective eyewear and mask should suffice to reduce transmission of herpesviruses to staff.

MEASURES TO REDUCE THE RISK OF TRANSMISSION OF BLOOD-BORNE VIRUS INFECTION TO DENTAL STAFF

Measures to minimize occupational transmission are discussed in other chapters in this book and elsewhere (Scully *et al.*, 1990b, 1994; Scully and Samaranayake, 1992; Ippolito *et al.*, 1993; CDC, 1993a; Miller, 1996).

Every member of the dental staff has a legal duty to ensure that all necessary steps are taken to prevent cross-infection to protect the patient, colleagues, themselves and their families or partners. All staff must be educated about the possible dangers of hepatitis, HIV and the other infections, their modes of transmission and the precautions necessary to prevent cross-infection, particularly vaccination against HBV. It is recommended that all members of the dental team should be fully immunized appropriately for the job they do and all should be confirmed as satisfactorily immunized against hepatitis B. A formal follow-up procedure should be introduced for the re-calling of all members of the team.

Routine safe working practice will ensure that both staff and patients are protected from transmission of infections from blood and body fluids. Immunocompromised staff should probably be absolved from the responsibility of treating virus-infected patients.

The most effective measure to prevent transmission is extreme precaution against accidental cuts and pricks from instruments or needles. Handwashing and hand care are important, and protective gloves must always be worn when in contact with body fluids (Molinari, 1995). Protective clothing, namely surgical gown, gloves, mask and eye protection, should be worn at all times during the treatment of all patients and also during the disinfection and cleaning of instruments and dental surgery. Despite suggestions to the contrary, the evidence is that gloves do not significantly reduce tactile discrimination (Solovan *et al.*, 1984; Brantley *et al.*, 1986; Wilson *et al.*, 1986; Hardison *et al.*, 1988; Matthews and Scully, 1989). Surgical gloves, however, do not provide reliably protection. They are readily perforated, often microscopically, and therefore must be changed between patients. Heavy domestic rubber gloves must be worn during instrument cleaning. Gloves must always be worn for clinical treatment by all personnel when touching blood, saliva, teeth or mucous membranes or items which have been in contact with them. Hands should be washed thoroughly when entering or leaving clinical areas and prior to eating or drinking. Soap dispensers and taps should be operated by the elbows or wrists, not gloves or hands. Open cuts and fresh abrasions to the skin should be covered with a waterproof dressing. Between each patient the gloves should be removed, hands washed and a new pair of gloves put on prior to proceeding. When wearing gloves contact with inanimate objects should be avoided as far as possible. When gloves are torn, cut or punctured, they must be removed immediately, hands thoroughly washed and regloving accomplished before completion of the dental procedure.

Eye protection should be worn by staff treating patients whenever dental aerosol or tooth fragments are generated. Patients should always be given eye protection during treatment. Eye washing facilities should be available in all clinical and laboratory areas in the event of an accident. Masks should be worn whenever dental aerosol or tooth fragments are generated.

There should be no eating, drinking, hair combing, teeth brushing, or application of cosmetics in clinical areas, including clinical laboratories (cloakrooms should be used). Any accident at work, however small, may be important. Re-sheathing of needles is to be avoided where possible:

- local anaesthetic syringe needles should be re-sheathed only when using an appropriate sheath-holding device;
- phlebotomy needles must not be re-sheathed but placed in a sharps container for disposal;
- all surgical sharps, glass items, burs, wire, etc. must be discarded into sharps containers, while ensuring that this is no more than three-quarters full. These boxes must be available in all areas where sharps are being used.

Gloves, masks and other protective clothing must not be worn or taken elsewhere, unless in an impervious container clearly labelled as infective. Clothing should be autoclaved before laundering.

All specimens for laboratory tests should be placed in appropriate containers and sealed into plastic bags separate from the request form. The bagged container and request form should then be sealed in a second bag for transportation to the laboratory.

In essence, therefore, the important precautions are:

1. All staff are immunized and boosters given appropriately.
2. Gloves should be worn if there is any chance of contact with blood, tissues or body fluids.
3. Cuts and grazes should always be covered with waterproof dressings.
4. Protective spectacles or goggles should always be worn.
5. Masks should be worn if there is a risk of splashes of blood or body fluids.
6. Extreme care should be taken with sharp instruments.
7. Needles should not be re-sheathed.
8. Used needles, syringes, scalpels and injection trays should be disposed of immediately into sharps containers.
9. Consideration should be given to amending working practices to avoid or reduce the handling of needles and sharp instruments.

THE ESSENTIALS OF INFECTION CONTROL (SEE ALSO ADDENDUM TO THIS CHAPTER)

Equipment

Disposable instruments should be used wherever possible and local anaesthetic cartridges and needles must never be re-used for any other patient. Used equipment should be clearly identified as infected and always handled with gloves before discarding into an impervious container, or washing and autoclaving.

Sterilization

Instruments, needles, etc. must be placed in an impervious container before sterilization or incineration, and must be labelled as infective. Disposable instruments, dressings, etc. should be incinerated. Non-disposable instruments should be rinsed in an effective disinfectant and sterilized immediately by autoclaving (134°C for 3 min) or hot air (160°C for 1 h). In hospital practice, ethylene oxide gas (10% concentration in carbon dioxide) at 55–69°C for 8–10 h can be used. It should be emphasized that solutions of ethyl or isopropyl alcohol, quaternary ammonium compounds or chlorhexidine are unreliable for inactivating viruses. Boiling instruments in a dental boiling water bath for 30 min is also unreliable.

Non-disposable instruments and dental impressions that cannot be sterilized by heat should be disinfected by immersion for at least 1 h (preferably overnight) in a suitable disinfectant such as hypochlorite (10,000 ppm available chlorine).

All working surfaces should be covered with disposable material. Working areas are disinfected with hypochlorite (1000 ppm available chlorine). Since HBsAg remains stable in dried blood for up to 4 months at room temperature (ACDP, 1995) and the survival of the other viruses is uncertain, spillage of blood should be disinfected by sprinkling it with sodium dichloroisocyanurate granules, or placing disposable tissues over it and pouring hypochlorite, 10,000 ppm available chlorine, over them, and leaving the agent to act for 30 min (see Hoffman and Kennedy, and Collins and Kennedy, Chapters 17 and 18 of this volume).

MANAGEMENT OF NEEDLESTICK INJURIES

This is discussed in Chapters 5, 6 and 7 of this volume, and elsewhere (Stewart et al., 1994). Should the skin be punctured by an instrument that has been used on a patient, body fluid or tissue, the area of skin should be liberally rinsed in water and the advice of the nearest Public Health Laboratory or hospital microbiologist sought. Where appropriate, blood from the patient on whom the instrument was used, and from the wounded person should be tested for HBeAg, HCV, HDV and HIV antibodies to determine the possible risks.

Any human material may be infected but certain patients present high inoculation risk. These may include those who are:

- HBsAg positive and HBeAb negative;
- HBeAg positive and HBeAb negative;
- Established cases of AIDS or HIV disease;
- IV drug users.

BEFORE A CLINICAL SESSION

Water should be run through each of the dental unit water systems (3 in 1 syringe, air-rotor coolants, etc.), according to manufacturer's recommendations.

Working surfaces and dental equipment should be cleaned and disinfected with hypochlorite–detergent mixture containing 1000 ppm available chlorine and 0.1% detergent.

The number of items of equipment and instruments laid out ready for use should be reduced to a minimum. Only supplies of frequently used materials should be laid out and where possible covered with lids.

INSTRUMENTS

Wherever possible instruments should be sterilizable or disposable. All instruments, including handpieces, must have been sterilized prior to use and should be laid on a sterile or clean tray using the appropriate sterile or clean working technique.

Cling film should be placed across the dental chair control buttons, operating light handles, ultrasonic scaler handpiece and 3-in-1 syringe bodies. The film must be changed or decontaminated with hypochlorite-detergent solution (see above) after every patient.

DURING A CLINICAL SESSION

High and low contamination zones must be defined and the surgery arranged accordingly. A small area around the patient which includes the dental unit and extends to include the waste disposal bag (Zone A) should contain only essential equipment, instrument, and materials. Only essential personnel should remain in or enter this area of potential high contamination. An open yellow waste disposal bag should be attached to a worktop or wall close to the operator and ensure a sharps container is accessible. In the remainder of the treatment area (Zone B), the contents should be kept to a minimum.

A good working posture should be maintained to reduce facial contamination from the patient's mouth. High volume suction should be used to reduce dental aerosols.

Touching anything with contaminated gloves/hands other than essential items should be avoided.

Blood or body fluid spillage must be dealt with as described on p. 141 above.

CLEARING AND CLEANING AFTER EACH PATIENT

The greatest risk of injury is during the clean up and disposal stage. It is advisable to wear heavy duty rubber gloves and to wash them during clean up procedures to reduce the spread of infection. Sharps should be removed first and placed in sharps containers.

Waste should be disposed of in plastic bags as follows:

*   Black – domestic waste, paper and non-infected items
*   Yellow – clinical waste and potentially infected items.

Yellow bags should be labelled appropriately to indicate source. It is the responsibility of clinicians to ensure that impressions are microbiologically safe before they leave the clinic. Alginate impressions should be sprayed with an alcoholic microbicide (e.g. Microzid). Dimensionally stable impressions should be treated with an aldehyde-free disinfectant (e.g. Durr).

EQUIPMENT

Detachable handpieces, ultrasonic scalers, aspiration tips, and 3-in-1 tips should be sterilized.

The chair, bracket table including the body of the 3-in-1 syringe, slow motor and the holder, operating light, and spittoon should be disinfected. All surfaces should be wiped with hypochlorite-detergent solution.

Any residual cement or impression material should be removed from handles, etc. and wiped with hypochlorite-detergent solution.

Anything likely to be contaminated with blood should be wiped thoroughly with strong hypochlorite.

All linen used in surgical procedures should be placed in a *red* security-fastened alginate-stitched bags.

GENERAL NOTES ON THE USE OF DISINFECTANTS AND STERILIZATION

Sterilization

Autoclaving
Sterilization of all items is the aim and can be achieved with autoclave or hot air oven but is difficult to achieve with disinfectants. Autoclaving is the method of choice for most instruments. There are two main types:

1. *Small, non-porous load.* For example the Little Sister, which is suitable for unwrapped instruments. Unperforated metal boxes need to be upturned to allow steam penetration and must be closed on removal from the autoclave. Only distilled water should be used when re-filling. The autoclave must not be allowed to boil dry.

The approximate cycle time is 15 min; this includes 134°C for a minimum of 3 minutes. Indicators (e.g. Bowie Dick type test tape) must be used with each load.

2. *Post-vacuum, porous load.* These are suitable for dressings and wrapped instruments.

In order to ensure autoclave functioning correctly, sterilization indicators (e.g. Bowie Dick type test tape) must be incorporated into every load.

Autoclave faults If a fault is noted whilst operating a sterilizer, the main electricity supply should be switched off and a warning notice 'not to be used' placed by the instrument. The defect should be reported. No attempt should be made to force the door of a faulty autoclave.

Disinfectants
Items that cannot withstand autoclaving e.g. anaesthetic masks, may have to be disinfected.

Choice of agent Care must be taken in choosing the correct agent, e.g. chlorhexidine is ineffective against hepatitis B, and therefore hypochlorite solution should be used.

Concentration Effectiveness decreases with: (a) age, therefore the expiry date should be checked; (b) use: fresh solutions should be made up according to the manufacturer's instructions. Disinfectants *must* be used at the correct concentration.

Cleaning Cleaning reduces the amounts of blood, saliva, dental materials and oils, which will otherwise reduce the effectiveness of disinfectants.

Contact Instruments must be in full contact with the disinfectant for an adequate immersion time.

Care Many disinfectants are corrosive chemicals which may damage metal instruments and fabrics as well as the skin and eyes.

Inhalation of fumes can be harmful – disinfectants should be kept in covered containers.

For further information about disinfectants see Hoffman and Kennedy, Chapter 17 this volume.

Environmental disinfectants
Agents
1. Sodium hypochlorite 10,000 ppm of available chlorine (e.g. undiluted Milton). Use this for the disinfection of areas contaminated with blood. Allow at least 20 min contact time.
2. Hypochlorite–detergent solutions – sodium hypochlorite, 1000 ppm available chlorine with 1% detergent (e.g. Lapol's liquid detergent). Use for disinfection of work surfaces – undiluted for swabbing down. Allow to dry.
3. Durr 212 – an aldehyde-free disinfectant with cleaning agent. Mix 20 ml with 1 litre of water to make a 2% solution. Undiluted solution has a shelf life of 2 years; diluted solution should be changed daily. Use to disinfect articles which cannot be autoclaved; immerse for at least 20 minutes. In case of known inoculation risk patients, immerse for at least 3 hours.
4. Phenolic type solution, e.g. Orotol. Add 50 ml to 1 litre of water. Use to disinfect and clean off suction apparatus.

After immersion in disinfectants all items must be thoroughly rinsed, especially hollow items, before being used in the mouth.

Skin surface antiseptics
1. Povidone iodine 7.5% surgical scrub (e.g. Betadine – surgical scrub). Use for hand and glove washing before and after clinical procedures. Rinse hands and dry thoroughly with disposable hand towel.
2. Chlorhexidine 4% surgical scrub (e.g. Hibiscrub). NB: for staff who are allergic to iodine. Use as above but note that gloves become sticky after washing with Hibiscrub.
3. 70% alcohol impregnated swabs (e.g. Medi-swabs). Use for skin preparation prior to injection; allow to dry before commencing procedure.

ADDENDUM: A TO Z OF RECOMMENDED PROCEDURES

Amalgam carriers (plastic or metal). Ensure free of amalgam, clean, dismantle, autoclave, store dry.

Anaesthetic equipment (e.g. laryngoscope, props, masks). Follow manufacturers' instructions – autoclave where possible, otherwise immerse in Durr 212 2% solution for 60 minutes, rinse in water, store dry.

Autoclave. Damp-dust outer casing and door seal daily. Wipe handle and controls with suitable detergent. Check distilled water level before use. Run regular tests.

Beakers. Disposable types recommended.

Bibs. Disposable types recommended. Clean plastic bibs with water and hypochlorite-detergent, then allow to dry.

Broaches. Plain or barbed – dispose into sharps container.

Bur brush. Autoclave at end of each session.

Burs. Clean thoroughly, using an ultrasonic bath or bur brush if appropriate. Autoclave and dry thoroughly. Surgical cases use pre-packed sterile burs.

Butterfly sponge pack. Wash under running water then autoclave, store dry but moisten before use. Not disposable.

Cheatle forceps. Clean, autoclave, store in a clean and dry place.

Composition bath. Empty, discard any material remaining. Wipe with hypochlorite-detergent after each patient.

Containers. Clean with detergent, rinse and dry.

Crown and bridge remover. Autoclave.

Crown forms. Tried in but unused – disinfect in Durr 212 2% solution.

Curing light. Wipe tip and outer casing with surface disinfectant.

Dappens pot – glass or plastic. See 'Containers'. For inoculation risk use plastic disposable type.

Denture brushes. Autoclave.

Drawers and cupboards. Clean weekly using hypochlorite-detergent and allow to dry.

Endodontic RCT instruments. Use pre-sterilized, otherwise collect a set of root canal instruments and autoclave them immediately prior to use. After use, clean in ultrasonic bath, arrange items in endobox and sterilize.

Face bows. Remove wax and debris. Wipe with a suitable disinfectant.

Floors. General – a domestic duty. If blood spillage, clean as for work surfaces.

Forceps. As for instruments but use pre-sterilized packs where possible.

Gauze, e.g. swabs, throat packs. Use pre-sterilized packs and dispose of carefully in clinical waste bag.

Glass. Slab – wipe clean immediately after use then wash with water and detergent, rinse and dry. Tumblers – clean then autoclave. Disposable mouthwash cup preferred.

Glasses (protective). After use, clean with water and detergent and store dry.

Gloves. Non-sterile gloves are available for use when working with patients or handling dirty instruments. Sterile gloves should be reserved for use with surgical cases. Domestic gloves should be used for cleaning purposes.

Gowns. Paper – disposable. Cotton, to laundry.

Gutta Percha points. Disposable.

Handpieces. Faults – sterilize prior to returning to the workshop. Laboratory – wipe with surface disinfectant after use on patient's dentures or appliances. Bracket for handpieces – wipe all surfaces with disinfectant between patients.

Hand washing. Use wrist or elbow to operator water tap. Use liquid soap, antiseptic scrub, e.g. Betadine, Hibiscrub in warm water, rinse thoroughly and dry on paper towel. Surgical procedures special washing technique – cover all cuts and grazes.

Impression composition. Wrap in gauze (sufficient for one patient). Discard after use.

Impression trays. Metal – sterilize trays not used after trying in the mouth. Disposable – discard impression material together with disposable trays into yellow plastic waste bags.

Instrument trays. After clearing, wipe with hypochlorite-detergent. Re-lay with cling film.

Lead apron. Wipe with hypochlorite-detergent and allow to dry.

Local anaesthetic kit. Syringe (metal) – wash, place in ultrasonic bath, autoclave (store dry). Cartridges-store in manufacturers packages until loading. Do not touch 'needle end'. Check expiry date. After use dispose in sharps container. Needles – do not break seal until needed. Re-sheathing – see 'Needles' below. Dispose in sharps container.

Metal instruments (non-surgical) and matrix retainers. Autoclave, store in clean and dry container.

Mirror (handheld face). Wipe with water and hypochlorite-detergent.

Mixers. Dentomat/Silamat – use on tray, wipe with detergent at end of each session.

Nail brushes. Use prepacked, sterilized.

Needles. Disposable needles must *never* be re-used on another patient. Re-sheathing of any needles should be avoided wherever possible.

Needlestick injury. Rinse wound under running water, encourage bleeding and then protect. Immediately report injury to member of staff in charge and follow Sharps Policy.

Occlusal plane guide. Clean, disinfect with hypochlorite-detergent.

Operating light. Switch off light. Wipe with hypochlorite-detergent. Do not over-wet. Do not wet back of light when warm. Light switch – protect with cling film

Orthodontic items. Autoclave whenever possible – if not then immerse in 2% Durr 212 solution for 20 minutes, rinse thoroughly. Blow-dry pliers with box joints after rinsing. Store dry.

Paper points. Autoclave. Use pre-sterilized packs or use a suitable autoclave cycle.

Photographic items, cheek and lip retractors, intra-oral mirrors. Clean then disinfect.

Plastic items, i.e. instruments, spatulas, etc. Check if item is meant to be disposed of or not. If not, clean and immerse in Durr 212.

Pliers. Oil joint if necessary, then as for orthodontic items.

Pulp tester. Switch off. Clean and disinfect probe tip. Wipe casing with surface disinfectant.

Reamers. See endodontic items.

Repairs, relines, rebases. Appliances or dentures from inoculation risk patients which need to be repaired or relined – immerse in a 2% Durr 212 for half an hour and rinse before proceeding.

Resuscitation kit. As for anaesthetic equipment and mask.

Rubber dam equipment. Clamp, clamp forceps – clean and autoclave. Frame, metal or plastic, punch – wipe with hypochlorite-detergent.

Scalpel blades. Use pre-packed, sterile. Remove blade using artery forceps, and place immediately into sharps container.

Scissors. Clean, check for sharpness, oil if necessary, autoclave.

Shade guide. Check manufacturers' guidance. Unless otherwise indicated, wipe or dip in a suitable detergent, store dry.

Sharp items. After use these *must* be placed in the sharps containers provided, together with other discarded 'sharp' items such as scalpel blades, hypodermic and endodontic syringes, burs, root canal instruments.

Silver points. Autoclave.

Skin preparation for venepuncture. Wipe area with 70% alcohol swab (mediswab).

Spatulas. Physically clean, autoclave or disinfect.

Specimens. Identify containers with patient's name and Institute number prior to use. For high-risk patients both the specimen and request card should carry a recognizable hazard label, 'Danger of Infection'. Gloves must be worn at all times by all staff involved in the collection of specimens from any patient. The specimen bottle and request card should be placed in separate compartments of the double-sleeved plastic specimen bag. All patients' specimens should be double-bagged for transport to the laboratory.

Spittoon. After each patient – flush through with water, wipe outside of bowl with damp tissue, check no debris or blood spots. After each session – as for 'after each patient' plus put half cupful of Orotol solution in bowl and leave.

Stainless steel instruments. See metal instruments. These should not come into contact with hypochlorite.

Suction apparatus. Before each patient: routine cases – check that the system works by drawing through clean water and set up new or sterile tips. Surgical cases – set up new or sterile tubing and sterile tips. Mobile aspirator – quarter fill bottle with strong hypochlorite solution (or Orotol) but *not* hypochlorite-detergent which may froth.

After each patient: routine cases – remove tips for disposal or autoclaving. Wipe down tubing with hypochlorite-detergent. Surgical cases – as for routine cases then remove tubing for cleaning and sterilization or disposal. Mobile aspirator – carefully empty contents of bottle into main drainage system, flush away with running water.

End of day: suction tips – draw water through on unit them remove. Check metal or hard plastic for cleanliness inside and outside. Autoclave. Central suction – flush with 1 litre of water. Wearing rubber gloves, remove waste trap filter, clean and replace then run Orotol through tubing. Clean and disinfect suction equipment before sending for repair.

Sutures. Use pre-sterilized packs. Dispose of needle into sharps container.

Syringes (a) Cartridge – dismantle and autoclave. Chip-clean metal tube outside and inside (by removal of any debris with wire and then irrigation with Durr 212 solution), autoclave bulb and tube. (b) Hypodermic – pre-sterilized and disposable. (c) Impression: (i) disposable; (ii) metal with disposable tips – autoclave; (iii) automixing syringes – suitable disinfectant. (d) Three-in-one – autoclave tip. Handle and bracket holder – wipe with hypochlorite-detergent.

Thermometers. After use immerse in 2% Durr 212 for 20 minutes – rinse and store dry in safe place.

Toothbrushes. Preferably use patient's own brush – demonstration brushes used in a patient's mouth should be given to the patient to take home. Denture brushes should be cleaned and autoclaved.

Trolleys. Before use – wipe with hypochlorite-detergent.

Ultrasonic bath. Used to clean instruments prior to sterilization.

Ultrasonic scalers. As for metal instruments tips.

Venepuncture. First wipe the site with an alcohol swab, e.g. Mediswab. Take care not to spill any blood. If minor spillage does take place mop up immediately then wipe affected area with sodium hypochlorite 10,000 ppm. Dispose of needle with syringe directly into sharps container and other items into a clinical waste bag. Provide dressings to cover injection site immediately following withdrawal of the needle.

Wax knife. Clean off wax, wipe with hypochlorite-detergent and autoclave if possible.

Willis bite gauge. As for metal instruments – autoclave.

Work surfaces. Before clinical session – wipe with hypochlorite-detergent on fresh cloth or paper towel. After each patient – wipe with hypochlorite-detergent. For blood-contaminated surfaces apply liberal amount of hypochlorite, 10,000 ppm available chlorine and then dry.

X-ray equipment. X-ray machine – switch off machine, wipe with hypochlorite-detergent. Cassettes – wipe with hypochlorite-detergent after use. File holders – wash to remove saliva then autoclave. Alternative is immerse in 2% Durr 212 for 20 minutes. X-ray films – after removal from the mouth, dry the film. Discard the empty film packet as clinical waste. Wash hands. Inoculation risk patients – seal each intra-oral film in a plastic bag ready for use. After exposure – film can be removed from plastic bag (without it touching the contaminated outside) and then processed as normal.

REFERENCES

ACDP (1995) *Categorization of Biological Agents According to Hazard and Categories of Containment*, 4th edn. *Advisory Committee on Dangerous Pathogens.* HSE Books, Sudbury.

Abe, K. and Inchauspe, G. (1991) Transmission of hepatitis C by saliva. *Lancet* 337, 248.

Abe, K., Kurata, T., Shikata, T. *et al.* (1987) Experimental transmission of non-A, non-B hepatitis by saliva. *Journal of Infectious Disease* 155, 1078–1079.

Aldershvile, J., Brock, A., Dietrichson, O. *et al.* (1978) Hepatitis B virus infections in Danish dentists. *Journal of Infectious Diseases* 137, 63–66.

Allander, T., Gruber, A., Naghari, M. *et al.* (1995) Frequent patient-to-patient transmission of hepatitis C virus in a haematology ward. *Lancet* 345, 603–607.

Allen, A.L. and Organ, R.J. (1982) Occult blood accumulation under fingernails; a mechanism for the spread of blood-borne infections. *Journal of the American Dental Association* 105, 455–459.

Almeida, O.P., de, de Souza Filho, F.J., Scully, C. *et al.* (1997) HIV in dental out-patients in Brazil. *Oral Surgery*, in press.

Alsip, G.R., Ench, Y., Sumaya, C.V. *et al.* (1988) Increased Epstein-Barr virus DNA in oropharyngeal secretions from patients with AIDS, AIDS-related complex, or asymptomatic human immunodeficiency virus infections. *Journal of Infectious Disease* 157, 1072–1076.

Alter, M.J. (1991) Inapparent transmission of hepatitis C: footprints in the sand. *Hepatology* 14, 389–391.

Amerena, V. and Andrew, J.H. (1987) Hepatitis B virus: the risk to Australian dentists and dental health care workers. *Australian Dental Journal* 32, 183–198.

Anon. (1987) Transmission of HIV by human bite. *Lancet* ii, 522.

Archibald, D.W., Zon, L., Groopman, J.E. *et al.* (1986) Antibodies to human T-lymphotrophic virus type III (HTLV-III) in saliva of acquired immunodeficiency syndrome (AIDS) patients and in persons at risk for AIDS. *Blood* 67, 831–834.

Archibald, D.W., Barr, C.E., Totosian, J.P. *et al.* (1987) Secretory IgA antibodies to human immunodeficiency virus in the parotid saliva of patients with AIDS and AIDS-related complex. *Journal of Infectious Disease* 155, 793–796.

Atkinson, J.C., Yeh, C.K., Oppenheim, F.G. *et al.* (1990) Elevation of salivary antimicrobial proteins following HIV-1 infection. *Journal of Acquired Immune Deficiency Syndromes* 3, 41–48.

Barr, C.E., Miller, L.K. and Lopez, M.R. (1992) Recovery of infectious HIV-1 from whole saliva. *Journal of the American Dental Association* 123, 37–45.

Bass, B.D., Andors, L., Pierri, L.K. *et al.* (1982) Quantitation of hepatitis B viral markers in a dental school population. *Journal of the American Dental Association* 104, 629–632.

Beasley, R.P. and Hwang, L-Y. (1983) Postnatal infectivity of hepatitis B surface antigen-carrier mothers. *Journal of Infectious Diseases* 147, 185–190.

Ben-Aryeh, H., Ur, I. and Been-Porath, E. (1985) The relationship between antigenaemia and excretion of hepatitis B surface antigen in human whole saliva and in gingival crevicular fluid. *Archives of Oral Biology* 30, 97–99.

Bennet, N.W., Carson, J.A., Fish, B.S. *et al.* (1985) An assessment of the prevalence of hepatitis B among health care personnel in Victoria. *Medical Journal of Australia* 143, 495–499.

Bond, W.W., Favero, M.S., Petersen, N.J. *et al.* (1981) Survival of hepatitis B virus after drying and storage for one week. *Lancet* i, 550.

Boozer, C.H., Shopper, T.H. and Weinberg, R. (1986) The presence of anti-HBc in a dental school patient population. *Journal of the American Dental Association* 112, 854–856.

Bowden, J.R., Scully, C. and Bell, C.J. (1989) Cross-infection control; attitudes of patients towards the wearing of gloves and masks by dentists in the United Kingdom in 1987. *Oral Surgery, Oral Medicine and Oral Pathology* 67, 45–48.

Brantley, C.F., Heymann, H.O., Shugars, D.A. *et al.* (1986) The effect of gloves on psychomotor skills acquisition among dental students. *Journal of Dental Education* 50, 611–613.

Breuer, B., Friedman, S.M., Millner, E.S. *et al.* (1985) Transmission of hepatitis B virus to classroom contacts of mentally retarded carriers. *Journal of the American Medical Association* 254, 3190–3195.

Broderson, M. (1974) Salivary HBsAg detected by radioimmunoassay. *Lancet* i, 675.

Brooks, S.L. (1981) Prevalence of herpes simplex virus disease in a professional population. *Journal of the American Dental Association* 102, 31–34.

Cancio-Bello, T.P., de Medina, M., Shorey, J. *et al.* (1982) An institutional outbreak of hepatitis B related to a human biting carrier. *Journal of Infectious Disease* 146, 652–656.

Capilouto, E.I., Weinstein, M.C., Hemenway, D. *et al.* (1992) What is the dentist's occupational risk of becoming infected with hepatitis B of the human immunodeficiency virus? *American Journal of Public Health* 82, 587–589.

CDC (1993a) Recommended infection-control practices for dentistry. *Morbidity and Mortality Weekly Report* 42, 1–11.

CDC (1993b) *HIV/AIDS Surveillance Report* 5, 13.

CDC (1995) Case-control study of HIV seroconversion in health-care workers after percutaneous exposure to HIV-infected blood – France, United Kingdom, and United States, January 1988–August 1994. *Morbidity and Mortality Weekly Report* 44, 929–932.

Chaudhary, R.K., Perry, E., Hicks, F. *et al.* (1992) Hepatitis B and C infection in an institution for the developmentally handicapped. *New England Journal of Medicine* 327, 1953.

Cleveland, J., Lockwood, S., Gooch, B. *et al.* (1995) Percutaneous injuries in dentistry: an observational study. *Journal of the American Dental Association* 126, 745–751.

Corey, L. and Spear, P.G. (1986) Infections with herpes simplex virus. *New England Journal of Medicine* 314, 686–681 and 749–757.

Cottone, J.A. and Puttaiah, R. (1996) Hepatitis B virus infection. *Dental Clinics of North America* 40, 293–307.

Couzigou, P., Richard, L., Dumas, F. *et al.* (1993) Detection of HCV-RNA in saliva of patients with chronic hepatitis C. *Gut* 34 (Suppl), S59–S60.

Cummings, C.G., Peutherer, J.F. and Smith, G.L.F. (1986) The prevalence of hepatitis B serological markers in dental personnel. *Journal of Infection* 12, 157–159.

Davies, K.J., Herbert, A.M., Westmoreland, D. *et al.* (1994) Seroepidemiological study of respiratory virus infections among dental surgeons. *British Dental Journal* 176, 262–265.

Davis, L.G., Weber, D.J. and Lemon, S.M. (1989) Horizontal transmission of hepatitis B virus. *Lancet* i, 889–893.

Davison, F., Alexander, G.J.M., Trowbridge, R. *et al.* (1987) Detection of hepatitis B virus DNA in spermatozoa, urine, saliva, and leucocytes, of chronic HBsAg carriers. A lack of relationship with serum markers of replication. *Journal of Hepatology*, 37–44.

Dodson, T.B., Nguyen, T. and Kaban, L.B. (1993) Prevalence of HIV infection in oral and maxillofacial surgery patients. *Oral Surgery, Oral Medicine, Oral Pathology* 76, 272–275.

Doublog, J.H., Gerner, N.W., Hurlen, B. *et al.* (1988) HIV and hepatitis B infection in an international cohort of dental hygienists. *Scandinavian Journal of Dental Research* 96, 448–450.

Dusheiko, G.M., Smith, M. and Scheuer, P.J. (1990) Hepatitis C virus transmitted by human bite. *Lancet* 336, 503–504.

Ebbeson, P., Melbye, M., Scheutz, F. *et al.* (1986) Lack of antibodies to HTLV-III/LAV in Danish dentists. *Journal of the American Medical Association* 256, 2199.

Eisenburg, J., Holl, J., Kruis, W. *et al.* (1977) Hepatitis risk to dentists. Studies on the frequency of hepatitis in dentists in the Munich region and in the district of Upper Bavaria. *Fortschritte der Medizin* 95, 1249–1258.

Epstein, J.B. and Scully, C. (1993) Cytomegalovirus; a virus of increasing relevance to oral medicine and pathology. *Journal of Oral Pathology and Medicine* 22, 348–353.

Epstein, J.B., Bucher, B.K. and Bouchard, S. (1984) Hepatitis B and Canadian dental professionals. *Journal of the Canadian Dental Association* 7, 555–559.

Epstein, J.B., Porter, S.R. and Scully, C. (1992) Non-A, non-B hepatitis and dentistry. *American Journal of Dentistry* 5, 49–56.

Everhart, J.E., Bisceglie, A.M.D., Murray, C.M. *et al.* (1990) Risk for non-A non-B (type C) hepatitis through sexual or household contact with chronic carriers. *Annals of Internal Medicine* 112, 544–545.

Faoagali, J.L. (1986) Hepatitis B markers in Canterbury dental workers: a sero-epidemiological survey. *New Zealand Medical Journal* 99, 12–14.

Feldman, R.E. and Schiff, E.R. (1975) Hepatitis in dental professionals. *Journal of the American Medical Association* 232, 1228–1230.

Felix, D.H., Bird, A.G., Anderson, H.G. *et al.* (1994) Recent non-sterile inoculation injuries to dental professionals in the Lothian Region of Scotland. *British Dental Journal* 176, 180–184.

Figueredo, J.F.C., Borges, A.S., Martinez, R. *et al.* (1994) Transmission of hepatitis C virus but not human immunodeficiency virus type 1 by a human bite. *Clinics in Infectious Disease* 19, 546–547.

Flynn, N.M., Pollet, S.M. and Van Horne, J.R. (1987) Absence of HIV antibody among dental professionals exposed to infected patients. *Western Medical Journal* 146, 439–442.

Fong, T.L,, Charboneau, F., Valinluk, B. *et al.* (1993) The stability of serum hepatitis C viral RNA in various handling and storage conditions. *Archives of Pathology and Laboratory Medicine* 117, 150–151.

Fox, P.C., Wolff, A., Yeh, C.K. *et al.* (1988) Saliva inhibits HIV-1 infectivity. *Journal of the American Medical Association* 116, 635–637.

Fox, P.C, Wolff, A., Yeh, C.K. *et al.* (1989) Salivary inhibition of HIV-1 infectivity; functional properties and distribution in men, women and children. *Journal of the American Dental Association* 118, 709–711.

Fried, M.W., Shindo, M., Fong, T-L. *et al.* (1992) Absence of hepatitis C viral RNA from saliva and semen of patients with chronic hepatitis C. *Gastroenterology* 102, 1306–1308.

Friedland, G.H. and Klein, R.S. (1987) Transmission of the human immunodeficiency virus. *New England Journal of Medicine* 318, 86–90.

Fultz, P.N. (1986) Components of saliva inactivate human immunodeficiency virus. *Lancet* ii, 1215.

Gerberding, J.L. (1994) Incidence and prevalence of human immunodeficiency virus, hepatitis B virus, hepatitis C virus and cytomegalovirus among health care personnel at risk for blood exposure: final report from a longitudinal study. *Journal of Infectious Disease* 170, 1410–1417.

Gerberding, J.L., Bryant Le Blanc, C.E. and Greenspan, D. (1986) Risk of dentists from exposure to patients infected with the AIDS virus. 26th Interscience Conference on Antimicrobial Agents and Chemotherapy. Abstract 1015, p. 283.

Gerberding, J.L., Bryant-LeBlanc, C.E., Nelson, K. *et al.* (1987) Risk of transmitting the human immunodeficiency virus, cytomegalovirus and hepatitis B virus to health care workers exposed to patients with AIDS and AIDS-related conditions. *Journal of Infectious Disease* 156, 1–8.

Glenwright, H.E., Edmondson, H.D., Whitehead, R.I.H. *et al.* (1974) Serum hepatitis in dental surgeons. *British Dental Journal* 136, 409–413.

Goebel, W.M. and Gitnick, G.L. (1979) Hepatitis B virus infection in dental students: a two year evaluation. *Journal of Oral Medicine* 34, 33–36.

Gooch, B.F., Cardo, D.M., Marcus, R. *et al.* (1995) Percutaneous exposure to HIV-infected blood among dental workers enrolled in the CDC needlestick study. *Journal of the American Dental Association* 126, 1237–1242.

Goto, Y., Yeh, C-K., Notkins, A.L. *et al.* (1991) Detection of proviral sequences in saliva of patients infected with human immunodeficiency virus type 1. *AIDS Research and Human Retrovirology* 7, 343–347.

Goubran, G.F., Cullens, H., Zuckerman, A.J. *et al.* (1976) Hepatitis B virus infection in dental surgical practice. *British Medical Journal* 2, 559–560.

Groopman, J.E., Salahuddin, S.Z., Sarngadharan, M.G. *et al.* (1984) HTLV-III in saliva of people with AIDS-related complex and healthy homosexual men at risk for AIDS. *Science* 226, 447–449.

Gruninger, S.E., Siew, C., Chang, S-B. *et al.* (1992) Human immunodeficiency virus type 1 infection among dentists. *Journal of the American Dental Association* 123, 57.

Hardison, D., Scarlett, M.I., Lyon, E. *et al.* (1988) Gloved and ungloved; performance time for two dental procedures. *Journal of the American Dental Association* 116, 691–696.

Hardt, F., Aldershvile, J., Dietrichson, O. *et al.* (1979) Hepatitis B virus infection in Danish surgeons. *Journal of Infectious Diseases* 140, 972–974.

Hatherly, L.I., Hayes, K. and Jack, I. (1980) Herpes virus in an obstetric hospital; asymptomatic virus excretion in staff members. *Medical Journal of Australia* 2, 273–275.

Heathcote, J., Cameron, C.H. and Dane, D.S. (1974) Hepatitis B antigen in saliva and semen. *Lancet* i, 71.

Herbert, A.M., Walker, C.M., Davies, K.J *et al.* (1992) Occupationally acquired hepatitis C virus infection. *Lancet* 339, 305.

Herbert, A.M., Bagg, J., Walker, D.M. *et al.* (1995) Seroepidemiology of herpes virus infections among dental personnel. *Journal of Dentistry* 23, 339–342.

Ho, D.D., Byington, R.E., Schooley, R.T. *et al.* (1985) Infrequency of isolation of HTLV-III virus from saliva in AIDS. *New England Journal of Medicine* 313, 1606.

Hogan, B. and Samaranayake, L.P. (1990) The surgical mask unmasked; a review. *Oral Surgery, Oral Medicine and Oral Pathology* 70, 34–36.

Hollinger, F.B. and Lin, H.J. (1992) Community-acquired hepatitis C virus infection. *Gastroenterology* 102, 1426–1429.

Hollinger, F.B., Grander, J.W. and Nickel, F.R. (1977) Hepatitis B prevalence within a dental student population. *Journal of the American Dental Association* 94, 521–527.

Hsu, H.H., Wright, T.L., Luba, D. *et al.* (1991) Failure to detect hepatitis C virus genome in human secretions with the polymerase chain reaction. *Hepatology* 14, 763–767.

Hurlen, B. (1980) Salivary HBsAg in hepatitis B infection. *Acta Odontologica Scandinavica* 38, 51–55.

Hurlen, B., Iversen, S.B. and Jonsen, J. (1979) Frequency of hepatitis in dental personnel in Norway. *Acta Odontologica Scandinavica* 37, 189.

Ippolito, G., Petrosillo, N. and Puro, V. (1993) Il rischio di infezione da HIV in odontoiatria. In: Infezioni da HIV e sindrome da immunodeficienza acquisita. *Rapporti Istisan* 5, 3–8.

James, S.P. and Sampliner, R.E. (1978) Hepatitis B in the dental setting. *Journal of Maryland State Dental Association* 32, 183–189.

Jenison, S.A., Lemon, S.M., Baker, L.N. *et al.* (1987) Quantitative analysis of hepatitis B virus DNA in saliva and semen of chronically infected homosexual men. *Journal of Infectious Diseases* 156, 299–307.

Jones, D.M., Tobin, J.H. and Turner, E.P. (1972) Australia antigen and antibodies to Epstein-Barr virus, cytomegalovirus and rubella virus in dental personnel. *British Dental Journal* 132, 489–491.

Kakizawa, J., Ushijima, H., Morishitam Y, *et al.* (1996) Diversity of HIV type 1 envelope V3 loop region in saliva. *AIDS Research and Human Retroviruses* 12, 561–563.

Kameyama, T., Kabashima, M., Futami, M. *et al.* (1985) Isolation of herpes simplex virus from saliva in patients operated. *Journal of the Japanese Dermatology Society* 34, 397–405.

Kameyama, T., Futami, M., Nakayoshi, N. *et al.* (1989) Shedding of herpes simplex virus type 1 into saliva in patients with orofacial fracture. *Journal of Medical Virology* 28, 78–80.

Karayiannis, P., Novick, D.M. and Lok, A.S.F. (1985) Hepatitis B virus DNA in saliva, urine and seminal fluid of carriers of hepatitis B e antigen. *British Medical Journal* 290, 1853–1857.

Kawashima, H., Bandyopadhyay, S., Rutstein, R. *et al.* (1991) Excretion of human immunodeficiency type 1 in the throat but not in urine by infected children. *Journal of Pediatrics* 118, 80–82.

Kiyosawa, K., Sodeyama, T., Tanaka, E. *et al.* (1991) Intrafamilial transmission of hepatitis C virus in Japan. *Journal of Medical Virology* 33, 114–116.

Klein, R.S., Phelan, J.A., Freeman, K. *et al.* (1988) Low occupational risk of human immunodeficiency virus infection among dental professionals. *New England Journal of Medicine* 318, 86–90.

Klein, R.S., Freeman, K., Taylor, P.E. *et al.* (1991) Occupational risk for hepatitis C infection among New York City dentists. *Lancet* 338, 1539.

Komiyama, K.. Moro, I., Mastuda, Y. *et al.* (1991) HCV in saliva of chronic hepatitis patients having dental treatment. *Lancet* 338, 572.

Kuo, M.Y-P., Hahn, L.-J., Hong, C.-Y. *et al.* (1993) Low prevalence of hepatitis C virus infection among dentists in Taiwan. *Journal of Medical Virology* 40, 10–13.

Leichtner, A.M., Leclair, J., Goldman, D.A. *et al.* (1981) Horizontal nonparenteral spread of hepatitis B among children. *Annals of Internal Medicine* 94, 346–349.

Levy, J.A. and Greenspan, D. (1988) HIV in saliva. *Lancet* 2, 1248.

Lewis, D.L., Arens, M., Appleton, S.S. *et al.* (1992) Cross-contamination potential with dental equipment. *Lancet* 340, 1252–1254.

Liou, T.-C., Chang, T.-T., Young, K.-C. *et al.* (1992) Detection of HCV RNA in saliva, urine, seminal fluid, and ascites. *Journal of Medical Virology* 37, 197–202.

Lodi, G., Porter, SR., Teo, C.G. and Scully, C. (1996) Prevalence of HCV-infection in health care workers of a London dental hospital (submitted).

Lot, F. and Abiteboul, D. (1992) Infections professionelles per le VIH en France; le point au 31 Mars 1992. *Bulletin Epidemiologie Hebdomaire* 26, 117–119.

Lubick, H.A. and Schaeffer, L.D. (1996) Occupational risk of dental profession survey. *Journal of the American Medical Association* 113, 10–12.

Lucht, E., Albert, J., Linde, A. *et al.* (1993) Human immunodeficiency virus type 1 and cytomegalovirus in saliva. *Journal of Medical Virology* 39, 156–162.

Lyman, D., Winkelsein, W., Ascher, M. *et al.* (1986) Minimal risk of transmission of AIDS-associated retrovirus infection by oral-genital contact. *Journal of the American Medical Association* 255, 1703.

Macaya, G., Visona, K.A. and Villarejos, V.M. (1979) Dane particles and associated DNA-polymerase activity in saliva of chronic hepatitis B carriers. *Journal of Medical Virology* 4, 291–301.

McQuarrie, M.B., Forghlani, B. and Wolochow, C.A. (1974) Hepatitis B transmitted by human bite. *Journal of the American Medical Association* 230, 723–724.

Matthews, R.W. and Scully, C. (1989) Can dental practitioners change established habits and wear disposable latex gloves? *Dental Practice* 27, 1–2.

Matthews, R.W., Hislop, W.S. and Scully, C. (1986) The prevalence of hepatitis B markers in high-risk dental out-patients. *British Dental Journal* 161, 294–296.

Mele, A., Ferrigno, L., Stazi, M.A. *et al.* (1990) SEIEVA Sistema Epidemiologico Integrato dell Epatite Virale Acuta. Rapporto annuale 1989; e risultati 1985–1989. Istisan 90/92.

Miller, C.H. (1996) Infection control. *Dental Clinics of North America* 40, 437–456.

Molinari, J.A. (1995) Handwashing and hand care; fundamental asepsis requirements. *Compendium of Continuing Education in Dentistry* 16, 834–836.

Molinari, J.A. (1996) Hepatitis C virus infection. *Dental Clinics of North America* 40, 309–325.

Mori, M. (1984) Status of viral hepatitis in the world community: its incidence among dentists and other dental personnel. *International Dental Journal* 34, 115–121.

Morris, R.E. and Turgut, E. (1990) Human immunodeficiency virus; quantifying the risk of transmission of HIV to dental health care workers. *Community Dentistry and Oral Epidemiology* 18, 294–298.

Mosley, J.W., Edwards, V.M., Casey, G. *et al.* (1975) Hepatitis B virus infection in dentists. *New England Journal of Medicine* 293, 729–734.

Nicholas, N.K. (1977) Viral hepatitis among practising dentists. *New Zealand Medical Journal* 588, 413–416.

Niederman, J.C., Miller, G., Pearson, M.A. *et al.* (1976) Infectious mononucleosis; EBV shedding in saliva and the oropharynx. *New England Journal of Medicine* 294, 1353–1359.

Noble, M.A., Mathias, R.G., Gibson, G.B. *et al.* (1992) Hepatitis B and HIV infections in dental professionals; effectiveness of infection control procedures. *Journal of the Canadian Dental Association* 57, 55–58.

O'Shea, S., Cordery, M., Barett, W.Y. *et al.* (1990) HIV excretion patterns and specific antibody responses in body fluids. *Journal of Medical Virology* 31, 291–296.

Ogasawara, S., Kage, M., Kosai, K-I. *et al.* (1993) Hepatitis C virus in saliva and breast milk of hepatitis C carrier mothers. *Lancet* 341, 561.

OSHA (1991) US Occupational Safety and Health Agency. Occupational exposure to blood-borne pathogens, final rule. *Federal Register* 56, 64175–64182.

Panis B., Ronmeliotou-Karayannis, A., Papaevangelou, G. *et al.* (1986) Hepatitis B virus infection in dentists and dental students in Greece. *Oral Surgery* 61, 343–345.

Piazza, M. (1989) Passionate kissing and microlesions of the oral mucosa: possible role in AIDS transmission. *Journal of the American Medical Association* 261, 244–245.

Piazza, M., Guadagnino, V., Picciotto, L. *et al.* (1987) Contamination by hepatitis B surface antigen in dental surgeries. *British Medical Journal* 295, 473–474.

Piazza, M., Borgia, G., Picciotto, L. *et al.* (1995) Detection of hepatitis C virus-RNA by polymerase chain reaction in dental surgeries. *Journal of Medical Virology* 45, 40–42.

Polakoff, S. (1989) Acute viral hepatitis B; laboratory reports 1985–8. *Communicable Disease Reports* ?, 26–29.

Porter, S.R. (1991) Infection control in dentistry. *Current Opinion in Dentistry* 1, 429–435.

Porter, S.R. and Lodi, G. (1996) Hepatitis C virus (HCV) – an occupational risk to dentists. *British Dental Journal* 180, 473–474.

Porter, S.R. and Scully, C. (1990) Non-A, non-B hepatitis and dentistry. *British Dental Journal* 168, 257–261.

Porter, S.R. and Scully, C. (1994) HIV; the surgeons' perspective. *British Journal of Oral and Maxillofacial Surgery* 32, 222–230.

Porter, K., Scully, C., Theyer, Y. *et al.* (1990) Occupational injuries to dental personnel. *Journal of Dentistry* 18, 258–262.

Porter, S.R., Scully, C. and Samaranayake, L. (1994) Viral hepatitis: current concepts for dental practice. *Oral Surgery, Oral Medicine, Oral Pathology* 78, 682–695.

Porter, S.R., El-Maaytah, M., Afonso, W. *et al.* (1995) Cross infection compliance of UK dental staff and students. *Oral Diseases* 1, 198–200.

Porter, S.R., Scully, C. and El-Maaytah, M. (1996) Compliance with infection control procedures in dentistry. *British Medical Journal* 312, 705.

Powell, E., Duke, M. and Cooksley, W.G.E. (1985) Hepatitis B transmission within families: potential importance of saliva as a vehicle of spread. *Australian and New Zealand Journal of Medicine* 15, 717–720.

Puro, V., Aspan, C., Olivi, G. *et al.* (1992) Sieroprevalenza di infezione da HIV nei pazienti di un ambulatorio odontoiatrico ospedaliero pubblico. *Ospedale S Camillo*, 208–211.

Reed, B.E. and Barrett, A.P. (1988) Hepatitis B virus carrier groups and the Australian community. *Australian Dental Journal* 33, 171–176.

Richman, K.M. and Rickman, L.S. (1993) The potential for transmission of human immunodeficiency virus through human bites. *Journal of Acquired Immuno-deficiency Syndromes* 6, 402–406.

Roy, K.M., Bagg, J., Follett, E.A. *et al.* (1996) Hepatitis C virus in saliva of haemophiliac patients attending an oral surgery unit. *British Journal of Oral and Maxillofacial Surgery* 34, 162–165.

Rozenbaum, W., Gharakhanian, S., Cardon, B. *et al.* (1988) HIV transmission by oral sex. *Lancet* i, 1395.

Sartori, M., La Terra, G., Aglietta, M. *et al.* (1993) Transmission of hepatitis C via blood splash into conjunctiva. *Scandinavian Journal of Infectious Disease* 25, 270–271.

Savage, C.M., Christopher, P.J., Murphy, A.M. *et al.* (1984) The prevalence of hepatitis B markers in dental care personnel at the United Dental Hospital of Sydney. *Australian Dental Journal* 29, 75–79.

Sawada, T., Taguchi, H., Miyoshi, I. *et al.* (1995) HTLV-1 proviral DNA in oral aspirates of newborns born to seropositive mothers. *Journal of the American Medical Association* 273, 284.

Scheutz, F., Mellbye, M., Esteban, J.I. *et al.* (1988) Hepatitis B virus infection in Danish dentists; a case control study and follow-up study. *American Journal of Epidemiology* 128, 190–196.

Schiff, E.R., de Medina, M.D., Kline, S.N. *et al.* (1986) Veterans administration cooperative study on hepatitis and dentistry. *Journal of the American Dental Association* 113, 390–396.

Schiff, E.R., De Medina, M.D., Hill, M.A. *et al.* (1990) Prevalence of anti-HCV in the VA dental environment from 1979–1981. *Hepatology* 12, 849 (abstract 45).

Scully, C. (1985) Hepatitis B; an update in relation to dentistry. *British Dental Journal* 159, 321–328.

Scully, C. (1989) Orofacial herpes simplex virus infections; current concepts on the epidemiology, pathogenesis and treatment, and disorders in which the virus may be implicated. *Oral Surgery, Oral Medicine and Oral Pathology* 68, 701–710.

Scully, C. (1995) Infectious diseases. In: Millard, H.D. and Mason, D.K. (eds) *1993 World Workshop on Oral Medicine*. University of Michigan Press, Michigan.

Scully, C. (1996) New aspects of oral viral diseases. In: Seifert, G. (ed.) *Oral Pathology: Actual Diagnostic and Prognostic Aspects*. Springer Verlag, Berlin, pp. 30–97.

Scully, C. and Bagg, J. (1992) Viral infections in dentistry. *Current Opinion in Dentistry* 9, 8–11.

Scully, C. and Matthews, R. (1990) Uptake of hepatitis B immunisation amongst United Kingdom dental students. *Health Trends* 22, 92.

Scully, C. and Mortimer, P. (1994) Gnashings of HIV. *Lancet* 344, 904.

Scully, C. and Porter, S. (1991) The level of risk of transmission of human immunodeficiency virus between patients and dental staff. *British Dental Journal* 170, 97–100.

Scully, C. and Samaranayake, L.P. (1992) *Clinical Virology in Oral Medicine and Dentistry*. Cambridge University Press, Cambridge.

Scully, C., Almeida, O.D.P. and Jorge, J. (1990a) Dental staff in Brazil and immunisation against hepatitis B. *British Dental Journal* 168, 184.

Scully, C., Cawson, R.A. and Griffiths, M.J. (1990b) *Occupational Hazards to Dental Staff*. British Dental Journal, London.

Scully, C., Porter, S.R. and Epstein, J.B. (1992) Compliance with infection control procedures in a dental hospital clinic. *British Dental Journal* 173, 20–23.

Scully, C., Blake, C., Griffiths, M. *et al.* (1994) Protective wear and instrument sterilisation/disinfection in UK general dental practice. *Health Trends* 26, 21–22.

Scully, C., Griffiths, M.J., Levers, H. *et al.* (1993) The control of cross-infection in UK clinical dentistry in the 1990s: immunisation against hepatitis B. *British Dental Journal* 174, 29–31.

Shapiro, E.D. (1987) Lack of transmission of hepatitis B in a day care center. *Journal of Paediatrics* 110, 93.

Shopper, T., Boozer, C., Lancaster, D. *et al.* (1995) Presence of anti-hepatitis C virus serum markers in a dental school patient population. *Oral Surgery, Oral Medicine, Oral Pathology* 79, 659–664.

Siegel, M.A. (1996) Syphilis and gonorrhea. *Dental Clinics of North America* 40, 369–384.

Siew, C., Gruninger, S.E., Mitchell, E.W. *et al.* (1987) Survey of hepatitis B exposure and vaccination in volunteer dentists. *Journal of the American Dental Association* 114, 457–459.

Siew, C., Chang, S-B., Gruninger, S.E. *et al.* (1992) Self-reported percutaneous injuries in dentists: implications for HBV, HIV transmission risk. *Journal of the American Dental Association* 123, 36–44.

Smith, H.M., Alexander, G.J.M., Birnbaum, W. *et al.* (1987) Does screening high risk dental patients for hepatitis B virus protect dentists? *British Medical Journal* 295, 309–310.

Solovan, D.F., Uldricks, J.M., Caccamo, P. *et al.* (1984) Evaluation of oral procedures performed with gloves; a pilot study. *Dental Hygiene*, March, 122–124.

Spitzer, P.G. and Weinger, N.J. (1989) Transmission of HIV infection from a woman to a man by oral sex. *New England Journal of Medicine* 320, 251.

Spruance, S.L. (1984) Pathogenesis of herpes labialis; excretion of virus in the oral cavity. *Journal of Clinical Microbiology* 19, 675–679.

Stewart, C.M., Jones, A.C., Bates, R.E. *et al.* (1994) Percutaneous and mucous membrane exposure: protocol in a southeastern dental school. *Oral Surgery* 78, 401–407.

Sugimara, H., Yamamoto, H., Watabiki, H. *et al.* (1995) Correlation of detectability of hepatitis C virus genome in saliva of elderly Japanese symptomatic HCV carriers with their hepatic function. *Infection* 23, 258–262.

SyWassink, J.M. and Lutwick, L.I. (1983) Risk of hepatitis B in dental care providers: a contact study. *Journal of the American Dental Association* 106, 182–184.

Takamatsu, K., Koyanagi, Y., Okita, K. *et al.* (1990) Hepatitis C virus RNA in saliva. *Lancet* 336, 1515.

Thomas, D.L., Factor, S.H., Kelen, J.D. *et al.* (1993) Viral hepatitis in healthcare personnel at the Johns Hopkins Hospital. *Archives of Internal Medicine* 153, 1705.

Thomas, D.L., Gruninger, S.E., Siew, C. *et al.* (1996) Occupational risk of hepatitis C infections among general dentists and oral surgeons in North America. *American Journal of Medicine* 100, 41–45.

Tullman, M.J., Boozer, C.H., Villarejos V.M. *et al.* (1980) The threat of hepatitis B from dental school patients. *Oral Surgery* 49, 214–216.

Vaglia, A., Nicolin, R., Puro, V. *et al.* (1990) Needlestick hepatitis C seroconversion in a surgeon. *Lancet* 336, 1315.

Vaglia, G., Calabrese, G., Pratesi, G. *et al.* (1985) Non-A non-B hepatitis in a dialysis population: spread by dental surgery? *Clinics in Nephrology* 268, pp. 000.

Vidmar, L., Poljak, M., Tomazic, J. *et al.* (1996) Transmission of HIV-1 by human bite. *Lancet* 347, 1762–1763.

Wahn, V., Kramer, H.H., Voit, T. *et al.* (1986) Horizontal transmission of HIV infection between two siblings. *Lancet* ii, 694.

Wang, J-T., Want, T-H., Lin, J-T. *et al.* (1991) Hepatitis C virus RNA in saliva of patients with post-transfusion hepatitis C infection. *Lancet* 337, 48.

Wang, J-T., Wang, T-H., Sheu, J-C. *et al.* (1992) Hepatitis C virus RNA in saliva of patients with post-transfusion hepatitis and low efficiency of transmission among spouses. *Journal of Medical Virology* 36, 28–31.

West, D.J. (1984) The risk of hepatitis B infection among health professionals in the United States: a review. *American Journal of Medicine* 287, 26–33.

Wilson, M.P., Ground, S., Tishk, M. *et al.* (1986) Gloved versus ungloved dental hygiene clinicians. *Dental Hygiene* 60, 310–315.

Yeung, S.C.H., Kazai, F., Randle, C.G.M. *et al.* (1993) Patients infected with human immunodeficiency virus type 1 have low levels of virus in saliva even in the presence of periodontal disease. *Journal of Infectious Disease* 167, 803–809.

Young, K-C., Chang, T-T., Liou, T-C. *et al.* (1993) Detection of hepatitis C virus RNA in peripheral blood mononuclear cells and in saliva. *Journal of Medical Virology* 41, 55–60.

Zuckerman, J., Clewley, G., Griffiths, P. *et al.* (1994) Prevalence of hepatitis C antibodies in clinical health-care workers. *Lancet* 343, 1618–1620.

CHAPTER 10
Nursing care

G. Griffiths

The risk of transmission of blood-borne pathogens to nurses, although small, cannot be denied. It will depend on the nature and frequency of exposure to blood or body fluids; the risk of transmission of infection after a single exposure to blood and the prevalence of infected patients. Most accidental exposure incidents are preventable, but there is a need for infection control programmes and continued emphasis on the employee's responsibility to ensure employee compliance with such procedures designed to protect them.

Waldron (1992) reported that despite every effort to educate staff, the number of needlestick injuries was high and continuing to rise. He estimated the costs of the management of needlestick injuries as between £7000 and £10,000 per annum depending on the type of injury, the infectivity of the source patient and, in the case of hepatitis B, the immune status of the recipient.

Occupational health department staff and infection control teams in health-care establishments have a key role in providing information and suitable training material. This must include nurses of all grades in all health-care establishments in addition to night duty staff and nursing agency staff as well as those who visit patients at home.

Training should start during induction courses so that staff are made aware of identifiable risks, and to ensure that they have a clear understanding of the action to be taken in dealing with situations in which exposure to blood-borne viruses and other microorganisms may occur. Nurses need to know how to avoid or to minimize these risks and be aware of the local codes of practice and how and when to report accidental exposure. Review and analysis of data from exposure incident reports will help to increase the effectiveness of the prevention programme. Codes of practice must include the safe disposal of clinical waste for all health-care workers, wherever they work, whether in hospital or other health-care establishments or in the patient's home.

Pre-employment medical examination of nurses can establish history of immunity to communicable diseases and possible immunity to hepatitis B, tuberculosis, etc. Hepatitis B vaccination is offered to nurses in most cases, but is particularly important for midwives, operating theatre and accident and emergency nursing staff and those working in psychiatric care.

The US Centers for Disease Control and Prevention (CDC, 1988) emphasized the routine use of Universal Precautions, a barrier method of minimizing exposure to blood and body fluids (see Hunt, Chapter 19 and Kibbler,

Chapter 20 this volume). In 1990 and again in 1995, guidance on safe working practices for clinical workers has been produced by the Advisory Committee on Dangerous Pathogens (ACDP, 1990, 1995). The 1995 document encourages a task-related approach to the selection of controls according to the varying degree of exposure to blood and the likelihood of penetrating injury.

By the nature of their work, surgeons and physicians are exposed to greater and more frequent risks, but nurses are now extending their roles to perform more procedures previously carried out by doctors. In particular, they take blood and set up intravenous infusions. Nurses and midwives taking blood in the patients' homes or clinics must not put themselves or others at risk of blood spillage, but must ensure safe transport of the specimens to the pathology laboratory (Royal College of Nursing, 1987; Health Services Advisory Committee (HSAC), 1991).

BLOOD-BORNE INFECTIONS

When blood containing infectious agents is transferred into the body of another person that person may also become infected.

As with acquisition of any infection there are three major factors which affect the outcome:

- the identity of organism, its virulence and mode of transmission;
- the dose of the infecting organism;
- the susceptibility of the host.

As carriers of blood-borne pathogens may be asymptomatic, spillage of any blood must be treated with caution (see Hoffman and Kennedy, Chapter 17 this volume) and every attempt made to avoid blood entering another person's bloodstream. In particular, it is imperative that all carers avoid needlestick or sharps injuries.

Although there are some bacteria and endoparasites which may cause infection if transmitted directly into the bloodstream in sufficient quantities (see Collins, Chapter 2 this volume), it is the viruses which are of most concern to the nursing profession: the human immunodeficiency virus (HIV) which causes AIDs and three of the hepatitis viruses – hepatitis B (HBV), C (HCV) and D (delta virus) (HDV) (see Jeffries, Chapter 1 and Hunt, Chapter 3 this volume).

Human Immunodeficiency Virus (HIV)

Although this book is concerned with blood-borne infections it should be noted that other body fluids may be contaminated with the virus, e.g.

Cerebrospinal fluid
Pericardial fluid
Synovial fluid
Breast milk
Semen

Pleural fluid
Peritoneal fluid
Amniotic fluid
Vaginal secretions
All unfixed tissues or organs
Any other body fluid containing visible blood.

Urine, faeces, saliva, sputum, tears, sweat and vomit, present a minimal risk of blood-borne infection unless contaminated with blood (although they may be hazardous for other reasons).

World-wide there are now 64 documented cases of occupationally-acquired HIV infections in health-care workers with a further 118 cases presumed (Heptonstall *et al.*, 1993). In April 1993, 4.7% of the 214,686 AIDs patients reported to the CDC, Atlanta, Georgia, had been employed in health-care settings. Only 6% of these patients had no risk factor for acquiring HIV infection other than by occupational exposure to blood or body fluids (Robert and Bell, 1994). Because most percutaneous exposures do not result in transmission of HIV it is evident that the level of viraemia in the source patient is significant, as is the volume of blood transferred. All HIV infected persons are considered to be capable of transmitting HIV to a susceptible person, but the further advanced the disease has progressed, the greater the amount of virus in the bloodstream (CDC, 1988).

Testing for HIV infection should be for the purpose of improving medical care or by specific request of the patient and must never be performed without the patient's permission and only then after he/she has received counselling. The adoption of Universal Precautions and the usual infection control procedures obviates the need for screening of patients for HIV as the assumption is that all human blood and specified body fluids are potentially infectious.

Many AIDs patients are being cared for in hospices or by community nurses at home. Other than giving injections it is unlikely that invasive procedures will be necessary or of benefit to the terminally ill. There must be clearly written guidelines, however, for nursing staff and carers, on safe disposal of clinical waste (HSAC, 1992) and dealing with blood spillage, e.g. from haemorrhage (see also Hoffman and Kennedy, Chapter 17, and Collins and Kennedy, Chapter 18 this volume).

Exposure to HIV infection
If the source individual has AIDS, is known to be HIV positive, or has refused to be tested, blood from the exposed person should be tested as soon as possible after exposure and repeated 12 weeks and six months later. If the source person is seronegative and is not in the at-risk group or not recently exposed to infection no further follow-up is necessary.

Hepatitis B

Guidelines for the UK Department of Health (DH, 1994) make it mandatory for surgeons and other health-care workers undertaking invasive procedures to show immunity to hepatitis B virus. In most health districts all nursing and midwifery staff are offered vaccination for hepatitis B. Routine immunization and monitoring of immunity status ensures that nurses are protected against the risks of hepatitis B transmission from their patients. Non-responders continue to be offered specific immunoglobulin in the event of inoculation injury (British Medical Association: BMA, 1995; DH, 1996).

Hepatitis B virus is more easily transmissible than HIV (Canadian Communicable Disease Report, 1992; see also Hunt, Chapter 3 this volume). It is much more stable and may survive for some time in the environment even in dried blood (Bond et al., 1981; ACDP, 1995). In the non-immune the transmission rate after percutaneous exposure to HBeAg-positive blood may be greater than 30% (Werner and Grady, 1982).

The cleaning of instruments and equipment must therefore be scrupulous. Rinsing blood off under running water should be the first step. Cleaning instruments with a nail brush is no longer recommended as it increases the risk of splashes into the eyes. Ultrasonic devices which are used in central sterile supply departments are the safest way of ensuring the removal of blood and body fluids. Caution is required in the use of these instruments and rubber gloves should be worn to protect the hands as petechial lesions may develop if they are immersed in the bath during its operation. There is also the risk of hearing impairment; operation in a sound-proof cabinet or wearing ear defenders is recommended (Kennedy, 1988).

Nurses are most at risk from needlestick injury but the virus may be transmitted as a consequence of blood splashing onto the mucous membranes or conjunctivae. Midwives are particularly at risk since they commonly experience splashes of blood and amniotic fluid. Although wearing of surgical face masks is no longer customary except in the operating theatre, their use is advised when splashing is likely to the lips and should be supplemented with goggles to prevent splashes to the eye. Alternatively face visors are available which protect the eyes and mouth (ACDP, 1995). Wearers of spectacles will have sufficient protection and need not wear goggles.

Hepatitis C

As many as 2% of Americans (including health-care workers) may be infected with hepatitis C virus (HCV) and it is conservatively estimated that there are 150,000 new infections each year. Currently 50–60% of patients acquire HCV in association with intravenous drug use (Sodeyama et al., 1991). The mode of transmission in the other 40% is unclear although sexual transmission is suspected.

A study in New York (Jochen, 1992) looked at serum samples of 1033 hospital employees and 2113 voluntary blood donor controls. Anti-HCV was found in 0.58% of the hospital employees and 0.24% in the controls. This was considerably lower than those found in New York dentists in 1991 (Klein et al., 1991).

Management guidance of occupationally acquired exposure to hepatitis C is available (Public Health Laboratory Service, 1993; BMA, 1996).

Hepatitis D

Little is known about the true extent of the infection with the hepatitis D virus (HVD) in the population except that the prevalence is higher in injecting drug users, as with other blood-borne viruses.

Routine use of the test for HBV in blood donors obviates the need to test for HDV (ACDP, 1995), as the latter is found only in the presence of the former.

Other Hepatitis Viruses

Hepatitis A and E viruses (HAV and HEV) are present briefly in the blood but are not transmitted in this way as there is no chronic carrier state. Infection arises principally from the faeces of the infected person or from contaminated food or water and produces symptoms of gastrointestinal infection.

There are other hepatitis viruses (see Jeffries, Chapter 1, Hunt, Chapter 3, and Morgan, Chapter 8 this volume) but they are of no concern here.

Cytomegalovirus

The majority of cytomegalovirus (CMV) infections are asymptomatic, but a small percentage of healthy young adults develop a mononucleosis syndrome with prolonged fever and malaise.

As with other blood-borne viruses CMV can be sexually transmitted, probably accounting for the rise in seroprevalence in young adults. A small percentage is accounted for by parenteral transmission of blood products and organ transplantation. By 35 years of age more than 60% of the population will have acquired the antibody to cytomegalovirus. Provided there is a normal immune system re-infection is improbable – unless from another strain.

The number of health-care workers acquiring CMV infection as a result of exposure to blood or body fluids appears to be very low. In the majority of people acquiring CMV the virus persists in a non-replicating state (latency) and may become reactivated only if the patients becomes immunocompromised (e.g. by HIV infection). Many investigators believe that a person with latent CMV infection will not transmit the virus even to recipients of blood transfusions, unless they are immunocompromised (Hibberd, 1995).

It is likely that the majority of health-care workers are infected with CMV at some time during their lives. Intrauterine and perinatal transmission of CMV is important because there is a 1–5% risk of severe fetal abnormalities and 5–20% risk of hearing disability and mental retardation.

That there is a small risk of acquiring the infection should be borne in mind by nurses, many of whom are in the child-bearing age group. It is difficult to achieve total prevention as patients are often symptomless and therefore unidentifiable. It is probably unwise for nurses who know or suspect that they are pregnant to care for immunocompromised patients who may have re-activated the latent virus (see Tookey and Packham, 1991; Tookey and Logan, 1994). (See also Jeffries, Chapter 1 this volume.)

RISKS OF ACQUIRING INFECTION

Operating Theatre Nurses

It is the surgeons and dentists who are most at risk in the operating theatre but the nurses assisting at operations (scrub nurses) also frequently handle instruments and sharps.

The identification of risk factors, depending on the type of operation performed and the length of time taken is vital in planning methods to limit accidental exposure to blood-borne viruses. Any variation in strategies for risk reduction needs to be based on the procedures involved and the equipment and instruments used rather than the degree of known or suspected infectivity of the patient (see also: Precautions involving use of equipment in operating theatre, etc. below).

Midwifery

The risks to midwives are similar to those to nurses but in addition they will also perform episiotomies and carry out suturing procedures, all of which have their attendant risks. As in the operating theatre, it is necessary to develop risk strategies and infection control procedures for all cases and not only for known or suspected infected cases.

Placentas must be included in arrangements for the disposal of clinical waste. Where home deliveries are carried out the attending midwife should send placentas to the nearest clinical waste disposal site for incineration.

Hepatitis B screening at early appointments in antenatal clinics ensures that risks from HBV are minimized. Those who are found to be seropositive will be checked at regular intervals during the pregnancy. The patient's notes will warn staff, but the woman will also carry a card giving her hepatitis B status which she should present if delivery occurs in another hospital.

Community Nursing and Health Visiting

The numbers of general practitioners has increased from 18,000 in 1949 to over 32,000 in 1994, with a further 2700 general practitioner trainees and assistants and approximately 16,000 practice nurses. In the UK general practitioner surgery premises are subject to health and safety legislation and inspection by the Health and Safety Executive. Notification of accidents occurring on the premises is a statutory obligation for all employees, as is the keeping of up-to-date records. This includes needlestick and sharps injuries. It applies also to Community Nurses who must report and record accidents that occur in a patient's own home.

Mental Health-care Environment

Mental illness accounts for approximately 14% of NHS in-patient costs but mental health-care services in the UK are provided in a variety of settings (Leggett, 1994), including the client's home, day hospitals, acute wards and hospital long stay wards, secure units and prisons (DH, 1993).

Upon admission clients may be highly stressed and poorly nourished and particularly susceptible to infection. Clients attending or admitted to centres for alcohol or drug dependency may have engaged in unsafe behaviour associated with high risk of infection with blood-borne pathogens, particularly HIV and hepatitis B, C and D. Suicidal tendency and self-mutilation may result in haemorrhage but self biting and scratching sometimes also occurs. Biting and scratching of carers has minimal risk since the client's saliva will have negligible amount of virus unless there is blood present (but see Collins, Chapter 14 this volume).

Downs' syndrome patients are more likely to be carriers of hepatitis B virus but may be asymptomatic. In a blood screening investigation for hepatitis B markers of 2239 people with learning difficulties, 5.5% were found to be carriers of hepatitis B e-antigen (Clarke *et al.*, 1984).

SHARPS INJURIES

Sharps are any item which can cause laceration or puncture wounds, presenting a special hazard if there is contamination by blood or body fluids, even if not visibly soiled. Examples include used hypodermic needles, any instruments used in invasive procedures, including such articles as disposable lancets for testing blood sugar, and splinters of bone which may be produced during orthopaedic surgery.

Many inoculation incidents occur as a result of re-sheathing of hypodermic needles; this is a dangerous practice unless it is performed with one hand (BMA, 1990). Needlestick injuries account for approximately 80% of all accidental exposures to blood. About 20% occur during use of needle, about 80% are associated with needle re-sheathing and the remainder occur during or after disposal (Jagger *et al.*, 1988; see also Jagger and Bentley, Chapter 5 this volume).

Accidental inoculation includes:

- all penetrating sharps/needlestick injuries;
- contamination of lesions with blood or body fluids;
- scratches or bites involving broken skin, i.e. bleeding or visible puncture wound;
- splashes into the eyes or mouth.

District nurses, midwives and health visitors carry needles, syringes and sharps containers in their cars but the health authority employer is ultimately responsible for providing the means for the safe disposal of sharps. This usually involves the nurse taking the full container to the nearest hospital incinerator or leaving it at a health centre for collection by arrangement with the hospital or local council.

SHARPS CONTAINERS

Sharps containers must conform to the British Standard (BSI, 1990) and one essential requirement is a line to indicate the 3/4 full mark above which it must not be filled. Accidents have been reported where staff have tried to push overflowing needles down and this can also lead to puncturing of the plastic by downward pressure on the needles. Once 3/4 full the bins must be securely sealed and taken directly to the nearest clinical waste collection point and never left in corridors to be collected. Sharps containers are described in detail by Collins and Kennedy, Chapter 18 this volume).

It is essential in busy areas such as intensive care units and accident and emergency departments where their usage is very high that sufficient numbers of these containers are placed in strategic positions so that the users of sharps do not have to move far to dispose of them. Sharps should be disposed of immediately by the user, wherever possible, and not left for disposal with others or handed from person to person.

REPORTING OF NEEDLESTICK AND SHARPS INJURIES

In the UK *The Management of Health & Safety at Work Regulations 1992 (Health & Safety at Work Act, 1974)* and the *Control of Substances Hazardous to Health Regulations 1994* (COSHH) require an assessment of risk to health, provision of health surveillance (where appropriate) and information for employees. Under the *Reporting of Injuries, Infections and Dangerous Occurrences Regulations 1994* these must be reported to the Health and Safety Executive if an infection results but not otherwise. But all such incidents should be recorded.

PROTECTIVE CLOTHING

Protective clothing was originally instituted to protect the patient but the need for protecting surgical staff has become more important recently. Manufacturers have introduced a number of modifications to aid the prevention of contamination.

Gowns

Gowns with impervious plastic reinforcements are available but are found to be hot and uncomfortable to wear even for short periods. Some surgical staff prefer to wear a waterproof apron beneath the gown and an extra pair of sleeves over it. The latter can be changed frequently before 'strike through' occurs to the inner gown.

Gloves

Wearing two pairs of gloves has been found to confer additional protection to surgeons and scrub nurses resulting in 60–80% decrease in inner glove perforation and visible glove contamination (Jackson and McPherson, 1992). Surgical gloves should conform to the appropriate British Standards (BSI, 1984/1994, 1992).

A recent simulation study in San Francisco General Hospital measured the amount of blood from a needle that penetrated a piece of filter paper under two conditions: a direct puncture, and through a vinyl or latex glove. There was significant reduction (up to 50%) in the amount of blood that reached the filter paper after the needle had passed through the glove (Gerberding *et al.*, 1990).

Cut-resistant glove liners have been developed to reduce the risk of penetrating injuries although in some cases this compromises dexterity and comfort.

Masks and Visors

These were originally designed to protect the patient from airborne pathogens, but are now also worn to protect staff from splashing of blood and body fluids onto the mucosa. Some manufacturers have added an extra waterproof layer to reduce the risk of fluid penetration.

Eye Protection

Those who ordinarily wear spectacles are considered to be adequately protected but goggles and visors or full face shields are available. Some include special 'fog filters' to reduce the problem of misting. They are of benefit during orthopaedic procedures and others where extensive spraying and spattering may occur. Non-disposable goggles may be washed with warm soapy water before being re-used.

Overshoes

These have not proved to be of benefit in preventing spread of infection and have the disadvantage of transferring bacteria to the hands when taking them off and on (Carter, 1990).

LAUNDRY

Linen soiled with blood and/or body fluids should not be rinsed or treated in clinical areas but sent directly for laundering in a red alginate-stitched bag inside a red linen bag. This allows the inner bag to be placed directly into the machine without further handling of the linen as the stitching will dissolve in warm water. Red denotes risk of infection so that no further labelling is necessary (DH, 1995).

EXPOSURE TO SUSPECTED BLOOD-BORNE INFECTION

Immediately after accidental percutaneous exposure, ocular or mucous membrane exposure or contact with damaged skin (abraded or with dermatitis):

- the injured part should be encouraged to bleed and the site washed thoroughly under running water;
- blood splashed into the eyes or mouth should be washed off with large volumes of water;
- the incident should be reported to the immediate superior; the type of injury and source patient, if known, should be recorded;
- the Occupational Health Department or appropriate medical adviser should be informed, according to local policy – as early as possible in order to:
  (a) assess the likelihood of blood-borne virus infection in the source case;
  (b) ascertain the vaccination history of the injured person;
  (c) obtain the injured person's consent for blood testing to establish existence of any pre-existing blood-borne infection (and therefore a baseline);
  (d) arrange counselling of the injured person should it be necessary;
  (e) administer hepatitis B immunoglobulin, preferably within 24 hours, as well as the first dose of the vaccination in a different site of the body *if the source patient is HBsAg positive.*

If the source patient is unknown or his/her hepatitis B status is unknown the exposed person should immediately commence a vaccination programme if not already so protected (see McCloy, Chapter 16 this volume).

GENERAL RECOMMENDATIONS

Prevention of Infection

- Assess the risk to health arising from the type of nursing and develop precautions to be applied;
- inform, instruct and train employees about the risks and precautions to take and introduce appropriate measures to prevent or to control risks;
- ensure continued compliance of these measures;
- inform on correct reporting of accidental exposure and keep records of reports;
- renew and analyse data from these reports to help increase effectiveness of the programme;
- ensure safe decontamination of spillages of blood/body fluids.

TO PREVENT PUNCTURE WOUNDS, CUTS AND ABRASIONS IN THE PRESENCE OF BLOOD AND BODY FLUIDS

- Where possible avoid or reduce the use of sharps but take extreme care when their use is unavoidable;
- use a re-sheathing device for needles if re-sheathing is unavoidable;
- protect all breaks in the skin with a waterproof adhesive dressing in addition to wearing gloves;
- protect the eyes and mouth with a visor or goggles and a face mask when splashing of blood or body fluids is likely;
- avoid contamination of the person or clothing with waterproof or water-resistant protective clothing;
- use good basic hygiene practices and avoid hand to mouth/eye contact;
- ensure safe environment with appropriate decontamination of hard surfaces;
- ensure safe disposal of all clinical waste, including sharps, by incineration.

PRECAUTIONS INVOLVING USE OF EQUIPMENT IN OPERATING THEATRE, ETC.

- Use a magnetic pad or emesis bowl to pass instruments to avoid placing sharps instruments directly into the hands; if this is not possible provide at least time to establish visual contact when sharp instruments are passed;
- use a thimble to protect the index finger of the non-dominant hand;
- staples have been found to cause fewer puncture injuries than sutures; 'superglue' may be used in some instances;
- use a knife blade holder which allows mechanical removal of the blade with one hand;
- needles on a needle holder placed in readiness for use must stand with the point down to prevent snagging of gloves.

For operations of long duration, routine changing of gloves should be carried out at intervals in addition to when there are signs of puncture.

SAFE DISPOSAL OF CLINICAL WASTE

The HSAC (1992) advises that suitable containers must be available to collect sharps and biological waste material. This includes provision of yellow plastic bags to the DH specification (National Health Service, 1993) and British Standard approved sharps containers (BSI, 1990) in the homes where injections are given – for which the general practitioner is also responsible. Under the *Environmental Protection Act 1990* a duty of care has been placed on those handling controlled waste, clinical waste is one of the categories of controlled waste (see Collins and Kennedy, Chapter 18, this volume).

SPILLAGES

The following precautions are recommended:

- wear disposable gloves and aprons (any lesion on exposed skin must first be covered with a waterproof adhesive dressing);
- use disposable cloth or paper towels soaked in freshly made hypochlorite solution (10,000 ppm available chlorine) to clean and decontaminate the area;
- discard into yellow plastic bag for incineration including the gloves and apron.

In the home, disposable tissue and a solution of bleach can be used and disposed of into the nearest WC. Gloves and aprons to be returned for disposal with any other clinical waste.

For larger spillages dichloroisocyanurate (NaDCC) granules are more effective and easier to use. They must be sprinkled onto the spillage and left for two minutes before clearing with disposable cloth or paper. Ensure that the room is well ventilated.

If these granules are unavailable the area can be diluted with hypochlorite solution (10,000 ppm available chlorine) and left for two minutes before clearing up (see Hoffman and Kennedy, Chapter 17, this volume).

Spillages of urine should be treated in this manner only if there is also visible blood present. Urine can promote the release of free chlorine from the solution which can present a hazard in poorly ventilated areas.

DISCUSSION

Accidental sharps injuries, although generally preventable, do not show signs of diminishing in frequency. The risk of acquiring HIV from such an injury is now known to be lower than that of hepatitis B but concern has also been growing recently regarding hepatitis C, particularly in the care of drug abusers.

The main risk to nurses is through needlestick injury, but blood splashes in operating theatres and labour wards, etc. and the risk of punctured gloves must be included when assessing the risks and developing effective measures to prevent transmission of blood-borne pathogens. Prominently displayed charts and posters such as those illustrated in Table 10.1, Figs 8.3 and 10.1 contribute to education and training.

There is still much controversy regarding the re-sheathing of hypodermic needles and Anderson et al. (1991) suggested four methods of doing this and several manufacturers have developed products to aid re-sheathing. Nevertheless, the British Medical Association (BMA, 1990), the World Health Organization (WHO, 1991), and the US Centers for Disease Control and Prevention (Richmond and McKinney, 1993) are agreed that the risks from re-

Table 10.1.   Reducing the risks of acquiring blood-borne infections.

Avoid or reduce use of sharps wherever possible	*If unavoidable* take extreme care in handling and disposal
Do not re-sheathe needles wherever possible	*If unavoidable* use a re-sheathing device or one hand only
Use good basic hygiene practice, especially adequate handwashing	Avoid hand to mouth\eye contact
Protect breaks in skin with waterproof adhesive dressings and disposable gloves	
Ensure safe disposal of all clinical waste and sharps by incineration	Where incineration is not available on site ensure safe handling and transport to incinerator
Where blood or body fluid splashes are likely	
Wear waterproof or water-resistant protective clothing	Protect eyes with goggles, mouth with mask or visor
Ensure safe decontamination of equipment and hard surfaces	
Precautions in use of equipment in operating theatre, labour ward, etc.	
In lengthy procedures routine change of gloves is necessary *in addition* to when they are punctured	
Place prepared needles and needleholder with points down to prevent snagging of gloves	
Use magnetic pad or emesis bowl to avoid placing instruments directly into hands or, at least, ensure visual contact when passing instruments	Use thimble to protect index finger of non-dominant hand Use knife blade holder for mechanical removal of blade with one hand
Use alternatives to sutures, e.g. 'superglue'	

sheathing needles are too great for the practice to be recommended. Every effort needs to be made to try to discourage this dangerous practice (see also Morgan, 1991).

Education, awareness and strict adherence to infection control practices cannot be over-emphasized in defending against occupational acquired blood-borne infections.

There is little likelihood that a vaccine for HIV and HCV will become available for many years, if at all, and the risk of occupational infection and cross-infection between patients may increase as the prevalence of infection builds up in the patient population, and where invasive techniques are employed.

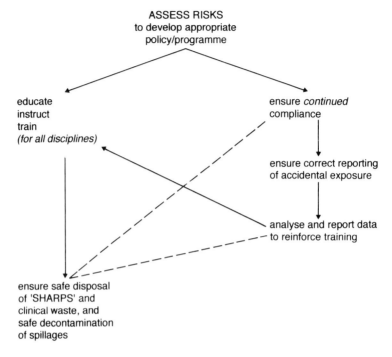

Fig. 10.1. Prevention of accidental exposure to blood-borne infection.

REFERENCES

ACDP (1990) *Categorization of Biological Agents According to Hazard and Categories of Containment.* Advisory Committee on Dangerous Pathogens. HSE Books, Sudbury.

ACDP (1995) *Protection Against Blood-borne Infections in the Workplace: HIV and Hepatitis.* Advisory Committee on Dangerous Pathogens. HMSO, London.

Anderson, D.C., Blowers, A.L., Packer, J.M.V. *et al.* (1991) Preventing needlestick injuries. *British Medical Journal* 302, 769–770.

BMA (1990) *A Code of Practice for the Safe Use and Disposal of Sharps.* British Medical Association, London.

BMA (1995) *A Code of Practice for Implementation of the UK Hepatitis B Immunisation Guidelines for the Protection of Patients and Staff.* British Medical Association, London.

BMA (1996) *A Guide to Hepatitis C.* British Medical Association, London.

Bond, W.W., Favero, M.S., Petersen, N.J. *et al.* (1981) Survival of hepatitis B virus after drying and storage for one week. *Lancet* i, 550–551.

BSI (1984) *BS4005: Specification for Single Use, Sterilized, Surgical Rubber Gloves.* British Standards Institution, London.

BSI (1990) BS 7320: Specification for Sharps Containers. British Standards Institution, London.

BSI (1992) *BS 455-1 Medical Gloves. Specification for Freedom from Holes.* British Standards Institution, London.

BSI (1994) *BS 455: Specification for Medical Gloves, Single Use.* British Standards Institution, London.

Canadian Communicable Disease Report (1992) Bloodborne pathogens in the health care setting: risk for transmission. *CCDR* 18, 177–184.

Carter, E. (1990) Ritual and Risk. *Journal of Infection Control Nursing: Nursing Times* 86, 63–4.

CDC (1988) Update: Universal Precautions for prevention of transmission of human immunodeficiency virus, hepatitis B virus and other blood-borne pathogens in healthcare settings. *Morbidity and Mortality Weekly Report* – 37, 377–382, 387–388.

Clark, S.K., Caul, E.O., Jancar, J. *et al.* (1984) Hepatitis B in seven hospitals for the mentally handicapped. *Journal of Infection* 8, 34–43.

Control of Substances Hazardous to Health Regulations (1994) HMSO, London.

DH (1993) *Health of the Nation: Key Area Handbook: Mental Illness.* Department of Health, London.

DH (1994) *AIDS/HIV Infected Health Care Workers: Guidance on the Management of Infected Healthcare Workers.* Department of Health. HMSO, London.

DH (1995) Hospital laundry arrangements for used and infected linen. *DH Circular HSG* (95)18. Department of Health, London.

DH (1996) *Immunisation against Infectious Disease.* Department of Health. HMSO, London.

Environmental Protection Act (1990). HMSO, London.

Gerberding, J.L., Littel, C., Tarkington, A. *et al.* (1990) Risk of exposure of surgical personnel to patient's blood during surgery at San Francisco General Hospital. *New England Journal of Medicine* 322, 1788–1793

Health & Safety at Work etc Act (1974). HMSO, London.

Heptonstall, J., Gill, O.N. and Porter, K. (1993) Healthcare workers and HIV surveillance of occupationally acquired infection in the United Kingdom. *Communicable Diseases Report Review* 3, R147–R153.

Hibberd, P.L. (1995) Patients, needles and healthcare workers. *Journal of Intravenous Nursing* 18, 65–76.

HSAC (1991) *Safe Working and the Prevention of Infection in Clinical Laboratories.* Health Services Advisory Committee. HMSO, London.

HSAC (1992) *Safe Disposal of Clinical Waste.* Health Services Advisory Committee. HMSO, London.

Jackson, M.M. and McPherson, D.C. (1992) Blood exposure and puncture risks of OR personnel. *Todays OR Nurse* 14, 5–10.

Jagger J., Hunt E., Brand-Elnagger J. *et al.* (1988) Rates of needlestick injury caused by various devices in a university hospital. *New England Journal of Medicine* 319, 284.

Jochen, A.B.B. (1992) Occupationally acquired hepatitis C infection. *Lancet* 340, 975.

Kennedy, D.A. (1988) Equipment-related hazards. In: Collins, C.H. (ed.) *Safety in Clinical and Biomedical Laboratories.* Chapman and Hall, London, pp. 22–23.

Klein, R.S., Freeman, K., Taylor, P.E. *et al.* (1991) Occupational risk for hepatitis C infection among New York City dentists. *Lancet* 338, 1540–1542.

Leggett, V. (1994) Infection control within mental healthcare environments: a community perspective. In: Worsley, M., Ward, K.A., Privett, S. *et al.* (ed.) *Infection Control.* ICNA, London.

Management of Health and Safety at Work Regulations (1992). HMSO, London.

Morgan, D. (1991) Preventing needlestick injuries. *British Medical Journal* 302, 1147.

National Health Service (1993) *NHS Performance Specification: Black Household Sacks, Medium Duty Clinical Waste Sacks, Heavy Duty Clinical Waste Sacks.* NHS Suppplies Authority, Alfreton.

Public Health Laboratory Service (1993) Hepatitis C virus, Guidance on the risks and current management of occupational exposure. *Communicable Disease Report Review*, R135–R139.

Reporting of Injuries, Diseases and Dangerous Occurrences Regulations 1995. HMSO, London.

Richmond, J.Y and McKinney, R.W. (Eds) (1993) *Biosafety in Microbiological and Bio-medical Laboratories.* Centers for Disease Control and National Institutes of Health, 3rd edn. Government Printing Office, Washington DC.

Robert, L.M. and Bell, D.M. (1994) HIV transmission in the healthcare setting. *Infectious Disease Clinics of North America* 8, 319–329.

Royal College of Nursing (1987) *Introduction to Hepatitis B and Nursing Guidelines for Infection Control.* London.

Sodeyama, T., Kyosawa, K., Urishihara, A. *et al.* (1991) Hepatitis C in hospital employees with needlestick injuries. *Annals of Internal Medicine* 115, 367–369.

Tookey, P. and Packham, C.S. (1991) Does cytomegalovirus present an occupational risk? *Archives of Diseases of Children* 66, 1009–1010.

Tookey, P. and Logan, S. (1994) Occupational risk of cytomegalovirus. *Review of Medical Microbiology* 5, 33–38.

Waldron, H.A. (1992) Needlestick injuries. *Lancet* 340, 975.

Werner, B.G. and Grady, G.F. (1982) Accidental hepatitis B surface antigen; positive inoculation. Use of 'e' antigen to estimate infectivity. *Annals of Internal Medicine* 97, 367–369.

WHO (1991) *Biosafety Guidelines for Diagnostic and Research Laboratories Working with HIV.* WHO AIDS Series 9. World Health Organization, Geneva.

CHAPTER 11
Clinical laboratories: mucocutaneous exposure

D.A. Kennedy and C.H. Collins

The analysis of blood, plasma and serum samples is a major activity of clinical laboratories, especially clinical chemistry and haematology laboratories. In the UK laboratory workers rarely collect blood; this task is performed by phlebotomists, nurses and doctors (see Bouvet, Chapter 7 this volume), although some haematology staff may be involved in the collection of bone marrow samples. The laboratory workers are thus more likely to be exposed to mucocutaneous contact, from the samples themselves, from contaminated surfaces and materials, and from aerosols (see Kennedy, Chapter 6 this volume). Percutaneous exposure, a risk in countries where laboratory staff perform phlebotomy is considered by Jagger and Bentley (Chapter 12, this volume).

Inevitably, some patients will harbour pathogenic microorganisms which can be transmitted to laboratory staff from their samples. Accordingly, there is a need to organize the work to minimize exposure to infection. A clinician who commissions a laboratory investigation is often able to determine whether the patient has a blood-borne infection and can label the blood sample or request form accordingly in order to alert laboratory staff to the risk involved. Several official publications have required or recommended that samples from patients known to be infected with, for example, a hepatitis virus or the human immune deficiency virus be labelled with the words 'High Risk' or 'Danger of Infection' (DH, 1979; Advisory Committee on Dangerous Pathogens (ACDP) 1990a; Health Services Advisory Committee (HSAC) 1991). But it does not follow that a patient or sample without such a label does not offer a blood-borne infection risk. Indeed, a laboratory report may be the first indicator of the presence of blood-borne pathogenic microorganisms in a patient under investigation. In addition, concern has been expressed, e.g. by Whale (1986), that by labelling some samples 'High Risk' a two-tier safety system would be created. The labelled samples would be treated with great care, while the unlabelled samples might be regarded as harmless and would be treated accordingly. It is therefore prudent to regard every single blood sample sent to the clinical laboratory, and every single patient for which sample collection has been requested, as an infection risk. It is against this background that the safety management strategy known as 'Universal Precautions' has developed, which is evolving into Standard Precautions (see Hunt, Chapter 19, and Kibbler, Chapter 20 this volume). In order to appreciate the logical basis of these precautions it is necessary to understand the mechanisms, i.e. portals of entry and routes of exposure, whereby blood-borne microorganisms can enter the body.

PORTALS OF ENTRY AND ROUTES OF EXPOSURE

The portals of entry for blood-borne pathogens, in considered order of importance, are: damaged skin, eyes, mucous membranes, lungs, mouth.

Damaged Skin

Apart from percutaneous exposure (Bouvet, Chapter 7 and Jagger and Bentley, Chapter 12 this volume) there is a potential in clinical laboratories for environmental surface contamination with blood (see Kennedy, Chapter 6 this volume). It is known that under experimental conditions hepatitis B virus (HBV) can survive in dried blood plasma for up to one week and possibly longer (Bond *et al.*, 1981). Thus, during the course of the work, unprotected skin lesions may come into contact with blood and transmission of viable pathogens is possible. Unfortunately very little investigational work has been done in this area.

Role of the Hand

Hands, including gloved hands, which handle blood and other samples are most likely to get contaminated during the course of work. Hand contamination may occur as a result of contact with blood-contaminated environmental surfaces. A contaminated hand may transfer blood to portals of entry (e.g. eye, mucous membrane – see below), other body surfaces and environmental surfaces. Box 11.1 illustrates the role of the hand in the transmission of contamination.

The Eye

The eye is a portal of entry for microorganisms (Papp, 1959). HBV has been transmitted experimentally in a chimpanzee by corneal inoculation (Bond *et al.*, 1982). A number of laboratory-acquired infections have been associated with sprays and droplets hitting the face or eye (Kennedy, 1988) thus demonstrating the potential for occupational blood-borne infection and the benefits of wearing spectacles. Recently Sartori *et al.* (1993) reported the transmission of hepatitis C virus by a splash into the eye of a nurse.

Mucous Membrane

HBV has been transmitted experimentally in human volunteers via the oral mucosa (Petersen, 1980) and mucous membranes are considered to be sites of entry for the human immunodeficiency virus (HIV) by the National Committee for Clinical Laboratory Standards (NCCLS, 1991). Other pathogens may be transmitted *via* mucous membranes.

Box 11.1. Role of the hand in transmission of contamination.

Hands and fingers are frequently injured
Hands get contaminated with workplace materials
Gloved hands get contaminated from external sources

Contacts*
hand to face in general
hand to mouth
hand to eye
hand to nose

Habits*
nose picking
nail biting
finger licking when page turning
pen and pencil chewing

*These may be involuntary.
After Kennedy (1995).

Lungs

A number of studies (see Kennedy, Chapter 6 this volume) have demonstrated that whole blood can become aerosolized by laboratory manipulations, probably by break-up of very thin sheets or ligaments. Plasma and serum can also become aerosolized, perhaps with greater ease than whole blood. Microorganisms that become airborne in aerosol clouds and enter the respiratory system by inhalation through the nose or mouth can give rise to systemic infection. The alveolar regions of the lungs are particularly vulnerable to infection and deposition here of fine particles in the range 3–5 µm aerodynamic diameter is maximal. However, the question remains whether at present blood aerosol amounts to a significant occupational health risk in clinical laboratories. Currently there appears to be general agreement that in occupational settings, and in the community, the risk of transmission of HBV and HIV by the airborne route is very low indeed. But other microorganisms that are naturally transmitted by the airborne route, e.g. brucellas, may be present in large numbers in blood samples (see Collins, Chapter 2 and Hunt, Chapter 3 this volume). *Mycobacterium tuberculosis* and *M. avium-intracellulare* are not infrequently isolated from the blood of patients with AIDS and the transmission of these tubercle bacilli to laboratory workers cannot be dismissed. The need for aerosol containment (see below) should always be considered when devising a prevention strategy.

Mouth

The mouth is a portal of entry for microbes and those which can survive a low gastric pH can enter the lower digestive system with potential for systemic infection. Transmission of pathogens may result if blood-contaminated objects enter the mouth or come into contact with its mucous membranes. Accordingly, mouth-pipetting, drinking, smoking and applying cosmetics and the like, are universally prohibited in clinical laboratories.

SAMPLE COLLECTION, METHODOLOGY AND EQUIPMENT

Some comments will be made on sample containers, inactivation of pathogens in blood or serum, serum and plasma separation, blood films, blood cultures, centrifuges and automated analytical equipment.

Sample Collection

Venous blood specimens are collected, either with a syringe and needle or by an evacuated blood collection system, e.g. the Becton Dickinson Vacutainer. There is evidence that both methods may disperse blood, thereby contaminating the environment (Lach *et al.*, 1983; Holton and Prince, 1986), although the syringe and needle method may give rise to more contamination than does the use of an evacuated blood collection system (Holton and Prince, 1986).

If the syringe and needle method is used the best type of container is one provided with a screw cap. Containers with rubber-bung-type closures are less satisfactory because blood tends to collect at the base of the closure or even between the lower part and the wall of the tube. When the closure is withdrawn small volumes of blood may be spattered with a risk of contaminating the operator's hands and surrounding environmental surfaces. Such dispersal can be minimized, if not prevented altogether, by placing a strip of absorbent paper around the cap and neck of the tube and then using it to withdraw the bung (Collins, 1993). This paper is then discarded into disinfectant. At least one manufacturer, aware of the problem, supplies disposable caps which may be placed over a bung so that it can be removed without contaminating the fingers.

Some evacuated tubes have rubber bung closures and others have screw caps. Holton and Prince (1986) compared the two types and concluded that there was less contamination of gloved hands when a screw-capped container was used. However, Tomlinson (1952) demonstrated that the removal of a screw cap also carries with it the risk of dispersal of droplets and aerosol (see Kennedy, Chapter 6 this volume).

An outbreak of hepatitis B has been attributed to re-use of a component of a capillary blood sampling device (Polish *et al.*, 1992).

Inactivation of Pathogens in Blood and Serum

Pre-treatment of plasma and serum samples to reduce the risk of HIV infection
All laboratory manipulations of samples containing viable HIV carry with them a risk of infection. Pre-treatment of samples in order to inactivate HIV, includ-

ing centrifugation and associated pipetting to collect plasma and serum, is no exception.

Nevertheless, pre-treatment remains an attractive proposition because if HIV (and other pathogens) can be reliably inactivated the risk to staff of becoming infected will, theoretically, be eliminated. In practice, however, it seems unlikely that any pre-treatment method, even with careful quality control, will guarantee total elimination of the risk of infection. It is clearly prudent, therefore, to handle all blood, serum and plasma samples, including those that have received pre-treatment, as though they were still capable of transmitting infection.

Clearly, there is no point in the pre-treatment of blood, serum and plasma if it affects adversely the results of a diagnostic investigation.

Heat treatment

Goldie *et al.* (1985) studied the effect of heat treatment on whole blood and serum before biochemical analysis. Clotted whole blood samples were placed in a water bath at 56°C for 30 min and then centrifuged. This treatment resulted in gross haemolysis of samples and since such samples are unreliable for biochemical analysis heat treatment of whole blood was not pursued further. However, similar treatment of serum samples produced encouraging results (with Na, HCO_3, urea, creatinine, glucose, uric acid, Ca, P, Mg, total protein, albumin, total bilirubin, immunoglobulins, cholesterol, triglycerides, T4, TSH, cortisol). With the exception of amylase and AST, enzyme results were significantly reduced. With a few differences, encouraging biochemical results for analytes other than enzymes after heat treatment of plasma or serum were reported by Lai *et al.* (1985), Van den Akker *et al.* (1985a) and Gow and Fallon (1985).

Evans and Shanson (1985) reported that heat treatment at 56°C for 30 min made no significant differences to the results with syphilis or hepatitis B serology, or for gentamicin, tobramycin and netilmicin assays. Holt *et al.* (1985) reported that sera spiked with aminoglycosides showed no change in concentration when measured spectrophotofluorimetrically after heat treatment. When measured microbiologically netilmicin and tobramycin assays showed no change after sera were heat treated for two hours. Gentamicin was not affected after sera were heat treated for one hour but levels fell after two hours' treatment. Other antimicrobials were less stable after heat treatment.

Van den Akker *et al.* (1985b) reported that heat treatment of sera did not affect the results of commercial micro-ELISA assay for HIV but that it did produce false-positive results in a different commercial assay.

β-propiolactone (BPL) treatment

β-propiolactone inhibits the reverse transcriptase activity of HIV and reduces the infectivity of other pathogens (Ball and Griffiths, 1985). Because of its high toxicity and demonstrated skin tumour production in animals, human contact should be avoided (Hathway *et al.*, 1991). Accordingly, BPL requires very careful handling and any risk reduction benefit from its use, in microbiology, will have to be weighed against the definite chemical risk.

Ball and Griffiths (1985) studied the effect on a range of biochemical tests of adding 25% BPL to samples. BPL was added to whole blood and the plasma was separated one hour later. There were significant differences in the results of many estimations carried out on this plasma, some due to haemolysis. When BPL was added to plasma after separation pH, bicarbonate and total CO_2 fell significantly. The α_1 band was reduced on protein electrophoresis and aspartate aminotransaminase (AST) and lactic acid dehydrogenase (LDH) results were 10–18% lower in treated plasma samples. Otherwise, analytical results (Na, K, Ca, total protein, glucose, amylase, urea, creatinine, phosphate, triglyceride) were not significantly affected. A limited study, by Ball and Bolton (1985) suggested that BPL does not interfere with the estimation of haemoglobin or total white cell or platelet counts.

Serum and Plasma Separation

Many investigations require the separation of serum or plasma from clotted or centrifuged blood. There is potential for percutaneous exposure, e.g. cuts and punctures from hypodermic needles or glass Pasteur pipettes. Plastic Pasteur pipettes are preferable to those made of glass, but are still able to cut the skin (Kennedy, Chapter 6 this volume). A syringe and needle assembly should never be used as a pipetting device (World Health Organization; WHO, 1993). Handling, and therefore exposure, may be reduced by using containers with additives, e.g. polystyrene beads or gel, that enable serum or plasma to be poured off, rather than pipetted. These containers may also be used in conjunction with analysers that are provided with sample probes which puncture the caps to aspirate a sample for analysis. Experiments using the 'double-tracer' method (Kennedy, Chapter 6 this volume) have demonstrated that serum and plasma separation is often associated with significant surface contamination caused by splashing and the deposition of small droplets.

Blood Films

Kohn (1976) noted that although extensive precautions had been proposed for work with liquid or clotted blood samples there was no mention of the risk associated with the handling of blood smears that might contain viruses. Dankhert et al. (1976) examined films made from blood and bone marrow from HBsAg-positive patients and found the marker on unstained films for up to 21 days after manufacture. Smears stained with Wright's stain using an automated slide stainer gave HBsAg-positive result for up to 14 days after staining. Dankhert et al. (1976) detected HBsAg in the staining fluid after the positive slides had been stained. Bond et al. (1981) showed that HBV remained viable for several days when blood on glass slides was kept under laboratory conditions. Bending and Maurice (1980) described a case of laboratory-acquired malaria associated with the making of a blood film. A method of decontaminating blood films has been described (ACDP, 1990b; see also Collins, 1993). Air-dried thick films are first fixed in methanol for 5 min and then placed, for 10 min, in the following mixture:

Formalin (37–40% w/v)   500 ml
Sodium hydrogen phosphate (NaH$_2$PO$_4$.H$_2$O)   22.75 g
Sodium dihydrogen phosphate (Na$_2$HPO$_4$)   32.50 g
Distilled water   4500 ml

The films are then washed three times in buffered water at pH 7.0 before being stained.

Blood Cultures

Risks associated with the collection of blood for blood culture, etc. are described by Bouvet (Chapter 7 this volume). In the laboratory, infection risks arise during subculturing, when 'venting' and from the probes of automated blood culture systems.

Until recently, the standard method of blood culture, which is still used in some laboratories, was to take 10–15 ml of venous blood from a patient using a syringe and needle and to inject it through the cap of a bottle containing liquid blood culture medium. A wide variety of automated blood culture systems is now available (Collins *et al.*, 1995), but subculturing must still be done at intervals. This entails either removing the closure in order to sample the contents with an inoculation loop or Pasteur pipette, or in the case of septum-capped bottles sampling with a syringe and needle. As indicated above, removing screw caps or other type of closure (see above) may generate aerosols and droplets when the film of liquid that is inevitably present between the cap and the rim of the bottle is broken (Tomlinson, 1952). Withdrawing a needle through a septum can also generate aerosols (Darlow, 1972).

All subculturing should therefore be conducted in a microbiological safety cabinet that conforms to and is operated in accordance with national standards such as BS 5726 (BSI, 1992). The additional precautions necessary to reduce dispersion of droplets within the cabinet are:

- loosing the closure a few seconds before removing it;
- if a syringe and needle must be used, wrapping an alcohol-soaked pledget of cotton wool around the needle before withdrawing it, and withdrawing it through the pledget (Hanel and Alg, 1955).

Evacuated blood culture bottles with septum caps have been offered by several commercial organizations. Apart from the obvious risk of needlestick injuries, small amounts of blood are occasionally left on the outside of the septum cap after the needle has been withdrawn (Gee, 1984). Unless this is wiped off with a disinfectant-soaked swab (see above) the residual blood remains an infection risk whoever handles the bottle. A further risk arises because of the need to vent the cap for the successful culture of some organisms, e.g. *Candida* spp. This is often done by inserting a hypodermic needle, the hub of which is plugged with cotton wool, through the septum. Again there is a risk of needlestick injuries.

Infection Risks Posed by Reagents Derived from Human Blood

Jones *et al.* (1985) examined 17 commercially-available blood clotting factor deficient reagents that were used in tests for bleeding disorders. Fifteen of them had been prepared from human plasma only, and two from human and bovine plasma. All factor VIII-deficient plasmas and three factor V-deficient plasmas gave anti-HIV positive results. It has been reported that other commercial factor VIII-and factor IX-deficient plasmas gave positive anti-HIV results (DH, 1985) and that all of five batches of immunoassay control sera of human origin tested gave anti-HCV results (Simmons *et al.*, 1990). A case was reported (Anon., 1989) in which a pregnant laboratory worker developed hepatitis B after suffering a cut while handling a quality control material which later gave a positive test result for HBV.

It is clear than any reagent for laboratory (including commercial quality control and calibration material, and material distributed by external quality assurance schemes) that is derived from human blood may possibly contain blood-borne viruses and other pathogens, and that they should be treated as potentially infectious in accordance with locally-agreed procedures. In addition, when purchasing reagents derived from human blood it is prudent, wherever possible, to specify that they are derived from donors who have been tested for markers of HBV, HCV and HIV infection, using methods of proven efficacy, and who have been found to be negative. Finally, bearing in mind that the greatest risk of occupational transmission is by the percutaneous route, e.g. as a result of cuts, it is best, whenever possible, to avoid the purchase of human blood-derived reagents that are packaged in vials with metal sealing rings. These are often found to be quite sharp and may become contaminated with the contents of the vial.

Sample Carry-over

Sample carry-over can increase infection risks. Carry-over of HBsAg-positive serum to a formerly HBsAg-negative serum was demonstrated when an automatic diluter with multiple-use tips was used. It was shown that even after wiping the tip about 20 nl of HBsAg-positive serum was transferred from the first tube in the series to the second, and even from this one there was a detectable carry-over to the third tube in the row (Centers for Disease Control; CDC, 1980). It has been reported by the Medical Devices Agency (MDA, 1996a) that, with reference to false positive hepatitis B test results, the use of an automated microplate analyser with a fixed metal probe resulted in sample carry-over and in some cases in contamination of primary samples.

The reports illustrate how it is at least theoretically possible for a non-infectious sample to be made infectious, thereby increasing the infection risk generally.

While carry-over of analytes has been of great interest to those who evaluate the performance of analytical systems, it does not appear to have received much consideration in the context of safety.

Centrifuges

Mechanical safety is the prerequisite of microbiological safety (Kennedy, 1988). Centrifuges should comply with the requirements of the International Electro-chemical Commission (IEC, 1992) and be regularly inspected and maintained by competent persons. Centrifugation of blood samples in unstoppered containers, and breakage of containers in centrifuges can produce blood aerosols and droplets resulting in environmental surface contamination (Whitwell *et al.*, 1957). In addition to stoppered tubes, sealed centrifuge rotors or sealed centrifuge buckets should be used where there is a risk of blood-borne infection. The standard (IEC, 1992) includes a microbiological test for sealed rotors and buckets. Breakage of, or leakage from, glass capillaries in microhaematocrit centrifuges can produce an aerosol (Rutter and Evans, 1972), and blood-stained glass presents a percutaneous infection risk. The small trays of sealing putty tend to become blood stained in use and thus may become a reservoir of infection. The presence of small glass fragments embedded in the sealing material increases the infection risk.

UNIVERSAL PRECAUTIONS

In the USA the adoption of Universal Precautions is a requirement of the Occupational Safety and Health Administrations's (OSHA) Blood-borne Pathogens Standard (OSHA, 1991) (see Hunt, Chapter 19 this volume). Similar precautions are recommended by other governments and the World Health Organization (see Kibbler, Chapter 20 this volume). Universal Precautions have the following basic elements.

- Training of staff.
- Reduction of sharps and management of essential sharps.
- Reduction of surface and air-borne blood contamination.
- Personal protective measures.
- First-aid measures.
- Quality assurance.

Box 11.2 (Kennedy, 1995) lists Universal Precautions that are recommended for a clinical haematology laboratory and which are generally applicable to most clinical laboratories. Some comments on the content of this box are pertinent:

Training of Staff

Staff should be trained in the basic principles of risk assessment and risk management. In particular, they should be encouraged to understand why they are at risk from blood-borne pathogens and in particular the relevant portals of entry, routes of exposure and how exposure can be controlled. Training should be kept under review, and regularly updated with regard to changes in work practices and range of pathogens likely to be encountered.

Box 11.2. Universal Precautions for haematology laboratories (Kennedy, 1995).

Control of Contamination
Require handwashing after blood exposure, after removing gloves and before leaving the laboratory.

Prohibit eating, drinking, smoking, application of cosmetics and lip balm, handling of contact lenses and storage of food or drink.

Use robust and leakproof sample containers and centrifuge in sealed buckets if the patient is known or suspected to be a carrier of HBV, HCV, HIV or other blood-borne pathogen.

Laboratory work – require the wearing of gloves, eye protection and face shields where there is a risk of direct contact with splashes, spray, spatter or droplets of blood, or where there are blood contaminated surfaces.

Sample collection – do not discourage the routine wearing of gloves and require that they be worn when a patient is known to be a carrier of HIV, HCV, HIV or other blood-borne pathogen; when a member of staff has cuts, scratches or other breaks in the skin; when there is extra risk of blood contamination, e.g. an un-cooperative patient and when the member of staff is receiving training.

Require that all splashes and spills be dealt with immediately and then decontaminate the area.

Require regular cleaning and decontamination of all environmental surfaces and all devices after use.

Sharps
Provide training in safe use and disposal.

Require that a syringe and needle should never be used as a pipetting device and avoid the use of sharp objects or glass wherever possible.

Require that contaminated needles and other sharps should never be bent, recapped, or removed unless there is no other option. If contaminated sharps must be so handled, it must be by mechanical means or a one-handed technique which minimizes the risk of injury.

Provide as close to the site of use as possible, adequate puncture-resistant, leakproof and adequately labelled sharps disposal containers. Institute measures to ensure that they are not over-filled, and require their safe use and disposal.

Immunization
Offer and make arrangements for HBV immunization and follow-up.

Offer and make arrangements for other appropriate immunization and follow-up.

First Aid Provisions
Make arrangements for immediate cleaning and dressing of cuts and other skin lesions.

When there is a risk of transmission of HBV, ensure that medical or other trained staff are available to administer combined active immunization and passive immunization with hepatitis B immunoglobulin and to provide follow-up.

When there is a risk of transmission of HIV, ensure that medical or other trained staff are available to provide counselling and to administer AZT prophylaxis if considered necessary.

When there is a risk of transmission of other pathogens, ensure that medical or other trained staff are available to administer appropriate prophylaxis or therapy.

Box 11.3. Sharps causing skin injury in clinical laboratories.

Blood cell counter component	Centrifuge component
Stapling machine	Scalpel
Glass fragments	Unbroken glass slide
Broken glass slide	Glass Pasteur pipette
Plastic Pasteur pipette	Glass test tube
Glass capillary tube	Edge of locker
Autoclave tin	Inside of sharps bin
Metal sealing ring of vial	Metal tube rack
Edge of seat	Edge of paper

After Kennedy (1988).

Control of Sharps

Every effort should be made to avoid the risk of sharps injury and it should be remembered that it is not only needles and lancets that can breach the skin. There is evidence that a wide range of devices found in laboratories are capable of cutting and puncturing skin (Box 11.3).

Needles should be tolerated only where their use is essential, e.g. in blood collection, and it cannot be over-emphasized that the use of syringes and needles as pipetting devices is a dangerous practice and that it should be positively discouraged (WHO, 1993). A plentiful supply of sharps containers (see Collins and Kennedy, Chapter 18 this volume) should be available wherever blood collection devices are used and these should be carefully managed to ensure safe disposal. Whether or not a needle should be re-sheathed after re-use is a continuing debate. If needles must be re-sheathed, steps should be taken to avoid needlestick. Steps should also be taken to avoid the risk of transmission of infection to a patient by the accidental re-use of re-sheathed blood-contaminated needles (Kennedy, 1995).

Surface Contamination

Surface contamination is often the result of spillage of blood or escape from damaged or leaking containers. Accordingly, it is recommended that robust and leakproof blood sample containers should be used, e.g. those that comply with the requirements of the International Standard Organization (ISO, 1995). On the assumption that the more that a sample is handled, the greater is the opportunity for contamination, when designing operating procedures every effort should be made to reduce the number of operating steps to a minimum. The prohibition of mouth pipetting, smoking, drinking, application of cosmetics and the like cannot be over-emphasized. Frequent hand washing and regular disinfection of environmental surfaces is essential.

Air-borne Contamination

Air-borne contamination is the result of production of blood aerosol (see above). Where there is a risk of air-borne transmission of pathogens, centrifugation of blood should be carried out in sealed centrifuge buckets that meet the requirements of the International Electrotechnical Commission (IEC, 1992). The sealed bucket should be opened in a microbiological safety cabinet of proven efficacy, e.g. one that complies with and is operated in accordance with a national standard (e.g. BSI, 1992). It must be stressed, however, that centrifuges should never be operated in open-fronted microbiological safety cabinets because they can disrupt airflow patterns leading to escape of potentially infectious aerosol (Collins, 1993). Wherever possible, operational procedures should minimize the need to pipette blood, pour blood and remove the closures of blood sample containers. Such manipulations are capable of producing an aerosol (see Kennedy, Chapter 6 this volume), and where an infection risk is possible they should be carried out in a microbiological safety cabinet. Automated equipment that avoids the need to remove closures is available.

Personal Protective Measures

These consist of two basic elements.

• Personal protective equipment.
• Immunization.

Personal protective equipment

Laboratory coats should always be worn in the workplace. When choosing a design, the object should be to give maximum protection from contamination with blood to the skin and the clothing worn under the coat. Accordingly, the degree of protection afforded should be the main determinant when making a purchasing decision. The traditional laboratory coat, which fastens at the front, offers limited protection to the neck, upper chest and wrist area. In contrast, the Dowsett and Heggie pattern overall (Dowsett and Heggie, 1972) offers good protection to these areas, and to the lower part of the body, if properly fastened. If at any time an overall becomes contaminated with blood, it should be carefully removed to prevent personal contact with the blood and placed in a clear or light blue coloured plastic bag for autoclaving (see Collins and Kennedy, Chapter 18 this volume) or in a laundry bag designated 'hot wash'. During work in which there is a higher risk of blood contamination, the overall should be supplemented with a disposable plastic apron which is used once only, and then discarded into an appropriate plastic bag for autoclaving or incineration.

Protective gloves should be worn and provision should be made for staff who have a sensitivity to latex and other glove materials (Berky *et al.*, 1992; MDA, 1996b). Wearing two pairs of gloves may be considered in some cases (see Morgan, Chapter 8 this volume).

Although ordinary sight-correction spectacles offer some protection it is much safer to wear protective spectacles with side-shields or full-face visors to give maximum protection to eyes from blood splashes. Cuts, grazes and other skin lesions should be protected with a water-proof adhesive dressing.

Immunization and follow-up procedures

Unfortunately, no HIV vaccine is available at present. However, staff should be offered, and arrangements should be made for HBV immunization and other appropriate immunization where available. Arrangements should be made to follow-up the efficacy of immunizations (see Morgan, Chapter 8 and McCloy, Chapter 16 this volume).

First Aid Measures

All blood splashes and spills should be mopped-up immediately, and this should rapidly be followed by decontamination of the affected area. Cuts and skin injuries should be cleaned immediately and covered with a waterproof dressing. Where there is a risk of transmission of HBV, trained staff should be available to administer combined active immunization and passive immunization with immunoglobulin and to provide follow-up (see Waldron, Chapter 15 and McCloy, Chapter 16 this volume). Where there is a risk of transmission of HIV, trained staff should be available to provide counselling. Where there is a risk of transmission of other pathogens, trained staff should be available to administer appropriate prophylaxis or therapy.

Quality Assurance

Quality has been defined as 'fitness for purpose'. In the context of Universal Precautions, quality assurance is all those planned and systematic actions necessary to provide adequate confidence in the efficacy of risk management procedures.

Quality assurance can be built within a framework that envisages that a series of searching questions can always be asked, at any time, about a procedure, e.g.

- How is it done?
- Why is it done that way?
- When is it done?
- How do you know that it works?
- Who is responsible for doing it?
- Have they been trained to do it
- Can you show me the evidence?

This approach can also be used when carrying-out safety audits in clinical laboratories.

REFERENCES

ACDP (1990a) *Categorization of Pathogens According to Hazard and Categories of Containment*, 2nd edn. Advisory Committee on Dangerous Pathogens. HMSO, London.

ACDP (1990b) *HIV – the Causative Agent of AIDS and Related Conditions*. Advisory Committee on Dangerous Pathogens. Health and Safety Executive, London.

Anon. (1989) Hepatitis B. *Hospital Hazards Material Management* 3, 5.

Ball, M.J. and Bolton, F.G. (1985) Effects of inactivation of HTLV-III on laboratory tests. *Lancet* ii, 99.

Ball, M.J. and Griffiths, D. (1985) Effect on chemical analysis of beta-propiolactone treatment of whole blood and plasma. *Lancet* i, 1160–1161.

Bending, M.R. and Maurice, P.D.L. (1980) Malaria: a laboratory risk. *Postgraduate Medical Journal* 56, 344–345.

Berky, S.T., Luciano, W.J. and James, W.D. (1992) Latex glove allergy: a survey of the US Army Dental Corps. *Journal of the American Medical Association* 268, 2695–2697.

Bond, W.W., Favero, M.S., Petersen, N.J. *et al.* (1981) Survival of hepatitis B virus after drying and storage for one week. *Lancet* i, 550–551.

Bond, W.W., Favero, M.S. and Petersen, N.J. (1982) Transmission of type B viral hepatitis via eye inoculation of a chimpanzee. *Journal of Clinical Microbiology* 15, 533–534.

BSI (1992) *British Standard 5726. Microbiological Safety Cabinets, Parts 1–4.* British Standards Institution, London.

CDC (1980) Hepatitis B contamination in a clinical laboratory – Colorado. *Morbidity and Mortality Weekly Report* 29, 459–465.

Collins, C.H. (1993) *Laboratory Acquired Infections,* 3rd edn. Butterworth-Heinemann, Oxford.

Collins, C.H., Lyne, P.M. and Grange, J.M. (1995) *Collins and Lyne's Microbiological Methods,* 7th edn. Butterworth-Heinemann, Oxford.

Dankhert, J., Uitentuis, J., Postma, A. *et al.* (1976) HBsAg hazard in blood and marrow smears. *Lancet* i, 1083–1084.

Darlow, H.M. (1972) Safety in the microbiological laboratory. In: D.A. Shapton and R.G. Board (eds) *Safety in Microbiology.* Society for Applied Bacteriology Technical Series 6. Academic Press, London.

DH (1979) *Code of Practice for the Prevention of Infection in Clinical Laboratories and Post-mortem Rooms.* HMSO, London.

DH (1985) Possible infection risk to laboratory staff from human blood-based coagulation factor deficient plasmas. *Safety Action Bulletin. SIB (85)30.* Department of Health, London.

Dowsett, E.G. and Heggie, J.F. (1972) A protective pathology laboratory coat. *Lancet* i, 1271.

Evans, R.P and Shanson, D.C. (1985) Effect of heat on serological tests for hepatitis B and syphilis and on aminoglycoside assays. *Lancet* i, 1457–1458.

Gee, B. (1984) An assessment of some hazards associated with the collection of venous blood. *Journal of Hospital Infection* 5, 102–103.

Goldie, D.J., McConnell, A.A. and Cooke, P.R. (1985) Heat treatment of whole blood and serum before chemical analysis. *Lancet* i, 1161.

Gow, A. and Fallon, R.J. (1985) Effect of inactivating HTLV-III on laboratory tests. *Lancet* ii, 99.

Hanel, E. and Alg, R.L. (1955) Biological hazards in common laboratory procedures. II. The hypodermic needle and syringe. *American Journal of Medical Technology* 21, 343–346.

Hathway, G.J., Proctor, N.H., Hughes, J. *et al.* (1991) *Proctor and Hughes' Chemical Hazards in the Workplace*, 3rd edn. Van Nostrand Reinhold, New York. pp. 491–492.

Holt, H.A., Bywater, M.J. and Reeves, D.S. (1985) Effects of inactivating HTLV-III on laboratory tests. *Lancet* ii, 99.

Holton, J. and Prince, M. (1986) Blood contamination during venepuncture and laboratory manipulation of specimen tubes. *Journal of Hospital Infection* 8, 178–183.

HSAC (1991) *Safe Working and the Prevention of Infection in Clinical Laboratories.* Health Services Advisory Committee, Health and Safety Commission. HMSO, London.

IEC (1992) Safety requirements for electrical equipment for measurement, control and laboratory use. EC 1010, Part 2–020: Particular requirements for laboratory centrifuges. International Electrotechnical Commission, Geneva.

ISO (1995) International Standard 6710. Single use containers for venous blood specimen collection. International Standards Organisation, Geneva.

Jones, P., Hamilton, P.J., Oxley, A. *et al.* (1985) Anti-HTLV-III positive laboratory reagents. *Lancet* i, 1457–1458.

Kennedy, D.A. (1988) Equipment-related hazards. In: Collins, C.H. (ed.) *Safety in Clinical and Biomedical Laboratories.* Chapman and Hall, London, pp.10–46.

Kennedy, D.A. (1995) Blood samples and reagents: hazards and risks. In: Lewis, S.M. and Koepke, J.A. (ed.) *Haematology Laboratory Management and Practice.* Butterworth-Heinemann, Oxford.

Kohn, J. (1976) Blood smears and hepatitis. *Lancet* ii, 226.

Lach, V.H., Harper, G.J. and Wright, A.E. (1983) An assessment of some hazards associated with the collection of venous blood. *Journal of Hospital Infection* 4, 57–63.

Lai, L., Ball, G., Stevens, J. *et al.* (1985) Effect of heat treatment of plasma and serum on biochemical indices. *Lancet* i, 1457–1458.

MDA (1996a) Labotech automated microplate analyser: risk of sample contamination and carry-over. Medical Devices Agency. Safety Notice MDA SN 9631. October 1996. MDA, London.

MDA (1996b) Latex sensitisation in the health care setting (use of gloves), MDA DB 9601. Medical Devices Agency, London.

NCCLS (1991) *Protection of Laboratory Workers from Infectious Disease Transmitted by Blood,* 2nd edn. NCCLS Document M29-T2, Vol. 11, No. 14. National Committee for Clinical Laboratory Standards, Villanova.

OSHA (1991) Occupational Health and Safety Administration. Bloodborne Pathogens Standard. Final rule. *Federal Register* 56 (235), 64175–64182.

Papp, K. (1959) The eye as a portal of entry of infections. *Bulletin of Hygiene* 34, 969–971.

Petersen, N.J. (1980) An assessment of the airborne route in hepatitis B transmission. *Annals of the New York Academy of Sciences* 353, 157–166.

Polish, L.B., Shapiro, C.N., Bauer, F. *et al.* (1992) Nosocomial transmission of hepatitis B virus associated with the use of a spring-loaded finger-stick device. *New England Journal of Medicine* 326, 721–725.

Rutter, D.A.K. and Evans, C.G.T. (1972) Aerosol hazards from clinical laboratory apparatus. *British Medical Journal* 1, 594–596.

Sartori, M., La Terra, G., Aglietta, M. *et al.* (1993) Transmission of hepatitis C via a blood splash into conjunctiva. *Scandinavian Journal of Infectious Disease* 25, 270–271.

Simmons, P., McOrmish, F., McCullough, P. *et al.* (1990) Contamination of immunoassay controls with hepatitis C virus. *Lancet* 338, 1539–1542.

Tomlinson, A.J.H. (1952) Infected airborne particles liberated on opening screw-capped bottles. *British Medical Journal* 2, 15–17.

Van den Akker, R., Hekker, A.C. and Osterhaus, A.D.M.E. (1985a) Effects of heat treatment of plasma and serum on biochemical indices. *Lancet* i, 1457–1458.

Van den Akker, R., Hekker, A.C. and Osterhaus, A.D.M.E. (1985b) Heat inactivation of serum may interfere with HTLV-III serology. *Lancet* ii, 672.

Whale, K. (1986) Is it time to rethink 'high risk' labelling? *Journal of Clinical Pathology* 39, 41.

Whitwell, F., Taylor, P.J. and Oliver, A.J. (1957) Hazards to laboratory staff in centrifuging screw-capped containers. *Journal of Clinical Pathology* 10, 88–91.

WHO (1993) *Laboratory Biosafety Manual.* 2nd edn. World Health Organization, Geneva.

CHAPTER 12
Clinical laboratories: percutaneous exposure

J. Jagger and M. Bentley

INTRODUCTION

In the past decade occupational risks to laboratory workers have become more serious than ever with the advent of the AIDS epidemic. One of the most disturbing statistics on the occupational transmission of the immunodeficiency virus (HIV) is the large proportion of clinical laboratory workers among the documented and probable cases of occupationally-transmitted HIV in the United States. Clinical laboratory workers accounted for 20% (32/163) of all cases reported through December 1996, ranking second only to nurses despite the fact that nurses are far more numerous in the health-care work force than clinical laboratory workers (Centers for Disease Control and Prevention: CDC, 1996). One explanation for this is that most of the cases attributed to clinical laboratory workers involved phlebotomists who sustained needlestick injuries from blood-drawing needles. It has become clear that the types of exposures that are most likely to result in HIV transmission are those that involve the direct inoculation of significant quantities of blood (CDC, 1995).

In addition to HIV, numerous studies have documented an impressive array of pathogens transmitted to clinical laboratory personnel, foremost among them hepatitis B (Levy *et al.*, 1977; Wruble *et al.*, 1977; Grist and Emslie, 1985: Jacobson *et al.*, 1985; Collins and Kennedy, 1987; Vesley and Hartman, 1988; see also Jeffries, Chapter 1 and Collins, Chapter 2 this volume). In one study hepatitis B virus (HBV) seropositivity rates were highest among laboratory workers whose clothing was routinely contaminated with blood, those who frequently performed blood gas analysis, and those working on multi-channel autoanalysers (Pattison *et al.*, 1974). These observations suggest that transmission of some blood-borne pathogens can also be associated with frequent contact and skin contamination. The potential for exposure to laboratory specimens contaminated with blood-borne pathogens remains high. One study in a hospital chemistry laboratory found that 6% of serum or plasma specimens received were HBV-contaminated, and 3% were HIV-contaminated (Handsfield *et al.*, 1987).

Nevertheless, during the past ten years great strides have been made to reduce the risk of laboratory worker exposures to, and infections by, blood-borne pathogens. The two most important landmarks in prevention in the United States have been the availability of an effective hepatitis B vaccine, and the implementation of the Occupational Safety and Health Administration (OSHA) Bloodborne Pathogen Standard, which was enacted in its final form in December of 1991 (OSHA, 1991). The Standard requires employers to provide the hepatitis B vaccine at no cost to employees, to provide an adequate supply of appropriate personal protective equipment, and to clean, maintain, and/or replace this equipment regularly. Puncture-resistant, leak-proof disposal containers must be provided in convenient locations, and must be replaced before becoming overfilled. If an employee is exposed to blood or body fluids, the employer must provide post-exposure follow-up, including employee and source patient testing for blood-borne pathogens, and records of reported exposures must be maintained. Finally, the employer must provide training in Universal Precautions, safe work practices, and the employer's obligations under the Standard to its employees (the OSHA Standard is reviewed by Hunt, Chapter 19 this volume; see also Kibbler, Chapter 20).

Although it is widely believed that these measures have resulted in significant improvements, there is little documentation of the effects of the Standard on the frequency of exposures and infections in the clinical laboratory environment. This is because there were no standardized exposure surveillance programmes in place before the implementation of these measures that would have made before-and-after comparisons possible. Another reason is that the number of exposure incidents reported from clinical laboratories in a single institution is likely to be too small to draw general conclusions about exposure and risk patterns. One approach for obtaining enough data to draw meaningful conclusions is to combine data from numerous institutions, provided they all use a standard surveillance system. The present analysis, including data from 64 hospitals, was carried out to determine exposure patterns in clinical laboratory settings in order to identify high-risk equipment and activities, and to accurately focus prevention initiatives in the areas of greatest need and opportunity.

METHODS

All percutaneous injuries or mucocutaneous exposures to blood or body fluids that occurred in clinical laboratory settings and were reported to the 64 participating EPINet12 hospitals (Jagger *et al.*, 1994; see also Jagger and Bentley, Chapter 5 this volume) during a two-year period, beginning in September 1992, were included in this analysis (306 cases). In order to identify the unique characteristics of the clinical laboratory environment that distinguish it from other areas of the hospital, these cases were compared with exposures occurring in patient rooms (3468 cases). Incidents that occurred to clinical laboratory personnel outside of the clinical laboratory environment and incidents that occurred with non-contaminated items were not included in this analysis.

RESULTS

More than 80% of the 306 exposures occurring in clinical laboratories were to laboratory technicians, technologists or phlebotomists. A small proportion were to nurses, physicians, and housekeepers. Figure 12.1 shows the types of devices causing sharp-object injuries in clinical laboratories. Needles caused the greatest number of injuries, accounting for 38% of all injuries. Of needle injuries, 51% were caused by blood-drawing needles in clinical laboratories, while in patient rooms only 23% of needle injuries were caused by blood-drawing needles ($\chi^2 = 32.8$, $P < 0.0001$). This is significant because blood-drawing needles are among the devices most commonly associated with the transmission of blood-borne pathogens. Another unique finding was that 22% of needle injuries in clinical laboratories were associated with needles that were used as tools, while only 3% of needle injuries in patient rooms were attributed to needles used as tools ($\chi^2 = 85.5$, $P < 0.0001$). It is a common practice in clinical laboratories to use needles and syringes for purposes other than what they were designed for. The lack of safer equipment specifically designed for laboratory applications often leaves laboratory workers no alternative but to use available equipment that puts them at unnecessary risk of injury.

Of particular interest was the finding that glass injuries (specimen tubes, capillary tubes, pipettes, and slides) accounted for 39% of injuries as opposed to only 0.3% of injuries in patient rooms ($\chi^2 = 918.9$, $P < 0.0001$). These are especially serious injuries because glass items are often containers for blood, so that when a laceration occurs, there is often the potential for the introduction of a large blood inoculum into the wound. Figure 12.2 shows how glass injuries happened. The largest number of incidents occurred during use of the glass item, which means that glass items frequently broke as they were being handled. Many plastic products that are safer alternatives than conventional

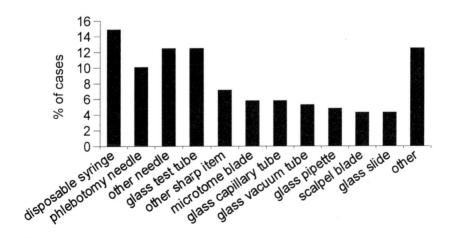

Fig. 12.1. Devices causing sharp-object injuries in clinical laboratories. 64 hospitals, 2 years, 208 cases.

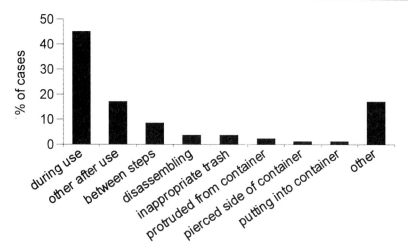

Fig. 12.2. Mechanism of glass injuries in clinical laboratories. 64 hospitals, 2 years, 82 cases.

glass equipment are commercially available. Vacuum tubes, specimen tubes, and capillary tubes, the devices most commonly used as blood receptacles, are now available in plastic and are highly resistant to breakage. Plastic pipettes are now more common in most laboratories than glass ones. The use of glass pipettes should be strictly limited to procedures for which plastic cannot be used.

A total of 76 mucocutaneous exposures was reported in clinical laboratories during the two-year interval. In 74% of these incidents the body fluid involved was blood, while in patient rooms only 54% of mucocutaneous exposures involved blood ($\chi^2 = 10.9$, $P < 0.001$). However, the quantity of blood or body fluid involved in exposures was smaller in clinical laboratory settings, where only 1% of exposures exceeded 5 ml of biological fluid, whereas in patient rooms 12% of exposures exceeded 5 ml of biological fluid ($\chi^2 = 8.4$, $P < 0.01$). Smaller amounts of biological fluids pose less risk to health-care workers. In clinical laboratories, the amount of biological fluid in a single exposure usually does not exceed the capacity of a specimen container, which is relatively small. However, risk could be even further reduced by limiting the amounts of blood or body fluids collected from patients to the minimum required for testing. Often only a few drops are needed for testing although several ml are collected from patients.

A notable characteristic of blood and body fluid exposures in clinical laboratories was the high percentage of cases in which the exposure was due to specimen containers leaking or breaking, or other product-related failures, as shown in Fig. 12.3. In clinical laboratories 97% of mucocutaneous exposures involved a product-related failure, whereas in patient rooms, 48% of cases were product-related, in comparison to direct patient contact ($\chi^2 = 69.2$. $P < 0.0001$). The design and integrity of specimen containers is an issue requiring more attention. In addition to leakage and breakage, the need to uncap specimen tubes to access contents creates a significant problem. Methods of specimen

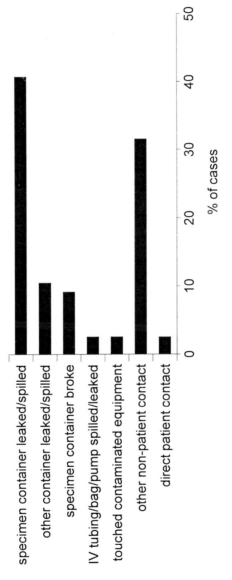

Fig. 12.3. Mechanism of blood and body fluid exposures in clinical laboratories. 64 hospitals, 2 years, 76 cases.

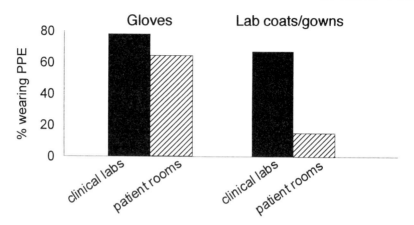

Fig. 12.4. Personal protective equipment worn at the time of exposure: clinical laboratories vs. patient rooms. 64 hospitals, 2 years, 909 cases.

access and transfer that do not require uncapping of specimen tubes would constitute an important advance in safety.

Another finding of interest is shown in Fig. 12.4. Clinical laboratory personnel more frequently wore personal protective garments at the time of mucocutaneous exposures than health-care workers exposed in patients' rooms. In particular, 78% of laboratory workers were wearing gloves at the time of exposure, as opposed to 64% of health-care workers exposed in patients' rooms ($\chi^2 = 5.9$, $P < 0.05$). Sixty-seven per cent of laboratory workers were wearing cloth laboratory coats or gowns at the time of their exposures, while only 15% of workers in patients' rooms were wearing laboratory coats or gowns ($\chi^2 = 123.2$, $P < 0.0001$). Clinical laboratory personnel appear to be more consistent in their use of personal protective equipment than other health-care workers. This may be explained in part by the controlled environment of the clinical laboratory as contrasted with patient care, where contact with blood or body fluids can occur without warning.

On the other hand, previous studies we have conducted provide evidence that cloth laboratory coats provide only an illusion of protection without reducing the risk of skin contact with biological fluids (Jagger et al., 1992; Jagger and Balon, 1995). The cloth laboratory coat remains the most common garment for covering the torso and arms of clinical laboratory personnel. If its purpose is to serve as a work uniform, or to prevent the soiling of clothes, the cloth laboratory coat can meet these goals. If its purpose is to prevent skin contact with biological fluids, it is irrelevant. At a minimum, fluid-resistant gowns or coats should be available in locations where there is a risk of splashing or spraying of biological fluids, although our data do not indicate that all laboratory workers would need this degree of protection at all times.

There is another reason, however, which may explain why laboratory coats and gloves did not prevent mucocutaneous exposures. We found that 74% of reported mucocutaneous exposures were to the eyes, nose, mouth, or other areas of the face. There is surely a reporting phenomenon contributing to this finding. We have noted similar selective reporting patterns among emergency department workers (Jagger *et al.* 1994; Jagger and Balon, 1995). A blood exposure to the eyes, mouth or other areas of the face is a shocking experience and more likely to be reported than blood contact with arms or hands. Face exposures are relatively infrequent, but it is clear that laboratory personnel rarely wear protective eyewear or face shields. These findings point to the need to identify situations and procedures that may result in splashing or spraying in which face exposure is a risk, and to selectively implement the use of protective face and eyewear under those circumstances.

CONCLUSIONS

The most serious risk of exposure to blood-borne pathogens among clinical laboratory workers is from needles used to draw or transfer blood samples, and from glass specimen tubes and glass capillary tubes used for collecting and storing blood samples. Plastic specimen and capillary tubes should be substituted for glass tubes wherever possible. For blood drawing, protective devices that shield or blunt phlebotomy and syringe needles after their use should be implemented not only among clinical laboratory personnel but hospital-wide. The use of needles and syringes as laboratory tools, which place workers at unnecessary risk, should be eliminated to the extent possible. The use of the common cloth laboratory coat as a fluid barrier should be eliminated and replaced, in at-risk locations and for at-risk procedures, by fluid-resistant gowns. Protective face and eyewear should be mandatory under circumstances in which face exposure is a risk. Manufacturers should improve the designs of specimen containers to minimize risk of leakage, breakage, and spillage, and should introduce features that would allow the transfer of fluid contents without the need to open containers. Quantities of blood and body fluid samples collected from patients should be limited to the minimum required for testing. Finally, every effort should be made to achieve a 100% hepatitis B vaccination rate among clinical laboratory personnel.

ACKNOWLEDGEMENT

Acknowledgment of all hospitals participating in the EPINet data-sharing network is provided at the end of Chapter 5 this volume.

REFERENCES

CDC (1995) Case control study of HIV seroconversion in health care workers after percutaneous exposure to HIV-infected blood (France, United Kingdom, and United States, January 1988–August 1994). *Morbidity and Mortality Weekly Report* 44, 929–933.

CDC (1996) *HIV/AIDS Surveillance Report*. 8, 21.

Collins, C.H. and Kennedy, D.A. (1987) Microbiological hazards of occupational needlestick and 'sharps' injuries. *Journal of Applied Bacteriology* 62, 385–402.

Grist, N.R. and Emslie, J. (1985) Infections in British clinical laboratories. *Journal of Clinical Pathology* 38, 721–725.

Handsfield, H.H., Cummings, M.J. and Swenson, P.D. (1987) Prevalence of antibody to human immunodeficiency virus and hepatitis B surface antigen in blood samples submitted to a hospital laboratory. *Journal of the American Medical Association* 258, 3395–3397.

Jacobson, J.T., Orlob, R.B. and Clayton, J.L. (1985) Infections acquired in clinical laboratories in Utah. *Journal of Clinical Microbiology* 21, 486–489.

Jagger, J. and Balon, M. (1995) EPINet Report: Blood and body fluid exposures to skin and mucous membranes. *Advances in Exposure Prevention* 1, 1–9.

Jagger, J., Detmer, D.E., Cohen, M.L. *et al.* (1992) Reducing blood and body fluid exposures among clinical laboratory workers: meeting the OSHA standard. *Clinical Laboratory Management Review* 6, 416–424.

Jagger, J., Cohen, M. and Blackwell, B. (1994) EPINet: A tool for surveillance and prevention of blood exposures in health care settings. In: Charney, W. (ed.) *Essentials of Modern Hospital Safety*, 3rd edn. Lewis Publishers/CRC Press Inc., Ann Arbor.

Levy, B.S., Harris, B.C. and Smith, J.L. (1977) Hepatitis B in ward and clinical laboratory employees of a general hospital. *American Journal of Epidemiology* 106, 330–335.

OSHA (1991) US Occupational Safety and Health Administration. Occupational exposure to bloodborne pathogens, final rule. *Federal Register* 56, 64004–64182.

Pattison, C.P., Boyer, K.M. and Maynard, J.E. (1974) Epidemic hepatitis in a clinical laboratory. *Journal of the American Medical Association* 230, 854–857.

Vesley, D. and Hartman, H.M. (1988) Laboratory-acquired infections and injuries in clinical laboratories: a 1986 survey. *American Journal of Public Health* 78, 1212–1215.

Wruble, L.D., Masi, A.T. and Levinson, M.J. (1977) Hepatitis-B surface antigen (HB Ag) and antibody (anti-HB) prevalence among laboratory and nonlaboratory hospital personnel. *Southern Medical Journal* 70, 1075–1079.

CHAPTER 13
Blood transfusion services

A.D. Kitchen and J.A.J. Barbara

INTRODUCTION

The key function of blood transfusion services (BTSs) is to collect and provide safe blood and blood products. This is achieved by specifically selecting 'safe' donors and by the subsequent laboratory screening of the blood collected. Thus, unlike clinical laboratories, the bulk of the pathological material present generally has a relatively low infection risk. However, there are certain risks which are uniquely associated with the collection and testing of donated blood. These centre around: (i) the fact that donors may be asymptomatic yet infectious; (ii) blood is screened only for a limited number of infectious agents; and (iii) a blood donation is a very large volume of potentially infectious material.

All four of the current infectious agents for which blood screening is mandatory in the UK (the hepatitis B and C viruses, the human immunodeficiency virus (HIV), and the agent of syphilis (*Treponema pallidum*)) give rise to asymptomatic infections, and at the time of donation, infected, and thus potentially infectious, donors declare themselves to be well and without any signs or symptoms of any illness. Thus BTSs will invariably collect and process potentially infectious donations from infected, but outwardly well, donors. The incidence of such donations will obviously depend upon the prevalence of the particular agent in the donor population and the effectiveness of donor selection procedures.

Because of their core function BTSs screen only for those infectious agents which are most likely to be transmitted by transfusion. Thus, although these are therefore also those most likely to be transmitted occupationally, this may not always be the case: a donor may be potentially infectious due to the presence of an agent not screened for by the BTS. Whether this would be transmitted to a patient is one issue, but in the context of this chapter the issue is whether a member of staff working with the donation could become infected if he or she were to come into direct contact with the blood.

Lastly, a blood donation is a large volume of potentially infectious material which may contain a high titre of infectious particles. A lot of manipulations may be performed on each blood pack, not only giving the potential of a spillage from the pack but also resulting in the splitting of a pack into component parts which would increase the number of potentially infectious products being handled.

This chapter, therefore, will concentrate mainly on two issues: identifying specific risks associated with specific activities within transfusion services, and the measures taken to minimize any risks to staff of occupationally acquired infections. In general there are two ways of preventing infection: preventing or limiting exposure to an infectious agent, and preventing or minimizing the extent of clinical disease in an exposed individual by appropriate clinical intervention. Clearly the two are very different: the first being achieved by the careful design and development of working procedures, and the second by active intervention, either by vaccination before staff become involved in activities likely to result in exposure to infectious agents, or by post-exposure intervention – passive immunization or anti-microbial therapy.

Finally, it is assumed that in all areas the principles and procedures of good laboratory practice (GLP) and good working practice are implemented. The sensible application of such procedures alone is central to the prevention of occupationally-acquired infection in all transfusion service staff.

INFECTIOUS AGENTS INVOLVED

An important issue is to identify the types and specific agents likely to be a risk to transfusion service staff. In addition, the distinction between risk of significant clinical disease and largely insignificant clinical disease needs to be considered. Any infectious agent likely to be present in donated blood is potentially, at least, capable of engendering infection; this would include viruses, bacteria and protozoa. In reality, however, and especially in those countries with good donor selection procedures and a low prevalence of infectious agents, the number of such organisms is small and the risk from most infectious agents is generally very low. None the less, the presence of some organisms does present an actual risk to staff; these include the parenterally transmitted viruses hepatitis B virus (HBV), hepatitis C virus (HCV) and the human immunodeficiency virus (HIV), of which the most infectious, and hence of greatest risk, is HBV. In addition, some specific individuals, notably pregnant women, may be at risk from viruses that can affect fetal development such as human cytomegalovirus (HCMV). Any risk from bacteria or protozoa is extremely low but none the less still exists; specific agents would include *Treponema pallidum* (the causative agent of syphilis), *Plasmodium* species (the causative agents of malaria) and *Trypanosoma cruzi* (the causative agent of Chagas' disease). More information about these viral, bacterial and protozoal agents is given by Jeffries, Chapter 1, and Collins, Chapter 2 this volume.

VACCINATION AND VACCINATION MONITORING

For many years staff in transfusion services were generally not considered to be at risk of occupationally-acquired infections. Although HBV is just one of a number of infectious agents that may pose a threat to BTS staff, and for which a vaccine is available, there are no vaccines or reliable prophylaxis for many of the other agents, or other appropriate pre-exposure protective measures. In the UK, in the last few years, some BTS staff have now been included in those groups of staff considered to be at increased risk of exposure to infectious agents, and HBV vaccination recommended for staff specifically coming into contact with blood and other pathological material. In most English transfusion centres vaccination programmes have been put in place, and staff are monitored regularly to assess their immune status and determine if and when further doses of vaccine are needed.

Although vaccination is widely used in the prevention of infection, especially when there is a high risk of infection, it is not without drawbacks. A major failing with many vaccination programmes is that of monitoring the antibody responses in vaccinated individuals. Unfortunately many individuals, including health-care workers, consider that once an individual has been vaccinated, he or she is/are immune for life from infection with that particular agent. That this may not be the case has been clear for many years. Although vaccination against some infectious agents does indeed appear to give virtually life-long protection, other vaccines may give only medium- to short-term immunity. Also, there is tremendous variation in the response to the same vaccine by different individuals. This may vary from long-term immunity in one individual to no detectable response in another individual given the same vaccine and at the same time. Such a variation in response has been found commonly with HBV vaccines, although the recombinant vaccines do appear to have a higher response rate than the native HBsAg-based products. Thus, depending upon the reasons for vaccination and relative risk of exposure, vaccinated individuals should be monitored to ensure that they do develop, and maintain, protective antibody levels.

For more information about hepatitis B vaccination see McCloy, Chapter 16 this volume.

INDIVIDUAL CRITICAL WORK AREAS WITHIN TRANSFUSION SERVICES

Within transfusion centres there is a whole variety of activities that involve different groups of staff. Generally only staff that come into direct contact with pathological material are deemed to be at risk of acquiring infection, but even office-based clerical staff may still come into contact with blood on some occasions. For example, there have been many instances where paperwork, donor records, etc. have become contaminated with blood spilt at donor sessions and have then been returned with the other records from the particular donor session for office staff to process, thereby exposing these staff to potentially infectious material. However, for the purposes of this text only those categories of staff obviously coming into contact with pathological material during the normal course of their work will be considered. Fortunately the use of computerized information transfer has reduced the risk of exposure in office staff.

Thorough training is an essential feature of prevention of occupational infection. All staff need to understand the actual risks involved, how to minimize exposure – to themselves and to others – and what needs to be done if exposure occurs.

Blood Collection Teams

The collection teams are the front line in that they are generally the only staff to come into direct contact with donors. However, some medical and technical staff will come into direct contact with those donors who need follow-up, including those who have been identified as being infected with a particular agent.

This direct contact involves initially close physical (e.g. skin-to-skin) contact and the possibility of exposure to aerosols. At this stage it is always hoped that any donors who have symptoms indicating possible active infection are identified and deferred. The effectiveness of donor screening systems varies tremendously, in global terms, depending upon the prevalence of infectious agents in the particular donor population and the rate of acquisition of infection within the general population from which the donors are drawn. Unfortunately, however, infectious, and apparently asymptomatic, donors may still be accepted for donation, and it is these donors who present a risk to collection team staff, other donors at the session, and ultimately recipients of the blood. However, it would be expected that such donors would be identified through the laboratory screening of the donations before they are issued for clinical use.

Following acceptance of a donor, venepuncture is performed and the donation collected. At this stage the risks are far easier to identify, and come from the possibility of needlestick injury from the needle used to collect the donation, spillage of the blood collected, or continued bleeding from the venepuncture site after collection has finished and the needle removed. Staff, and indeed other donors, could be exposed to infectious agents present in the blood of an asymptomatic donor, *via* the used needle or any spilt blood from that donor. Finally, even after the donation there is still the possibility of infection through post-donation bleeding at the venepuncture site.

A number of measures are taken to minimize any risk of infection of the team staff and other donors at the donor session. All staff are made aware of the risks of infection, no matter how small these risks really are, and well trained in the basic visual assessment of donors, the safe handling of the needles, blood packs, and sample tubes, how to dress properly the venepuncture site, and how to deal with any spillages of blood or other body fluids, no matter how large or small. The re-sheathing of needles is a significant cause of accidental needlestick injuries, especially with the large numbers of needles used at a blood transfusion centre and is avoided by immediately placing the used needles into the disposal containers now available (see Morgan, Chapter 8 this volume). In addition, personal responsibility is expected, staff are trained to cover any areas of broken skin, to report accidents and to report immediately any episodes of acute febrile illness. The use of disposable gloves, whilst clearly providing a physical protective barrier, can decrease dexterity (but see Scully, Chapter 9 this volume) and could in some cases possibly increase the risk of needlestick; latex is no barrier to any needle.

Although, eventually, the microbiological status of most of the donated blood will be known and any incidents at the session involving contact with blood can thus be resolved appropriately, there may be occasions when blood is collected from donors but is never microbiologically tested. There are a number of reasons for this including: the cessation of collection because of donor illness or clotting of the bleed line, or the realization that, for some reason, the donor should not have been accepted and allowed to donate. Whatever the reason, in such circumstances any risk to staff may be compounded by the failure to perform microbiological tests on the donations.

Blood Processing Activities

The majority of donations collected are processed at the Transfusion Centres to prepare the blood components needed as part of modern transfusion medicine. Most of this processing involves manipulations of the blood within the closed systems of the blood packs into which the donations have been collected.

Centrifugation is most commonly used to separate the individual components, the separated components then being isolated inside the different packs making up the standard blood collection sets which are used today. Once individual components have been prepared and separated, the individual packs are separated by sealing the connecting tubing with external metal clips and/or heat seals. A greater understanding of the viability and stability of the individual blood components prepared has meant that the processing of blood is now carried out as soon as possible after collection, and the procedures used are also a lot quicker. Invariably this means that the processing of the donations is often finished long before the laboratory screening of the donations has even started, resulting in the handling of large numbers of untested, potentially infectious, donations.

Significant risks occur at each centrifugation stage and during the transfer steps. The bursting or leaking of a pack under centrifugation results in the generation of spray, very large numbers of droplets, and aerosols which may result in surface and air-borne contamination across the work area. This is difficult to prevent because the workload and the centrifugation regimen used generally preclude the use of sealed rotors. The transfer of blood or plasma between packs is generally performed by pressure applied to the surface of the packs to force products to move between packs; the pressure is quite high and is applied evenly to the flat surfaces of the blood packs. Any pin-holes, flaws, tears or other defects quickly become apparent, and can result in a 'jet' of blood/plasma being sent across the work area, possibly contaminating a large area. The use of gloves when handling blood packs can help to minimize exposure to spilt material, but there remains a risk of contact with other parts of the body, notably eyes and mouth. The use of full face protection systems would help to minimize any risks to processing staff, but may not always be compatible with the working practices used and can also be very restrictive for the staff. However, although any spillage could involve a large amount of material, the quality of the blood packs used and the design of the processing procedures have increased significantly over recent years, resulting in a much lower incidence of spillages, and hence fewer individual exposures to potentially contaminated material, during the processing of blood into blood products.

Large-scale Fractionation Plants

Many BTSs prepare blood products such as Factor VIII, albumin solutions and immunoglobulins by processing large plasma pools. Although only tested and cleared plasma is used as the raw material there is still potentially a small risk to staff handling the plasma. This risk arises from the large volumes involved and the need to thaw and open the packs to pool the plasma for processing. Thus workers are exposed to potentially infected plasma from thousands of different donors. Whilst it is hard to identify and quantify any specific risk, none-the-less exposure to an infectious agent not routinely screened for but present in the raw plasma cannot be discounted. But such an agent would subsequently be removed and/or inactivated at a later stage of the fractionation process.

Blood Testing Laboratories

Blood screening laboratories present an infection risk to the staff working in them. By the very nature of the work there is frequent handling of large numbers of blood samples in open systems, involving direct contact with them. There are various potential sources of infection, including the generation of droplets and aerosols during centrifugation, and skin contact from spilled blood. Although much of the testing is automated, sample handling, sample preparation, cleaning and maintenance of equipment and general laboratory disposal and decontamination procedures expose staff to risk of blood-borne infection. It is important that, wherever possible, the minimum amount of work is done with known infectious material, and that it is safely removed as soon as possible. However, all samples have to be treated as potentially infectious and all working practices must be designed so that if infectious material is unknowingly encountered, the standard operating procedures are sufficient to provide protection for staff.

There is a significantly higher risk to staff working in laboratories that handle directly samples from patients. Such laboratories are those that deal with cross-matching of donations prior to transfusion, and red cell or white cell/platelet antibody investigations. The samples that are examined in these laboratories will come from a broad cross-section of hospital patients, including those with blood-borne diseases. Again, good working practices and awareness are needed to minimize any risk of infection.

DISPOSAL OF WASTE

The disposal of waste material generated by transfusion centres is a subject that has engendered a lot of debate over the last few years. A number of different types of waste are generated, and these may need to be handled in different ways. Non-pathological waste such as waste paper and uncontaminated packaging material, etc. should be disposed of like any other general non-infectious waste and needs no special handling except to be kept separate from any waste containing even the smallest amounts of pathological material.

Collection-team Waste

Waste from collection teams contains little free liquid matter, but it comprises mainly soiled dressings and wipes, used collection needles and occasionally heavily soiled wipes when a significant spillage or other accident involving pathological material has occurred. Needles should be placed in sharps containers: all other waste in the appropriate biohazardous waste sacks. All this waste is then sent for incineration.

Laboratory and Processing Waste

Laboratory waste may be liquid, solid or semi-solid and may be in relatively large quantities. Depending upon the types of tests performed and the equipment in use, waste fluids usually make up the majority of laboratory waste; the residual donation samples (semi-solid waste), remaining at the end of testing are the next most significant part. Solid waste consists mainly of wipes used for routine cleaning and mopping up of spills. Apart from broken glassware, which may or may not present any infection risk, laboratories usually do not generate significant numbers of sharps. An additional waste material is blood that has passed its expiry date. Although such donations have been microbiologically tested (see above), the large volumes of blood that are sometimes disposed of may still present a potential risk to staff because of the presence of organisms for which tests were not performed.

The waste fluids generated in laboratories often contain only small amounts of pathological material diluted in large quantities of simple inorganic wash solutions. These solutions are usually treated with hypochlorite, or other appropriate disinfectant before disposal into the sewer. Waste disposal is considered by Collins and Kennedy, Chapter 18 this volume.

ACTION AND FOLLOW-UP OF ACCIDENTS INVOLVING PATHOLOGICAL MATERIAL

Any member of staff who is involved in an incident resulting in exposure to pathological material is immediately investigated to see if he or she had been infected as a result of the incident. A blood sample is collected immediately and a portion of serum is stored for reference. Depending upon local policy, the sample may then be tested for evidence of any previous or even current infection, most often just for HBV, but sometimes for other blood-borne viruses. The material involved in the incident, if it can be clearly identified, is also tested for the presence of infectious agents (if this has not already been done, e.g. in the case of blood donations). Depending upon the results of the initial investigations, treatment may be offered, such as vaccination for HBV, or counselling as appropriate for other infectious agents where no vaccine or other appropriate treatment exists, further samples may be taken over the next few months, or no further action is taken except that the reference sample is stored frozen indefinitely. In all such investigations it is important first of all to determine if exposure to infectious material has occurred and to decide what intervention is needed. Subsequently it is also important to determine whether any resulting infection was from the incident or was a result of some other completely unrelated activity. Generally cases of exposure to HBV or HIV are also reported to the PHLS Communicable Diseases Surveillance Centre for long-term monitoring (see also Waldron, Chapter 15, and McCloy, Chapter 16 this volume.)

CONCLUSIONS

Although in general BTS staff work with low risk material they must be considered to be potentially exposed to risks similar to those of staff working in general pathology laboratories, and the same precautions and advice and guidance provided. Where appropriate, prophylaxis for relevant infectious agents should be offered and strategies should be developed to facilitate response to any exposure of staff to pathological material.

CHAPTER *14*
Non-health-care occupations

C.H. Collins and D.A. Kennedy

Health-care workers, i.e. those who care for the sick and injured, are well provided with guidance on protection against blood-borne infections, but this is not necessarily the case with other occupations. Some service, professional and trades organizations offer advice to their members and employees and two small pocket books which give advice to employees have been published (Morris, 1993; Collins, 1994).

One may distinguish three groups of workers in occupations other than health care who are at risk:

1.   Those who expect that their duties *will* involve exposure to blood, for example funeral and mortuary staff.
2.   Those who expect that their duties *may* involve exposure to blood; these include police and scenes of crime officers, prison and other custodial staff, social workers and fire-fighters
3.   Those who would *not* normally expect to be exposed to blood, but who may occasionally experience exposure as a result of careless activities of other persons. This wide group includes public service, transport, postal, refuse collection and cleaning staff.

This chapter addresses the prevention of blood-borne infection, notably with the viruses of hepatitis B (HBV), hepatitis C (HCV), and the immunodeficiency virus (HIV), in certain of these occupations and activities. Post-exposure procedures are considered by Waldron, Chapter 15, and McCloy Chapter 16 this volume.

Assessing risk in any of these groups is difficult because of the absence of data and the tendency to include all workers, e.g. those in administrative and non-operation staff as well as those who are clearly at risk. Assessments should therefore be made only in respect of those who are, or may be, exposed to blood or blood products in their daily activities.

FUNERAL AND EMBALMING STAFF

Cadavers may be infected with a wide variety of pathogens (Healing *et al.*, 1995). This chapter is concerned only with blood-borne pathogens but funeral directors have a right to be informed in cases of deaths due to any infectious diseases and should expect any infectious cadavers to be in a body bag. Unnecessary bagging of bodies, however, does cause problems.

There is little risk of blood-borne infections from uninjured cadavers dead of natural causes other than hepatitis and AIDS, but cuts and abrasions should always be covered with waterproof dressings and protective clothing, at least gloves and gowns, should be worn when bodies are prepared for viewing by relatives, etc. prior to burial or cremation. This usually involves 'hygienic preparation' – washing the cadaver's face and hands, tidying the hair, and perhaps the application of some cosmetics.

Embalming, however, which is intended to delay decomposition when delay in burial or cremation is inevitable does expose morticians to blood. This involves infusion with formalin-containing preparation via a needle which is inserted into a main artery, with drainage of effluent from the heart into a closed vessel. Accordingly it gives considerable scope for sharps injuries and mucocutaneous exposure. Beck-Sague et al. (1991) found that 39% of 539 respondents to a questionnaire had at least one needlestick injury in the previous 12 months; skin contact and splashes of blood to mouth or eyes were also common occurrences among embalming staff. In Europe embalming is discouraged (unlawful in France) in cases of AIDS, hepatitis B and C, and the Group 4 infections as defined by the Commission of the European Communities (1990), e.g. the haemorrhagic fevers.

Precautions

If a cadaver is mutilated there is a risk of exposure to blood and staff should wear full protective clothing, including boots, full-length overalls, aprons and gloves (two pairs). Masks or visors should also be worn when there is the possibility of blood splashes. All cuts, scratches and other abrasions of the skin should be covered with waterproof dressings, even if the area will be covered by gloves and protective clothing.

A high standard of personal hygiene is essential: hands should be washed both before and after the removal of gloves. Staff should shower after completion of preparing mutilated bodies or an embalming. Eating, drinking and smoking should be banned while work is in progress.

Funeral staff who handle cadavers should receive hepatitis B vaccine.

See 'General Guidance on Handling the Dead', below.

MORTUARY AND POST-MORTEM ROOM STAFF

Cadavers received in mortuaries may have been victims of accidents or other violence and may be bloody.

The clothing of victims of accidents and assaults may also be contaminated with blood and should therefore be handled with care.

If the deceased has succumbed to an infectious disease the cadaver should be delivered to the mortuary in a body bag which should be adequately labelled, but body bags should not be used indiscriminately. It is important that mortuary staff, and funeral staff who may subsequently collect the body, are fully aware of the nature of the infection. Unfortunately, the cause of death is sometimes not known and may not even be immediately revealed by a post-mortem.

Post-mortem examinations are not usually carried out on persons who have died as a result of an AIDS-related, hepatitis B and C or a Group 4 biological agent (e.g. a haemorrhagic fever) infection. HIV may survive in cadavers for up to 14 days (van Roy, 1996); HBV may survive and remain infectious for longer periods (Bond, 1984; Bond et al., 1981). Refrigeration of bodies may not diminish the potential for the transmission of HIV and HBV.

Precautions

General precautions for the prevention of infection in the mortuary and post-mortem room are set out by the Health and Safety Executive (HSAC, 1991). Here we are concerned only with blood-borne infections.

In addition to the protective clothing recommended above, those who perform post-mortem examinations should wear waterproof aprons that overlap the boots, more than one pair of gloves (surgical or heavy duty over latex or neoprene).

Sharps injuries, from instruments, bone and bone splinters, and blood splashes (mucocutaneous exposure) are the main infection risks in post-mortem examinations.

Accidental cutting of the hand during a post-mortem examination poses a serious risk of such infection and the choice of gloves is of primary interest. Bickel and Diaz (1990) evaluated, for autopsy use, metal mesh gloves and filleting gloves of the type that are available in sporting goods stores in the USA. Among other considerations, they reported that metal mesh gloves are 'workable' but somewhat heavy and cumbersome. The filleting gloves, which were cheaper than metal mesh gloves, were made of a combination of stainless steel and a modern synthetic material. They were machine washable and re-usable and withstood liquid disinfection and autoclaving. They were also considerably lighter and more flexible than the metal mesh gloves. The authors were unable to cut the filleting gloves with a scalpel or an autopsy knife and felt that they could be highly recommended for autopsy use. They cautioned, however, that neither kind of gloves had been evaluated for resistance to needle punctures.

Bell and Ironside (1992) found that chain-mail gloves could be worn between two pairs of rubber gloves as an extra precaution against cutting injuries during autopsy work. They found them flexible and convenient but they did not prevent needlestick injuries.

Blood may be dispersed as small droplets or aerosols (see Kennedy, Chapter 6 this volume) by electrically-operated saws and by energetic washing of tissues and washing out body cavities: packing with absorbent material which is left in the cavity is a safer alternative. Washing down the table after the autopsy may also generate aerosols.

Blood should not be allowed to dry on instruments. Initial washing with warm, rather than hot, water containing disinfectant and/or detergent is recommended, to avoid coagulation of blood, etc. on their surfaces and in their crevices (Healing *et al.*, 1995). They should then be sterilized by autoclaving, or left immersed for 24 hours in a suitable disinfectant. Although hypochlorites are the preferred disinfectants for blood-contaminated materials they may attack metals. Some phenolics are alternatives (see Hoffman and Kennedy, Chapter 17 this volume).

All mortuary staff should be vaccinated against hepatitis B, as well as against the other diseases recommended by the Department of Health (DH, 1996).

WASTE FROM FUNERAL PREMISES AND MORTUARIES

Blood and body fluids may be flushed, gently, into the public sewer. Splashing and aerosol formation must be avoided. Spillages that cannot be flushed away may be absorbed with chlorine-releasing granules or one of the hygroscopic disinfectant powders or granules which are then placed in colour coded plastic contaminated waste bags for incineration (see Collins and Kennedy, Chapter 18 this volume).

Sharps should be placed in approved sharps containers (British Standard: BSI, 1990) and disposable articles, gloves, protective clothing, shrouds, dressings, swabs and tissue into colour coded plastic contaminated waste bags for incineration.

Clothing to be salvaged should be treated as 'infectious laundry' (see Griffiths, Chapter 10 this volume). Alternatively, before laundering, it may be soaked overnight in a suitable disinfection solution (e.g. Tergene), which does not damage fabrics.

GENERAL GUIDANCE ON HANDLING THE DEAD

General guidance on the prevention of infection of staff who handle the dead is given by the Department of Health (DH, 1988), Beck-Sague *et al.* (1991), Healing *et al.* (1995), Young and Healing (1995) and Gershom *et al.* (1995). Advice on health and safety in the funeral service is published by the National Association of Funeral Directors and associated organizations (1992), and in mortuaries and post-mortem rooms by the Health Services Advisory Committee (HSAC, 1991).

POLICE AND SCENE OF CRIME OFFICERS

Although there are no reliable data about occupationally-acquired blood-borne infections in police officers and civilians employed on operational duties by police forces it is obvious from reports in news media that some officers and civilians are at risk of exposure when they are dealing with violent criminals, intravenous drug abusers and investigating crimes where blood has been spilled. It also clear that higher ranking officers, who are engaged in administrative and supervisory duties, are less at risk than relatively junior and operational officers.

In the late 1970s, when the general public became aware of hepatitis B as a result of 'outbreaks' associated with hospitals, such as that in Edinburgh (Bone et al., 1971), Martin et al. (1981) reported that of 312 blood samples selected at random from over 3000 received at a forensic laboratory, five (1.6%) were positive for HBsAg; at that time the incidence among blood donors was 0.7–0.8%. Some years later Morgan-Capner et al. (1988) reported that of 284 police officers tested, 2.8% exhibited HBV markers and 1.1% a marker suggesting that they had suffered acute HBV infection. Welch et al. (1988) tested 425 officers and found that only three (1%) had markers indicating evidence of a past infection. But of 198 suspected source individuals, 33 (16.7%) were HBsAG positive, i.e. were carriers. Half of these were HBeAg positive, and therefore were highly infectious. Fagan (1990) considered that the risk to 'front line' police officers is similar to that of hospital personnel, but Anstad (1991) argued that although police officers probably face higher risks than members of the general public they are at a much lower risk than health-care workers.

Mantineo (1996) noted that in Italy there was a 1.5% seroprevalence in newly recruited police officers but in those with more than two years service it was 14%. None of the recruits were HBsAg-positive, against 3.4% of the older officers.

It is misleading to compare the results obtained from operational police officers with those from the general public (usually derived from blood donors) as the operational officers are involved with a population subset that is more likely to be infected, e.g. drug traffickers and users, and prostitutes, although no figures are available for the criminal fraternity as a whole.

The circumstances in which police officers may suffer puncture wounds and mucocutaneous exposure are:

- personal assaults – uncontrollable situations;
- detention of violent suspects;
- bites, not necessarily by violent persons (see under Custodial Services, below);
- needlestick injuries during searches of suspects' clothing, property or habitation;
- accidental inoculation during first-aid procedures;
- scenes of accidents involving personal injuries;
- scenes of crimes (including collection of evidence);
- investigating suspicious fires;

(Hu et al., 1991; Goldschmidt, 1995; see also Forensic Science Service, 1993).

Precautions Against Exposure and Infection

Universal Precautions, as applicable in clinical situations, are largely inappropriate in operational circumstances. The following precautions are recommended (adapted from various sources):

- cover all cuts and abrasions with waterproof dressings;
- wear leather or disposable gloves;
- place sharp objects in plastic clinical waste containers;
- place contaminated objects other than sharps in plastic bags or boxes;
- require suspects to empty their own pockets and bags and turn them inside out;
- do visual inspection and feel outside of pockets before putting hands inside them;
- do visual inspection of premises, cupboard drawers, bags, etc. before manual inspection;
- use barrier device (see below) for mouth-to-mouth resuscitation (or use mouth-to-nose method);
- bag contaminated clothing, bedding, etc., for machine 'hot wash';
- clean/disinfect hard surfaces with hypochlorite (bleach, diluted $1:1$) in water plus detergent.

Some forces issue officers with 'preventative kits' (Fig. 14.1) as part of their appointments. These small pouches contain: a pair of plastic disposable gloves; a yellow-coloured plastic waste disposal bag; a plastic disposable face shield (to prevent mucocutaneous contact between patient and first aider); an antiseptic wipe.

These might well be augmented with a pair of plastic disposable forceps for picking up hypodermic needles.

Immunization

Police officers who are at risk should have received the usual vaccines etc. before or immediately after recruitment. At present these do not include hepatitis B vaccine to which there appears to be opposition, probably based on risk assessments for the force as a whole (non-operational as well as operational) and probably based on cost (approximately £35 per officer). This may soon be resolved by European Community law. Meanwhile it would seem to be sensible to assess risk on an individual basis – the work done by the officer – and to give the vaccine to those who are exposed under the circumstances listed above. Guidance to police forces is issued by Leigh (1995).

PRISON AND CUSTODIAL STAFF: SOCIAL WORKERS

These workers are at risk of exposure to the blood of two groups of persons who are in their care: those of violent disposition, and intravenous drug abusers. Prison and other custodial officers are not infrequently assaulted by their charges, as are social workers, either in the course of their field work or in their

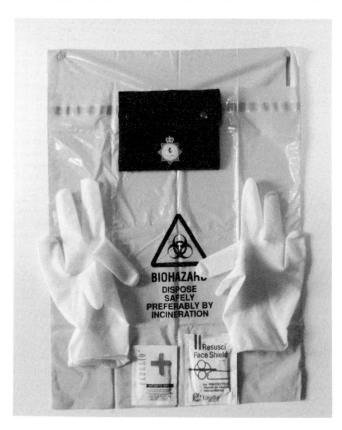

Fig. 14.1. 'Preventative kit' as issued to some police officers. (Courtesy of the Chief Constable of Kent and Selles Medical Ltd.)

offices while interviewing applicants for benefits. These assaults may include biting and scratching. There have been several reports of the transmission of hepatitis B and C viruses by human bites (McQuarrie *et al.*, 1974; Cancio-Bello *et al.*, 1982; Dushieko *et al.*, 1990; Figueredo *et al.*, 1994; Vidmar *et al.* 1996). There have also been newspaper reports of attempts by persons who know that they are HIV or HBV positive to transfer their own blood by threatening other people with contaminated sharps.

Individuals taken into custody or care may carry used hypodermic needles and syringes on their persons or in their personal belongings.

No data are available for UK prisoners, but Mathei (1996) reported that in Belgium HBV seroprevalence was no higher than that among the general population, although there was perception of major risk among prison staff.

Personal Precautions

Vigilance and foresight are necessary qualities in avoiding encounters and contacts. As with other persons at risk, all cuts and abrasions should be covered with waterproof dressings. Care should be taken in investigating the contents of pockets, bags and belongings of clients. Visual inspection should precede manual examination. Even wearing the thickest type of gloves will not guarantee protection against sharps.

Injuries and exposures should be reported as soon as possible to the occupational health department or a general medical practitioner.

There is a strong argument in favour of giving hepatitis B vaccine to these officers, even in the absence of statistics about exposure and infection.

FIREFIGHTERS

While firefighters may encounter dead and injured people in the course of their duties, and frequently enter burning or burned premises which have been used by drug addicts who have left behind their needles and syringes, their many other duties may also bring them into more direct contact with blood. The fire services deal with many accidents and emergencies where people have been injured and are bleeding and who have to be rescued and require resuscitation and first aid. Even after rescue, blood-contaminated materials and surfaces have to be cleared away.

Precautions

The official issue protective overalls must always be worn when casualties and badly burned or mutilated bodies are moved. Most brigades provide their crews with casualty handling packs. These contain: a pair of polythene coated overalls; a pair of plastic gloves; a pair of surgical gloves; a plastic disposable face shield (to prevent mucocutaneous contact between patient and first aider); and a yellow plastic bag to receive contaminated materials.

After handling such casualties the following decontamination procedures are carried out (Kent Fire Brigade):

* outer gloves washed, cleaned in disinfectant solution and removed;
* disposable gloves washed in disinfectant solution but NOT removed at this stage;
* goggles and headgear removed and washed in disinfectant solution; conspicuity coats removed and cleaned with a solution of disinfectant;
* boots and leggings liberally scrubbed with a solution of disinfectant;
* contaminated porous clothing removed and bagged for disinfection and laundering;
* disposable gloves washed in disinfectant solution, removed and bagged for disposal;
* hands and other areas that may have been contaminated washed in a solution of disinfectant;
* used equipment wiped over with a solution of disinfectant.

The disinfectant recommended by several Brigades is Trigene, a halogenated tertiary amine type with a surfactant. This is used as a spray for surfaces and large equipment. Contaminated clothing is fully immersed in it for at least

15 minutes before laundering. After use residual Trigene may be flushed into the public sewer.

Disposable waste should be delivered to the authorized waste disposal organization for incineration.

Injury or exposure to blood should be reported as soon as possible to the designated officer who will arrange for medical care and advice.

OTHER PUBLIC SERVICE PERSONNEL

The risk of exposure to blood in the various public services, such as transport, maintenance, cleaning and refuse collection and disposal, and leisure services, comes mainly from hypodermic needles and syringes discarded, sometimes with ingenuity by intravenous drug abusers. Of less importance are improperly discarded sanitary towels and tampons.

The sites where needles have been found include:

- underneath seats, in and behind upholstery in public transport;
- in telephone kiosks, even in the coin slots,
- in various parts of public conveniences, including those on trains and aircraft;
- in post boxes;
- in public and domestic rubbish bins;
- in parks and the grounds of public buildings;
- in 'squats', derelict buildings and building sites;
- in the ballast on railway lines.

Precautions

It is obviously impossible for staff to take 'universal precautions' in these areas. Even the wearing of gloves may cause problems, but in some circumstances, e.g. in open areas, heavy duty footwear will offer protection against puncture by needles.

Cleaning and maintenance staff should not pick up needles with their fingers, even when wearing gloves. In high risk areas they should be provided with forceps or tongs for this purpose, and small sharps containers for safe disposal.

Refuse collectors may be protected by heavy duty footwear, overalls and gloves. Their vehicles should carry yellow waste bags and disinfectant; and also disinfectant wipes as washing facilities may not be immediately available.

REFERENCES

Anstad, G. (1991) Should police officers be immunised against hepatitis B? *Journal of Occupational Medicine* 33, 845–846.

Beck-Sague, M., Jarvis, W.R., Fruehling, J.A. *et al.* (1991) Universal precautions and mortuary practitioners: influence on practices and risk of occupationally acquired infection. *Journal of Occupational Medicine* 33, 874–878.

Bell, J.E. and Ironside, J.W. (1992) How to tackle a possible Creutzfeldt-Jakob disease necropsy. *Journal of Clinical Pathology* 46, 193–197.

Bickel, J.Y. and Diaz, A. (1990) Metal mesh gloves for autopsy use. *Journal of Forensic Science* 35, 12–13.

Bond, W.W. (1984) Survival of hepatitis virus in the environment. *Journal of the American Medical Association* 251, 397–398.

Bond, W.W., Favero, M.S., Petersen, M.J. *et al.* (1991) Survival of hepatitis B virus after drying for one week. *Lancet* i, 550–551.

Bone, J.M., Tonkin, R.W. Davison, A.M. *et al.* (1971) Outbreak of dialysis-associated hepatitis in Edinburgh 1969–1970. *Proceedings of the European Dialysis and Transplant Association* 8, 189–197.

BSI (1990) *BS 7320. Specification for Sharps Containers.* British Standards Institution, London.

Cancio-Bello, T.P., de Medina, M., Shorey, J. *et al.* (1982) An institutional outbreak of hepatitis B related to a human biting carrier. *Journal of Infectious Disease* 146, 652–656.

Collins, C.H. (1994) *Blood-borne Diseases in the Workplace: a Pocket Guide.* (UK edition). Genium, Schenectady and H & H Scientific, Leeds.

Commission of the European Communities (1990) *Council Directive on the Protection of Workers from Risks Related to Exposure to Biological Agents at Work.* 90/679/EEC.

DH (1988) *Information to Undertakers – Infectious Diseases.* PL/CMO/88/8. Department of Health, London.

DH (1996) *Immunization against Infectious Diseases.* Department of Health. HMSO, London.

Dusheiko, G.M., Smith, M. and Scheuer, P.J. (1990) Hepatitis C virus transmitted by a human bite. *Lancet* 336, 502–504.

Fagan, E.A. (1990) *Hepatitis B: the Disease and its Consequences.* European Conference on Hepatitis B as an Occupational Health Hazard. Gower Medical Publishing, London.

Figueredo, J.F.C., Borges, A.S. and Martinez, R. (1994) Transmission of hepatitis C virus but not human immune deficiency virus type 1 by a human bite. *Clinics in Infectious Disease* 19, 546–547.

Forensic Science Service (1993) *Safety at Scenes of Crime Handbook.* The Forensic Science Service, London.

Gershom, R.M., Vlahof, D., Farzadegan H. *et al.* (1995) Occupational risk of human immune deficiency virus, hepatitis B virus and hepatitis C virus infections among funeral service practitioners in Maryland. *Infection Control and Hospital Epidemiology* 16, 194–197.

Goldschmidt, R. (1995) Exposure to blood (ambulance workers, first aid workers, police and firemen). *Proceedings of a Symposium on Bloodborne Infections – Occupational Risks and Prevention.* International Section of the International Social Security Association on the Prevention of Occupational Risks in Health Service, Paris, INRS.

Healing, T.D., Hoffman, P.N. and Young, S.E.J. (1995) The infection hazards of human cadavers. *CDR Review* 5, Review No 6, R61–R68 PHLS Communicable Disease Centre, Colindale.

HSAC (1991) *Safe Working and the Prevention of Infection in the Mortuary and Post-mortem Room.* Health Services Advisory Committee, Health and Safety Commission. London, HMSO.

Hu, D.J., Kane, M.A. and Hayman, D.L. (1991) Transmission of HI virus and other blood borne pathogens in health care settings; a review of risk factors and guidelines for protection. *Bulletin of the World Health Organization* 69, 623–630.

Leigh, A. (1995) *Hepatitis B and the Police Service.* Home Office Police Research Group. Home Office, London.

McQuarrie, M.B., Forghlani, B. and Wolochow, C.A. (1974) Hepatitis B transmitted by a human bite. *Journal of the American Medical Association* 230, 723–724.

Mantineo, G.A. (1996) HBV prevalence rate in the Italian police force. A retrospective study and analysis of the current situation. *International Conference on Communicable Diseases as Occupational Hazards.* Jerusalem, 18–26 February 1996.

Martin, P.D., D'Mello, L.Z. and Dulake, C. (1981) Incidence of HBsAg in blood samples submitted to the Metropolitan Police Forensic Science Laboratory. *Forensic Science International* 17, 1–3.

Mathei, C. (1996) Hepatitis B among prison staff in Belgium – survey on prevalence and interventional data. *International Conference on Communicable Diseases as Occupational Hazards.* Jerusalem, 18–26 February 1996.

Morgan-Capner, P., Hudson, P. and Armstrong, A. (1988) Hepatitis B markers in Lancashire police officers. *Epidemiology and Infection* 100, 145–151.

Morris, R.J. (1993) *Blood-borne Diseases in the Workplace: a Pocket Guide.* (US edition). Genium, Schenectady.

National Association of Funeral Directors (1992) *Health and Safety in the Funeral Service.* British Institute of Embalmers, Solihull.

Roy, J van. (1996) HIV infection – occupational risk assessment for health care workers handling human corpses. *International Conference on Communicable Diseases as Occupational Hazards.* Jerusalem, 18–26 February 1996.

Vidmar, L., Poljak, M., Tomazic, J. *et al.* (1996) Transmission of HIV-I by human bite. *Lancet* 347, 1762–1763.

Welch, J., Tilzey, A.Z., Bertrand, J. *et al.* (1988) Risk to Metropolitan Police officers from exposure to hepatitis B. *British Medical Journal* 299, 835–836.

Young, S.E.J. and Healing, T.D. (1995) Infection in the deceased: a survey of management. *CDR Review,* 5, Review No 6, R6–R73. PHLS Communicable Disease Centre, Colindale.

CHAPTER 15
Role of the occupational health department

H.A. Waldron

The occupational health department in a hospital has two principal duties in relation to infections acquired at work: prevention and treatment. Each will be discussed in turn below. It is not my intention that this chapter should be an instruction manual for occupational health departments, but rather that it should serve to indicate to other health professionals concerned with the prevention and treatment of hospital acquired infections how they might reasonably expect their own occupational health department to function.

Infectious diseases have always loomed large in the occupational hazards faced by what I suppose one now has to call health-care workers. (This misnomer is all part of a general conspiracy to convince the public that doctors and nurses are interested in health, when what they thrive on professionally is disease. They should truly be called disease-care workers but I doubt that this sobriquet is likely to find any sort of acceptance in our increasingly politically-correct society.) The old bugbear of the medical profession at least, was tuberculosis, and as one comforts oneself reading the obituaries of former colleagues, it is interesting to note how many of those who practised before the last war contracted the disease and were sent to the mountains of Switzerland to recuperate. Nowadays, the rate of tuberculosis for health-care workers as a whole is no greater than that of the general public although some groups, including pathologists and morticians, experience a higher rate. Given the socioeconomic status of health-care workers, however, one would expect the rate of tuberculosis to be lower than that of the general public suggesting that occupational factors are still of aetiological significance. In view of the modern recrudescence of the disease it is likely that rates in health-care workers will rise in the future unless attention to protection is vigorously observed.

The diseases which pose the greatest threat at present are those which can be contracted from infected blood and the most common means of exposing oneself to risk is from a so-called sharps injury, usually caused by a needle. Needlestick injuries are common in hospitals (Astbury and Baxter, 1990; McCormick *et al.*, 1991; Mallon *et al.*, 1992) and in my own hospital, we have well over 100 a year reported to us (among a staff of approximately 3000) and we know from some unpublished research which was carried out a couple of years ago, that only about one in seven of all needlestick injuries is reported to the occupational health department; the true prevalence is therefore more like 700–800 per year. Other workers report similar discrepancies between the apparent and true rates of needlestick injuries (Tandberg *et al.*, 1991; Mallon *et al.*, 1992).

Of the many diseases which it is possible to contract from infected blood, only two cause much alarm among those who injure themselves – AIDS and hepatitis B – but there is increasing concern about hepatitis C, particularly in those hospitals with large liver units although the evidence suggests that the risk of contracting this disease from needlestick injuries is relatively low (Hernandez *et al.*, 1992; Struve *et al.*, 1994).

THE PREVENTION OF BLOOD-BORNE INFECTIONS

It is a basic tenet of occupational health practice that the first duty of an occupational health department is to prevent workers suffering any ill effects from their work and this applies equally to blood-borne infections as to any other hazard. There are very simple ways of preventing blood-borne infections in hospital, the most important of which is the safe disposal of sharp instruments into containers specially designed for the purpose (Linnemann *et al.*, 1991; Weltman *et al.*, 1995; see also Collins and Kennedy, Chapter 18 this volume). This safe disposal includes not attempting to re-sheath used needles, a practice which anyone walking casually round a hospital will find still occurs and which is cited as a frequent cause of accidents (Choudhury and Cleator, 1992). All occupational health personnel have horror stories of needles sticking through the side of sharps containers, of giving sets being found in soiled linen, of needles being found on patients' lockers or in rubbish bags or in drains in various hospital departments. On this account, it is very often not the individual who is so careless with the needles who suffers the injury but a member of the portering, domestic or laundry staff and the prevalence of injury among these groups of staff may be higher than among nursing or medical staff (Waldron, 1985), even though the two latter groups account numerically for more accidents (Astbury and Baxter, 1990; Mallon *et al.*, 1992; Moss *et al.*, 1994).

There is absolutely no excuse for needles or other sharps to be disposed of in an unsafe manner. There are sharps bins on the market which comply with the relevant British Standard (BSI, 1990) and are adequate for the purpose so that it is (almost) impossible to poke a needle through the side while it is a simple matter to train staff to use them. One still finds that bins are overfilled and that staff will then push a syringe with its attached needle hard into the box (and with some containers, through the side); that staff will push down needles in a box in order to get a few more in; that phlebotomists and other blood takers do not have sharps boxes with them when they do their rounds – one phlebotomist of my acquaintance took her dirty needles round with her in a plastic bag – and so on. Sometimes the reason is that not enough sharps boxes are provided and one may have to remind managers that they have a personal legal duty to do all in their power to protect staff at work under the terms of the *Control of*

Substances Hazardous to Health Regulations 1994 but in these present cost-conscious times, that does not always overcome their fiscal scruples and, unfortunately, no manager has been prosecuted under the terms of these regulations to my knowledge.

The inculcation of safe working practices among medical staff is a task which is almost Herculean in concept and Sisyphean in performance. Medical staff are generally indifferent to their own health and attempts to bring them into the occupational health net are by no means always successful; I have colleagues who claim great success with doctors in this respect but I have never been able to emulate them. In my experience, and that of others (Weltman *et al.*, 1995) doctors are most likely to dispose of needles and other sharps in an unsafe way in hospital and when questioned about it, the most usual excuse is that they are so busy that they do not have time to change the way they work. In one informal study we asked nurses who had pricked themselves on needles on the wards of the hospital if they could identify the doctor who was responsible and about half said they could – or were willing to say they could. I wrote to those who had been identified, telling them what had happened as a result of their carelessness and offered to discuss safe working practices with them, but none chose to do so.

How do occupational health departments go about trying to prevent sharps injuries then? Mostly through intensive propaganda; by the distribution of leaflets and posters, by showing videos, by giving talks at inductions for different categories of staff, by giving advice to new members of staff as they are seen and by investigating accidents which occur to see how practices can be changed and by advising management on the need for more safety equipment. It should be noted that in many cases written or video material will have to be presented in different languages, the number of which will depend on the number of different nationalities represented among the hospital staff; this is likely to increase the cost and decrease the efficacy of the message. The reported effectiveness of campaigns to reduce sharps injuries varies a good deal. Some authors (Haiduven *et al.*, 1992) have found a good response with a substantial reduction in the number of injuries. Others (Linnemann *et al.*, 1991), by contrast, found no improvement, other than a decrease in the number of injuries associated with re-sheathing needles and concluded that 'new approaches are needed to reduce needlestick injuries'. Following efforts to reduce the incidence of needlestick injuries that rate may actually *rise*, due to better reporting, and one study concluded that a constant rate of needlestick injury following the increased provision of sharps boxes may 'paradoxically represent a modest preventive effort' (Smith *et al.*, 1992). My own experience tends to support the view that intensive preventive programmes will result, at best, in the injury rate remaining relatively stable but even this modest success is worth striving for.

When injuries have occurred to medical and nursing staff during their work, the occupational health department should use the occasion to reinforce advice on the safe disposal of needles, emphasizing the danger of re-sheathing needles and stressing the need for the appropriate use of protective equipment. Surgeons and others carrying out invasive procedures may splash themselves with blood and advice should be given about covering skin wounds, of wearing eye or face protection and of protecting the hands. There has been a great deal of research into the benefit to surgeons of wearing two pairs of gloves and there is some evidence that this may prevent injury. In one study, for example, blood contact with the skin was reduced by 70% compared with single glove use (Greco and Garza, 1995). Pathologists performing autopsies on patients known or suspected to be HIV positive may wear specially armoured gloves which will reduce their risk of injury and the occupational health department should be able to advise on all these matters (see also Morgan, Chapter 8, and Collins and Kennedy, Chapter 14, this volume).

In the matter of prevention it is obviously important for the occupational health department to liaise very closely with the hospital's control of infection committee and the health and safety committee (if there is one) and they must be represented on these bodies. Occupational health staff have no executive functions in relation to health and safety and their proposals often carry more weight when transmitted through a committee which has representatives of other specialties than when issued simply from the department.

Universal Precautions

The approach taken to reduce the risk of blood-borne infections, particularly by surgeons, is to take what are referred to as Universal Precautions. That is, all patients are treated as though they are highly infectious and protective equipment is worn consistent with this view (see Hunt, Chapter 19 and Kibbler, Chapter 20 this volume).

From an occupational health standpoint this method of dealing with risk has nothing to commend it. Protective measures here – as in any other risky job – should be commensurate with the degree of risk, which is to say, that the most stringent precautions are taken when the risks are greatest. To assume that *all* patients are highly infectious is nonsensical although it is certain that in some hospitals and in some surgical specialties the proportion of high-risk patients may be considerable, but even so, it will be considerably less than 100%.

The advice about what precautions are most suitable depends upon a valid assessment of risk and although it is now unfashionable to talk of high-risk individuals, in this instance, it is exactly apposite. The risks to health-care workers come from the fact that some of their patients are infectious and contact with their blood may lead to the health-care worker becoming infected. When dealing with infectious patients, therefore, measures must be taken to minimize the risk of blood contact; when dealing with patients who are not infectious, less stringent precautions will be satisfactory.

As part of the risk assessment, it is necessary to know whether the patient *is* infectious or not and I can see no reason why surgeons and their occupational health advisers should not have this knowledge. Employers are legally obliged to assess the risks which their employees are asked to take and surgeons might reasonably say that hospital managers or clinical directors are failing in their duty if information about the infectivity of a patient is kept from them. Thus I have sympathy when surgeons call for HIV testing prior to surgery, for example (Stotter *et al.*, 1990). (In passing, it is equally reasonable that patients should be given the same information about their surgeons, should they wish it.)

Arguments about testing patients for the benefit of their medical or nursing attendants will continue and it may never be possible to reconcile the views of clinicians treating infectious patients, who are jealous to maintain confidentiality at whatever cost to others, and occupational health professionals and those who are at risk whose preoccupations are with protecting themselves and others from contracting potentially fatal diseases. The matter does not seem greatly different, however, from the requirement to attach a biohazard label to blood taken from a patient with hepatitis B, for example, so that laboratory staff can take appropriate precautions to avoid contamination.

VACCINATION

The situation with respect to hepatitis B infection changed dramatically when an effective vaccine became available. The first vaccine was derived from human plasma and there was initially some resistance to it because of fears that it might carry a risk of contracting AIDS. Despite numerous assurances to the contrary, take up of the vaccine did not become widespread until a genetically-engineered product came on the market. The vaccine should be offered to all those who are in regular patient contact or with potentially infected material, and to others who are at risk of sustaining a sharps injury; this will include portering, domestic and laundry staff as discussed above (DH, 1993). Some employers now *require* all staff at risk to be vaccinated and this is a trend which seems likely to increase in the future although for reasons of protecting management rather than staff.

After a course of vaccination, antibody levels *must* be determined and if they are unacceptable, a booster should be offered. The timing of subsequent boosters is a matter for debate, but guidelines have been suggested and occupational health departments will wish to follow these until better information becomes available (Table 15.1). Most medical schools now require their students to have a course of vaccination, and nurses should also be vaccinated during their training. The occupational health department must ensure that adequate records are kept of the dates of vaccinations and of the results of any antibody studies. This information should be entered on some form of record to be given to members of staff so that they can provide evidence of their immunity to hepatitis B if required by their manager or by any subsequent employers.

Hepatitis B vaccination produces an acceptable level of antibodies in at least 90% of cases and a single booster will produce an adequate response in about half the non-responders (Waldron, 1990). Those who do not respond to a booster should be investigated to determine whether or not they have previously been infected with hepatitis B and, most importantly, whether or not they are themselves likely to be infectious; if they are, then they must not be allowed

Table 15.1. Surface antibody levels and action to be taken after hepatitis B vaccination.

Surface antibody levels (miu ml^{-1})	Result of vaccination	Action to be taken
<10	Non-responder	Immediate booster and re-assessment of antibody levels at 4–6 weeks
10–99	Unsatisfactory response	Immediate booster and re-assessment of antibody levels at 4–6 weeks
100–999	Acceptable response	Booster at 3 years
≥1000	Good response	Booster at 5 years

Based on BMA (1995).

to carry out invasive procedures until the matter has been discussed with an occupational health consultant (see below).

Despite booster doses of vaccine, a relatively small number of staff will not seroconvert for reasons which are still not obvious. They must be reassured that the fact that they have not seroconverted does not put them at any greater potential risk than they would otherwise have been; the importance of meticulous attention to safe working practices must be impressed upon them and they must be encouraged to report any sharps injuries promptly to the occupational health department.

There is often a tendency among those who have acquired adequate immunity from vaccination to assume that is the end of the matter. Anyone expressing this view should be firmly disabused; vaccination is a fall-back position and the only *sure* way of reducing the risk of contracting hepatitis B is to adhere to safe working practices; in this context it is well to remind staff of the possibility of contracting other diseases against which they may have no immunity at all. (Hepatitis B vaccination is also discussed by McCloy, Chapter 16 this volume.)

Staff Carrying out Exposure Prone Invasive Procedures

Following a number of well-publicized outbreaks of hepatitis B among patients who had been treated by e-antigen-positive surgeons, in August 1993 the Department of Health issued a guidance note entitled *Protecting health-care workers and patients from hepatitis B* (DH, 1993). The guidelines applied to all health-care workers who carried out what were called exposure prone invasive procedures. (This is usually abbreviated as EPIP and health-care workers who fall into this category are often inelegantly referred to as EPIP workers, or simply, EPIPs.) The procedures covered by these guidelines are all those in which there is a risk that the blood of the operator may gain entry into the body of a patient and the guidelines are all about protecting patients – and thus sparing the Department of Health and health authorities and trusts the expense of large scale look-back exercises – and protecting staff is of secondary importance. The British Medical Association has recently published a Code of Practice for the implementation of the DH guidelines (BMA, 1995).

All health-care workers who carry out these invasive procedures are required to produce evidence that they are not e-antigen positive. Categories of staff who fall within the remit of these regulations include all surgeons and dentists, midwives who may suture, and those who may insert intravenous or intra-arterial lines. It is important that the occupational health department, the control of infection team and the various directorates within the hospital determine which categories of staff are to be considered as carrying out invasive procedures and arrangements must be made to ensure that the immune status of all has been determined. As noted in the recent *Addendum* to the guidance note, the target date for testing staff has passed and the status of all those carrying out invasive procedures should now be known.

In general, members of staff have responded reasonably well to these requirements but there has been difficulty in a substantial minority of cases. For example, surgeons who have contracts with more than one hospital may slip through the net and it has proved difficult to persuade some dentists to comply. It is also difficult to ensure that the immune status of locum staff is known before they start work and it is not my experience that all locum agencies take their responsibilities under health and safety legislation very seriously. A number of trusts – perhaps now the majority – require members of staff who will carry out invasive procedures to provide evidence of the immune status before they will issue them with a contract.

Health-care workers who are found to be surface antigen positive must not carry out invasive procedures until their e-antigen status has been established, and the *Addendum* to the guidance note requires that this must be carried out in more than one laboratory, each using a different type of assay. Blood sampling for these tests must be carried out by the occupational health department or someone deputed to do so on their behalf. Members of staff found to be e-antigen positive will not be allowed to carry out invasive procedures and it is suggested that they should be either re-deployed or retrained; the Department of Health, however, does not provide funds for either of these eventualities. The most likely outcome for a surgeon who is found to be e-antigen positive is that he will become unemployed and thereby sustain a considerable financial loss for carrying a disease which he will in all probability have contracted from one of his patients. There is some provision for treatment with alpha interferon and an e-antigen health-care worker who is successfully treated and remains e-antigen negative for 12 months after the cessation of treatment *may* (my italics) be able to resume exposure prone procedures, according to the *Addendum*. There is, of course, nothing said about payment during the months during and following treatment, and the probability that the kindly trust will keep a job open during this time can well be imagined. Fortunately, there seem to have been very few health-care workers who have been found to be e-antigen positive and their ultimate recourse may be to go to law to sue the health authority in whose employ they were when they contracted the disease in the first place, assuming, that is, that they can ever determine where and when this was.

TREATMENT FOLLOWING SHARPS INJURIES

It is essential that all staff know that they must report to the occupational health department as soon as possible after they have sustained a sharps injury. If – as so often happens – the injury takes place at a time when the department is closed, then suitable alternative arrangements must be in place. It may be that the injured person goes to the Accident and Emergency Department or to one of the wards of the hospital designated for the purpose. There must also be a consultant on call who is available to give advice; this may be the consultant occupational physician or the control of infection consultant.

The occupational health staff must determine the severity of the injury and assess the degree of risk and an adequate assessment of risk requires that the status of the donor patient (that is, the patient who was passively involved in the procedure) is known. Specifically, the occupational health staff need to know if the donor patient is hepatitis B, hepatitis C or HIV positive and this will require information from the medical staff treating him or her; it may require further blood testing, which can only be carried out with the patient's consent. Based on this information, the occupational health staff will know whether the recipient is potentially at risk of contracting any of these infections. Sometimes there is difficulty in obtaining information about the donor patient's status. Where this happens it is either because the patient has gone home or because the clinicians are unwilling to take blood from the patient for this purpose (Oakley et al., 1992). In these cases, there is little alternative but to consider the patient to be 'high risk' and act accordingly.

The action to be taken will depend upon knowledge of the donor patient's status and the recipient's hepatitis B antibody levels. It the donor patient is 'low risk', that is, is not a carrier of either hepatitis B or C, and is not HIV positive, then no further action need be taken, except to obtain the circumstances of the accident, investigate further if necessary and advise on safe working practices if appropriate.

The risks to the recipient when the donor patient *is* infectious are considerable, certainly in terms of normal occupational health practice. For a non-immune recipient, the risk of contracting hepatitis B is approximately 1 in 200; the risk for HIV transmission following a single percutaneous exposure is about 1 in 300 and about 1 in 3000 for a single mucocutaneous exposure. These risks are at least two orders of magnitude *greater* than would normally be acceptable for an occupational risk which is potentially life-threatening and on no account should be taken lightly.

The greatest risk of infection comes from a deep penetrating injury with a hollow needle used to take blood, closely followed by any other injury in which the patient's blood is known to have entered the recipient's tissues. The greater the amount of blood inoculated, the greater the risk to the recipient. Splashes on mucous membranes or onto the skin are less hazardous, unless there are open wounds on the skin.

Hepatitis B

If the injury involves a patient who is a hepatitis B virus carrier, then the recipient's antibody levels must be determined, either from the notes or with an urgent assay. Where the antibody level is satisfactory (Table 15.1), no further action need be taken. Where the level is low, then a booster dose of vaccine should be given. The most critical circumstance is when the donor patient is infectious and the recipient is unprotected, either because of failure to react to a course of vaccination or because he or she has not been vaccinated at all. In this case, it is necessary to administer specific immunoglobulin and at the same time start a rapid course of vaccination. In the UK specific immunoglobulin is available *only* for at-risk recipients whose accident has involved a patient *known* to be infectious. Follow-up of these cases is mandatory and the occupational health department must have procedures in place which permit this to happen. In all cases of needlestick injury, blood should be taken for serum save so that in the event of recipients developing hepatitis or seroconverting, their status at the time of the accident can be determined.

Hepatitis C

The number of needlestick injuries in which donor patients with hepatitis C are involved appears to be increasing, certainly in hospitals with large liver units. There is no vaccine against hepatitis C and so nothing which can be done prophylactically to protect recipients. The best that can be done is to take blood for serum save and then monitor antibody levels subsequently to determine whether or not the recipient has been infected. Where there is any doubt at all in the minds of the occupational health personnel, advice must be obtained from the local consultant virologist or the local hepatologist. Fortunately, the evidence suggests that the hepatitis C virus is not as infectious as the hepatitis B virus and so the risks of contracting an infection are correspondingly less. This will be of little comfort to those who may develop the disease, however, and serves to underline the important role of the occupational health department in trying to prevent accidents occurring in the first place. The British Medical Association has recently published a guide to hepatitis C (BMA, 1996).

Human Immunodeficiency Virus (HIV)

Despite the very large number of sharps injuries which occur worldwide, a relatively small number of health-care workers is known to have seroconverted. Nevertheless, accidents which involve donor patients who are HIV positive or who have AIDS may cause profound distress to the recipient and usually the first task of the occupational health department following such an accident is to comfort and reassure the recipient. (I use the word 'comfort' here instead of the much overworked 'counsel' because I think that comforting is much closer to what actually takes place.) Until very recently it was widely felt that there was little which could usefully be done in the way of prophylaxis; some departments offered Zidovudine (AZT) in the hope – rather than the expectation – that it might be helpful, but in my own experience, very few of those who started a course of AZT following a sharps injury completed it. The report in

Table 15.2. Odds ratio of seroconversion following percutaneous exposure to HIV-infected blood

Risk factor	Adjusted odds ratio	95% confidence interval
Deep injury	16.1	6.1–44.6
Visible blood on device	5.2	1.8–17.7
Procedure involving needle placed directly in vein or artery	5.1	1.9–14.8
Terminal illness in donor patient	6.4	2.2–18.9
Post-exposure use of zidovudine (AZT)	0.2	0.1–0.6

Based on Table 1, CDC (1995).

the 22 December 1995 issue of *Mortality and Morbidity Weekly Report* (CDC, 1995), however, has changed thinking very abruptly. The report contained the findings of a case–control study of seroconversion after percutaneous exposure to blood infected with HIV which showed that a 4-week course of AZT beginning promptly after exposure reduced the likelihood of seroconversion by a factor of 5 (see Table 15.2). The situation has also been greatly affected by the evident success of the introduction of new drugs to treat HIV-positive patients, including the use of the protease inhibitor Indinavir. In a later issue of *MMWR*, the CDC (1996) has recommended post-exposure prophylaxis as shown in Table 15.3, and occupational health departments will almost certainly wish to follow these recommendations in future.

In Table 15.3 it will be seen that the type of exposure is categorized according to the presumptive degree of risk, and the decision whether or not to offer prophylaxis depends on the degree of risk. Prophylaxis is *recommended* when there is percutaneous inoculation of blood from a patient with AIDS or with a high titre of HIV. In this case three drugs should be taken, AZT, 3TC (Lamivudine) and Indinavir. Where the accident involves a donor of unknown HIV status but who is considered to be at risk of infection, then prophylaxis with AZT and 3TC should be offered.

Because of the complexity of this issue it is strongly recommended that each hospital should develop its own written policy about post-exposure prophylaxis and make sure that the procedures are in place by which it can be effectively implemented. One particular matter will need special attention.

For post-exposure prophylaxis to be effective, it is necessary that it begins as quickly as possible after the exposure, certainly within 1–2 hours. For this to be at all possible, those who are potentially at risk should already have read and understood that part of the post-exposure policy which details the risk and points out clearly when prophylaxis is required. Each individual should determine whether or not they will take the course of medication if they have an accident and they must be able to have quick and ready access to starter doses of it; they may be able to get the drugs from the occupational health department,

Table 15.3. Recommendations for chemoprophylaxis following occupational exposure to HIV-infected blood[1].

Type of exposure	Source material	Prophylaxis	Recommended regime[2]
Percutaneous	Blood[3]		
	Highest risk	Recommend	AZT+3TC+IDV
	Increased risk	Recommend	AZT+3TC±IDV
	No increased risk	Offer	AZT + 3TC
	Fluid containing visible blood, other infectious fluid or tissue	Offer	AZT+3TC
	Other body fluids	Not offer	
Mucous membrane	Blood	Offer	AZT+3TC±IDV
	Fluid containing visible blood, other infectious fluid or tissue	Offer	AZT±3TC
	Other body fluids	Not offer	
Skin, increased risk[4]	Blood	Offer	AZT+3TC±IDV
	Fluid containing visible blood, other infectious fluid or tissue	Offer	AZT±3TC
	Other body fluids	Not offer	

1. Based on Table 1, CDC (1996).
2. AZT, zidovudine 200 mg three times daily; 3TC, lamivudine 150 mg twice daily; IDV, indinavir 800 mg three times daily.
3. Highest risk: deep injury with large volume of blood *and* blood with high titre of HIV. Increased risk: *either* exposure to large volume of blood *or* blood with high titre of HIV. No increased risk: *neither* large volume of blood *nor* blood with high titre of HIV.
4. For skin, risk is increased for exposures involving a high titre of HIV, prolonged contact, contact over an extensive area or contact with broken skin.

their own ward, the genitourinary medicine department or the accident and emergency department, for example; but wherever the drugs are to be obtained must be well known around the hospital so that an injured member of staff can start taking them as soon as possible after an accident.

For this to happen, some way has to be found to ensure that every member of staff has a copy of the policy document and that starter doses of the prophylactic drugs are kept at strategic points around the hospital. When accidents occur at a time when the occupational health department is closed, the injured member of staff may decide to take the prophylactic drugs but he or she must report to the occupational health department at the earliest opportunity thereafter so that the circumstances of the accident can be recorded, blood taken and follow-up arranged. It is a relatively simple matter to see that every member of staff gets a copy of the policy but quite another to ensure that each reads and understands it and has coolly decided what to do in the event of an accident. Coolness is not the term which springs most quickly to mind to describe the events following a high-risk accident; panic is a rather better choice. Seldom has the recipient read the policy and he or she will turn to whoever is closest at hand to seek advice and under these circumstances, the advice given is frequently at odds with the official line. All those who are involved in developing and implementing a post-exposure prophylaxis policy – occupational health personnel, virologists, the control of infection team and infectious diseases consultants – will need to spend a great deal of time and effort informing (or reminding) their colleagues and others of the proper procedures to follow.

Those who elect to take post-exposure prophylaxis will need regular follow-up to monitor the side-effects of the medication and to advise on any remedial action and to determine whether any symptoms noted by the recipient are due to side-effects of the drugs or to the effects of the HIV. It is not very likely that any of the occupational health personnel will have the necessary experience or knowledge to be able to bring the required degree of expertise to bear on this task and it is much better that it be done by a consultant in genitourinary medicine albeit under the aegis of the occupational health department. One of the genitourinary medicine consultants could be deputed to take on the task of following up members of staff taking post-exposure prophylaxis and see them in the occupational health department, entering details in the occupational health records which are the most secure in the hospital. In this way the recipient will be assured that he or she is getting the best advice during the period of follow-up and that confidentiality is being rigidly observed.

HIV-infected staff
Although the risks of HIV-positive health-care staff infecting their patients is remote – indeed, only a single instance has been recorded to date – those who become HIV positive following an accident will need to adhere to the guidelines published by the UK Health Departments in 1994 (DH, 1994). Health-care workers who are HIV-positive must seek special occupational health advice and must not undertake invasive procedures until they have received that advice. The advice will normally be expected to come from a consultant in occupational health but occupational health – and other – staff may contact an advisory panel which the Department of Health set up for guidance. Occupational health professionals who are aware of infected health-care workers who have neither sought nor followed advice on their working practices should inform their local Director of Public Health of this fact.

CONCLUSION

In many hospitals the occupational health department is a much under-used resource and this must act to the detriment of the welfare of the staff as a whole. Many – perhaps the majority – of occupational health departments are nurse-based and nurse-led and do not have the services of a consultant occupational physician which tends to lessen their impact on the medical staff particularly.

The special expertise which occupational health personnel have is in the assessment of risk and in the prevention of blood-borne infectious diseases this expertise must be called upon when developing and implementing strategies with this aim in mind. It is well for staff in other specialties to recognize that although occupational health personnel may not be as knowledgeable as they are in their own specialist areas of virology, infectious disease and so on, techniques of risk assessment and risk communication are common to *all* risks and there will be no-one in the hospital better able to advise on these matters than those in the occupational health department. So far as treatment and follow-up of staff who sustain injuries, then it is likely that occupational health staff will not have the expertise necessary to advise those who elect to take post-exposure prophylaxis after an incident involving an HIV-positive donor patient, and this is best left to those who have.

REFERENCES

Astbury, C. and Baxter, P.J.(1990) Infection risks in hospital staff from blood: hazardous injury rates and acceptance of hepatitis B immunization. *Journal of the Society of Occupational Med*icine 40, 92–93.

BMA (1995) *A Code of Practice for Implementation of the UK Hepatitis B Immunisation Guidelines for the Protection of Patients and Staff.* British Medical Association, London.

BMA (1996) *A Guide to Hepatitis C.* British Medical Association, London.

BSI (1990) *BS 7320. Specification for Sharps Containers.* British Standards Institution, London,

CDC (1995) Case–control study of HIV seroconversion in health-care workers after percutaneous exposure to HIV-infected blood. *Morbidity and Mortality Weekly Report* 44, 929–933.

CDC (1996) Provisional Public Health Service recommendations for chemoprophylaxis after occupational exposure to HIV. *Morbidity and Mortality Weekly Report* 45, 468–472.

Choudhury, R.P. and Cleator, S.J. (1992) An examination of needlestick injury rates, hepatitis B vaccination uptake and instruction on 'sharps' technique among medical students. *Journal of Hospital Infection* 22, 143–148.

Control of Substances Hazardous to Health Regulations 1994. HMSO, London.

DH (1993) *Protecting Health Care Workers and Patients from Hepatitis B.* Recommendations of the Advisory Group on Hepatitis. HMSO, London.

DH (1994) *AIDS/HIV-Infected Health Care Workers: Guidance on the Management of Infected Health Care Workers.* Department of Health, London.

Greco, R.J. and Garza, J.R. (1995) Use of double gloves to protect the surgeon from blood contact during aesthetic procedures. *Aesthetics and Plastic Surg*ery 19, 265–267.

Haiduven, D.J., DeMaio, T.M. and Stevens, D.A. (1992) A five-year study of needlestick injuries: significant reduction associated with communication, education, and convenient placement of sharps containers. *Infection Control and Hospital Epidemiology* 13, 265–271.

influenza (Hib). Many countries are now adding hepatitis B to their routine infant immunization programmes. As vaccination has been, and is primarily, aimed at population control of infectious diseases many, if not the majority, of health-care workers are likely to commence their career having been vaccinated against the common infectious diseases of childhood which will, in the future, include hepatitis B.

PASSIVE IMMUNIZATION

Passive immunization is the administration of an immune serum, human or animal, which contains preformed antibodies. Antibodies can be produced naturally, actively induced, passively acquired by injection or transferred from mother to fetus/infant via the placenta or breast milk.

In respect of the prevention of blood-borne infections two product types are relevant – normal human immunoglobulin (IgG), and hyperimmune serum globulin. Antisera with high titres of antibody against specific microorganisms have been used to provide short-term protection following exposure to potentially serious infections. Passive and active immunization can be used in combination particularly if the infection has a long incubation period. The use of specific hepatitis B immunoglobulin (HBIG) will be discussed later in this chapter.

IMMUNIZATION PROGRAMMES

The Key Principles of Immunization Programmes for Health-care Workers

Before establishing any occupational vaccination programme, particularly for newly-developed vaccines, risk-benefit and cost-benefit evaluations will be required to answer questions which may arise with regard to the safety, efficacy, and cost of the vaccines on the one hand, against the severity of infection, available treatment, prevalence and incidence of the disease on the other. Acceptability, which will influence uptake and compliance is also important.

'Ownership' of the vaccination programme – agreement by the employer, employee and unions of its value – will make its implementation more effective, as will decisions at the outset on funding and administration. Complementary education programmes must include information not only about the vaccine but also the disease, its occupational aspects and the consequences of non-vaccination. Any preventive programme must be audited and evaluated at regular intervals with regard to its effectiveness and relevance to changing patterns of disease and developments in vaccine manufacture.

Who is eligible?

For any one particular vaccine, the answer to this key question will depend on the disease profile, age-specific attack rate, method of transmission, local epidemiological pattern, the definition of the 'at-risk' employee and country specific legislation, regulations or guidance. The 'at-risk' employee is identified by asking the simple question, 'What do you do?' and determining by work activities, not solely job title or place of work, if infection can be transmitted. Vaccination programmes may therefore include students, support workers such as technicians, morticians, instrument sterilizers, housekeeping and laundry staff. The potential for health-care workers to transmit infection to vulnerable patient groups, for example the immunosuppressed and elderly, must also be recognized. Eligibility for hepatitis B vaccine is discussed in detail later in this section.

Vaccination status must be reviewed on promotion, relocation, overseas posting, after accidental exposure to a biological agent and when epidemics occur in the local or national population.

Vaccination programmes must also address practical issues stipulating dose, schedule and site of vaccination in accordance with manufacturers' instructions and the actions to be taken in the case of non-response. Comprehensive records are essential.

Vaccination in relation to health and safety legislation

Vaccination of health-care workers exposed to infectious diseases as a direct consequence of their work is specifically provided for in legislation. In 1990, the European Council Health and Safety Directorate (DG V) issued, as part of a cluster of health and safety legislation, the Directive: *Protection of Workers from Risks Related to Exposure to Biological Agents at Work* (Commission of the European Communities, 1990). This required that effective vaccines, where they exist, should be made available to exposed workers who are not already immune.

But vaccination is an important element of the hierarchical approach to risk management and control of biological agents at work, adapting the principles already applied to managing the risk from hazardous substances in the workplace. The requirement actively to consider vaccination as part of the overall control strategy is set out in Annex VII of the Classification Directive (Commission of the European Communities, 1993). Adopted in 1993, this amends the Biological Agents Directive, classifies biological agents into one of four groups based on their ability to cause human disease by infection, denotes those agents where records must be kept for exposed workers and for which a vaccine exists. Annex VII, the Code of Practice on vaccination, requires:

- assessment of the risk of exposure to biological agents;
- the offer of vaccination to workers at-risk where effective vaccines exist;
- vaccination in accordance with national law and practice;
- the provision of a certificate of vaccination if necessary.

The United Kingdom has incorporated the EC Directives into the *Control of Substances Hazardous to Health Regulations 1994.* However, for many infectious diseases no vaccines are available and protection depends on safe working practice and strategies, for example, universal precautions, which are discussed further in relation to post-exposure prophylaxis.

ACTIVE IMMUNIZATION: HEPATITIS B

Hepatitis B is recognized as the most important infectious occupational disease. Based on data available in 1992 Van Damme and Tormans (1993) developed an epidemiological risk model for occupationally-acquired hepatitis B infection in European health-care workers which suggests that approximately 24,000 new hepatitis B infections occur each year in health-care workers in Europe if no vaccination is given. Using accepted projections of clinical outcomes and assuming vaccination coverage of 40% in traditional health-care workers and 25% in allied health-care workers they estimated that 502 health-care workers in northern Europe and 6423 workers in southern Europe become infected with the hepatitis B virus (HBV) each year. Furthermore, one European health-care worker would die each working day from the consequences of occupationally-acquired hepatitis B infection. Current estimates are that 837 health-care workers in northern Europe and 7463 in southern Europe contract hepatitis B each year (Van Damme et al., 1995).

Determination of Occupational Risk

Occupational risk of hepatitis B infection has traditionally been associated with health-care professionals who may be defined as: '... persons whose work activities involve regular physical contact with patients and/or their blood or body substances for the purpose of providing care and investigational and/or therapeutic intervention ...'

Well-designed epidemiological studies in the 1970s and early 1980s allowing for confounding factors such as age, sex, socioeconomic status, previous treatment with blood or blood products and membership of high-risk groups or participation in social high-risk activities showed that not only health-care professionals, as defined above and including for example, doctors, dentists, nurses and midwives, but also those without direct patient contact and those working outside health-care institutions had an increased risk of occupational HBV infection. It was also clear that risk is relative to and measured in terms of what people actually do and not their job title or place of work; risk increases with length of service.

In summary, those eligible for vaccination will include health-care workers and others, including students and trainees, who have direct contact with patients and their body fluids or who are likely to experience regular physical contact with blood or blood-contaminated secretions and excretions. The various other occupational 'at-risk' groups defined by risk assessment are listed in Box 16.1. In addition, the occurrence of regular splashing or contamination of protective clothing will also demonstrate that these workers are at occupational risk.

Box 16.1. Non health-care professionals at risk from occupational hepatitis B infection

Health Services
- housekeeping
- catering
- laundry
- sterile supplies service
- maintenance – equipment
- machinery

Public Services
- ambulance/paramedic
- police
- firefighters
- teachers (institutions)
- waste disposal

Education/Research
- university
- polytechnic
- pharmaceuticals

Embalmers/Morticians

Prison staff

Many European and North American countries, areas of mainly low ende-micity, introduced recommendations during the 1980s for the vaccination of those at high risk for HBV infection, most of whom were adults. The UK Health Departments (DH, 1996) adopts a hepatitis B vaccination strategy based on the identification of groups at risk which can be broadly categorized as those relating to patients, those at occupational risk and those at risk because of their lifestyle. The current recommendations of the DH for HB vaccination are summarized in Box 16.2.

General guidance on the protection of health-care workers from HIV and HBV was issued by the Health Departments in 1990 (DH, 1990). In 1993 the NHS Management Executive issued further guidance (DH, 1993) as a result of a small number of patients being infected with HBV by their health-care worker (almost exclusively a surgeon). The purpose was to ensure that health-care workers who may be at risk of acquiring hepatitis B from a patient are protected by immunization, and to protect patients against the risk of acquiring hepatitis B from an infected health-care worker. All health-care workers who perform 'exposure prone procedures' (EPP) which are described as: '. . . those where there is a risk that injury to the worker may result in the exposure of the patient's open tissue to the blood of the worker including those where the worker's gloved hands may be in contact with sharp instruments, needle tips and sharp tissues (spicules of bone or teeth) inside a patients open body cavity, wound or confined anatomical space where the hands or fingertips may not be completely visible at all times'. . . must be immunized against HBV. The professional groups identified in the guidance are summarized in Box 16.3.

Vaccine Development

Safe and effective HB vaccine has been available for over ten years. The first commercially available HB vaccine was prepared from the plasma of chronic

Box 16.2. Department of Health recommendations for hepatitis B immunization

Patient groups at risk:
- babies borne to mothers who are chronic carriers or who have had acute HB infection during the pregnancy
- close family contacts of a case or carrier
- families adopting children from countries with a high prevalence of HB
- haemophiliacs and others receiving regular blood transfusions/blood products and those who administer them
- patients with chronic renal failure

Occupational groups at risk:
- health-care workers including students and trainees who have direct contact with blood or blood-stained body fluids or patients tissues
- other occupational risk groups e.g. morticians, embalmers
- other occupational risk groups based on risk assessment of an individual's exposure
- police, ambulance, rescue services
- prison staff
- staff and residents of residential accommodation for those with severe learning disabilities (mental handicap)
- long-stay travellers/workers in endemic areas

Groups at risk through lifestyle:
- parenteral drug misusers
- individuals who change sexual partners frequently
- homosexual/bisexual men
- prostitute men and women
- those with multiple sexual partners
- prisoners

From DH (1996).

asymptomatic HBsAg carriers. Inactivation steps ensured the non-infectivity of the plasma-derived vaccine (PDV) which was introduced in the UK in the early 1980s. PDVs have an excellent record of safety and efficacy. They are now produced in Asia and used in a number of countries.

Several factors inhibited the widespread acceptance and use of the PDV but principally it was the recognition that the human immunodeficiency virus (HIV), the causative agent of AIDS, was predominant in the same populations with a high prevalence of HBV infection who were donating serum for vaccine manufacture. There were serious concerns about the potential for the PDV to transmit HIV. Nevertheless, no case of HIV or any other infection due to vaccination for HBV has been documented.

A new generation of HBV vaccines became commercially available in the mid-1980s, having been developed using genetic engineering techniques. At present, yeast and mammalian cells are used for the expression of HBsAg in these DNA recombinant vaccines. The vaccines do not contain live viral parti-

Box 16.3. Health-care workers who must be vaccinated against HBV.

- surgeons
- dentists
- doctors who carry out exposure-prone procedures
- midwives
- theatre nurses
- renal dialysis nurses
- independent contractors
  - GPs
  - dentists
  - midwives
  - podiatrists
- medical, dental nursing and midwifery students

From DH (1993).

cles. The purified antigen is formulated as a vaccine by adsorption onto aluminium hydroxide and contains thiomersal as a preservative. HB vaccine should be stored between +2 and +8°C. It should not be frozen as this reduces effectiveness.

Vaccination Schedule

The schedules for vaccination approved by most regulatory authorities are:

- 0, 1 and 6 months (an initial dose followed by two further doses at one and six months);
- 0, 1, 2 and 12 months (an initial dose and two further doses at one and two months after the first dose with a booster at 12 months).

Protective antibody titres, measured as anti-HBs, are achieved using a wide variety of schedules but, in general, the doses of vaccine in the primary course should be given at least one month apart. The two HB vaccines licensed for use in the UK are interchangeable.

Vaccination Site

The site of vaccination and administration technique are critical to achieving a maximum response. The vaccine is given by intramuscular injection into the deltoid in adults (or into the anterolateral thigh in neonates and infants). The intradermal route and buttock injections are not recommended (CDC, 1994).

Immune Response

HB vaccines are highly effective with seroconversion rates of around 95% in young healthy adults. Seroconversion is defined as the production of any anti-

HBs. The generally accepted antibody level for seroprotection against infection is 10 miu ml^{-1} (Frisch-Niggemeyer et al., 1986). In a heterogeneous population such as health-care workers seroprotection is achieved in approximately 85% of vaccinees following completion of the primary course.

Persistence of detectable antibody titres has been shown to be dependent on the peak antibody titre achieved following the primary vaccination series but protection against clinical disease or development of the carrier state lasts long after detectable antibody has disappeared (Leroux et al., 1994). Vaccinees whose antibody titres following the primary course of vaccine are between 10 and 100 miu ml^{-1} are classified by some authorities as poor responders. Where resources are available, they may be given further doses of vaccine as described below in an attempt to improve the response.

The response to vaccination is influenced by age, obesity, immunocompetence, genetics and smoking (Shaw et al., 1989; Hollinger, 1990). Poorer responses increase in frequency with increasing age and/or weight of the vaccinee. Males tend to respond less well than females (Westmoreland et al., 1990). There is a small minority of persons who appear to be genetically incapable of mounting an immune response to existing HB vaccines.

Safety

Several hundred million doses of HB vaccine have been distributed worldwide. Side-effects can in general be classified as local reactions at the site of injection or systemic reactions due to either direct irritating effects of the vaccine or from hypersensitivity to traces of allergens. The vaccines are well tolerated. The majority of side-effects which do occur are minor and transient and localized to the site of injection. They are commonly reported as mild tenderness or local irritation comprising erythema, swelling and induration and most likely to be caused by the aluminium hydroxide component of the vaccine. General symptoms are reported as fatigue, headache, dizziness and nausea and occasionally fever, diarrhoea, myalgia, arthralgia and rashes.

As a general principle the use of any drugs or vaccines is avoided in pregnancy. Female workers at high risk of occupational HBV infection should be advised to complete the course of HB vaccine before pregnancy begins but HB vaccination is not contraindicated in pregnant or lactating women. The genetically engineered HB vaccines do not contain live virus particles. Where pregnancy coincides with a course of HB vaccine the programme should be completed because the consequence of acquiring infection, both to the mother and the newborn infant, outweighs the discomfort of any possible minor side-effects.

Pre-and Post-vaccination Serological Screening

The cost-effectiveness of screening of workers prior to vaccination is determined by the national prevalence of hepatitis B. In countries of high endemicity it may not be cost-effective to screen for previous evidence of infection although, in these countries, endemicity may also be a reflection of living conditions and social status and there may be subsets of health-care workers whose prevalence rates more accurately reflect those in countries of low endemicity. These health-care workers would be at higher risk. Although it has been consistently shown that workers exposed to blood or body substances on a regular basis have higher prevalence rates than the control population, within countries of low endemicity it is not cost-effective to routinely screen workers prior to hepatitis B vaccination.

The routine measurement of antibody titres following hepatitis B vaccination in infants and adolescents is not recommended. However, health-care workers and others at occupational risk are not a homogenous group with respect to age, weight and gender. Routine measurement of antibody titres 6–8 weeks after the third dose of vaccine will identify those who have not responded and who, if pre-vaccination screening has not been undertaken, may have had previous subclinical infection.

Non-responders

The antibody level for seroprotection against infection is 10 miu ml^{-1} (Frisch-Niggemeyer *et al.*, 1986). Vaccinees with antibody titres less than 10 miu ml^{-1} are non-responders. Approximately 30–60% of non-responders will respond following three additional doses of vaccine. These may be given as a repeat vaccination course or as separate doses given 2–3 months apart. Antibody titres should be measured 6–8 weeks following the third dose (Grosheide and Van Damme, 1996). Vaccinees who continue to fail to attain an antibody titre greater than 10 miu ml^{-1} should be further investigated as non-response to vaccination may be due to either a genetic inability to respond or previous subclinical infection with HBV. The first stage in the investigation is to determine core antibody (anti-HBc) as a marker of natural infection. Those who are negative are classed as true non-responders and must receive clear instruction relating to the use of protective clothing and safe working practice and consideration of restriction of work activities, for example by prohibiting work with the virus or in departments where there is known to be a high percentage of patients who are carriers.

Persons who are anti-HBc positive should be further investigated, with counselling and consent, for markers of chronic infection. Not only is there the potential for them to transmit infection to family members or sexual partners but also, if they are undertaking invasive procedures, to their patients. HBeAg positive carriers must not carry out 'exposure prone procedures' (DH, 1993). Identification of carriers also allows the opportunity of referral for specialist clinical care and treatment if appropriate.

COMPLIANCE WITH HEPATITIS B VACCINE PROGRAMMES

Safe and effective vaccines against hepatitis B have been available for over ten years and yet, even amongst those at highest risk of occupational infection, the

Table 16.1. HB vaccination rates in health-care workers by country – 1990.

Country	Profession	% Vaccinated	References
Germany	Nurses	47–61	Vogt-Versloot, 1990
Spain	Nurses	38	Martos, 1990
UK	HCWs	42–65	Carruthers, 1990
UK	Doctors	49	Burden and Whorwell, 1991
UK	Nurses	25	Burden and Whorwell, 1991
Belgium	Doctors	44	Van Damme et al., 1990
Belgium	Dentists	71	Van Damme et al., 1990
France	HCWs	45	Abiteboul et al., 1990

uptake rates in many countries is of the order of 40–60% and highly variable as illustrated in Table 16.1 (Carruthers, 1990; Martos, 1990; Van Damme et al., 1990; Vogt-Versloot, 1990; Burden and Whorwell, 1991). The reasons for poor uptake of HB vaccine include:

- the first vaccine was plasma derived;
- congruity of risk groups for AIDS and HBV;
- potential for plasma to transmit AIDS and other viruses;
- low endemicity of HBV in NW Europe;
- cost of vaccine;
- attitude of administrators;
- attitude of HCWs: not at risk; no time/not convenient; too old.

Following a survey which estimated that, on average, only 45% of eligible health-care workers had been vaccinated (Abiteboul et al., 1990) the French government introduced legislation (*Journal Officiel*, 1991) making vaccination compulsory for all at-risk workers since when vaccination uptake has increased significantly. A survey of public sector hospitals conducted at the end of 1991 showed that 84% of nurses were covered by vaccination (Gouaille, 1992) and an additional 1 million doses of HB vaccine had been used in that 12 month period.

Universal HB Vaccination

Although countries have made recommendations for hepatitis B vaccination based on their epidemiological pattern of disease, the strategy to vaccinate those in high risk groups has failed to control HBV infection in countries of low endemicity. Those at risk through lifestyle are difficult to target and often are infected before presenting at clinics where vaccine is available. Although much effort has been expended on vaccinating occupational groups at risk (in practice, health-care workers) they contribute of the order of 2–9% to the total pool of HBV infection (Kane et al., 1991).

Control in these countries will be achieved only with a universal programme of infant or adolescent vaccination or both and continuing the vaccination of adults in high risk groups for a limited period.

PASSIVE IMMUNIZATION AND POST-EXPOSURE PROPHYLAXIS

Active vaccination of those at risk for HBV infection remains the mainstay of prevention. As there will always be a percentage of vaccinees who do not respond, vaccination must be part of an overall risk control and management strategy which must have a clear policy for the management of all contamination accidents. In all cases this will include immediate attention to the wound, the reporting of the accident to the responsible person, and determining the immune status of the employee and the source of the contamination for the principal blood-borne infectious agents hepatitis B and C viruses (HBV, HCV), and human immunodeficiency virus (HIV). The range of options will vary according to whether specific preventive measures are available or not.

Unlike hepatitis B there are no vaccines specific for HCV and HIV. Furthermore, it is unrealistic to anticipate developments in the near future. There are no specific immunoglobulins for passive post-exposure prophylaxis. The use of normal human immunoglobulin (IgG) has no proven efficacy (Public Health Laboratory Service; PHLS, 1993). Therefore, prevention of occupational HCV and HIV infection must depend on awareness and understanding as to how it can happen in the workplace, a consistently high standard of working practice, and the use of appropriate protective clothing. Strategies include:

- screening and treatment of blood products;
- universal precautions;
- safe systems of work including appropriate protective clothing;
- sterilization of re-usable medical equipment;
- correct use of multi-dose ampoules.

Hepatitis B

For HBV the specific actions post-exposure will range from vaccination and the use of hepatitis B immune globulin (HBIG) simultaneously in those workers who have been exposed to a carrier of the virus and who are unvaccinated or have an antibody titre less than $10\,\text{miu}\,\text{ml}^{-1}$ through to no action in those workers who have maintained a protective antibody titre (PHLS, 1992). The combination treatment of HB vaccine and HBIG is over 90% effective in preventing HBV infection following documented exposure. Maximum effect is obtained when HBIG is administered within 48 hours of the exposure accident but will retain some benefit if administered up to 14 days following the accident. The options for post-exposure prophylaxis for HBV infection are summarized in Table 16.2.

Hepatitis C

Studies from many countries including the United States (Forseter *et al.*, 1993; Thomas *et al.*, 1993), Australia (Bowden *et al.*, 1993), Spain (Perez-Trallero *et al.*, 1992), Northern Italy (Campello *et al.*, 1992), show that the prevalence of

Table 16.2.   Summary of post-exposure prophylaxis against hepatitis B.

Immune status of employee/HCW	Hepatitis B status of patient/source of exposure		
	Unknown*	Negative	S/E antigen positive
Unvaccinated	Vaccinate + or − HBIG	Vaccinate	Vaccinate + HIBG
Vaccinated responder: Anti HBS <10 miu ml^{-1} Anti HBS >100 miu ml^{-1}	Booster No action	Booster No action	Booster + HBIG No action
Non-responder to primary course (3 doses)	Booster doses +/− HBIG	No action or booster doses	HBIG + booster doses
Non-responder to primary course + booster (6 doses)	HBIG	No action	HBIG

+, with; −, without
*Risk assessment will aid decision making on requirement for HBIG.

HCV in health-care workers is not greater than in the local reference population and is of the order of 0.5–2%. Prevalence is associated with exposure to
recognized risk factors and older age, but not occupation. Cross-sectional studies which show a higher prevalence rate in specific groups of health-care workers cannot conclude that this is due to occupational transmission, rather that it
reflects the prevalence in the local population from which those health-care
employees are recruited.

However, occupational transmission does occur. Polish *et al.* (1993), using a
logistic regression model, showed that the principal risk activity is needlestick
injuries. The estimated risk of transmission of HCV following a needlestick
injury is 3% or less (Sodeyama *et al.*, 1993). Following an exposure accident the
following steps must be taken:

- immediate attention to the wound;
- report the accident to the responsible person;
- determine the HCV antibody status of the employee as a baseline reference;
- determine the HCV status of the source of the contamination;
- counsel employee and if the source of the contamination is positive for
  HCV undertake regular serological monitoring for 12 months.

Counselling requires that the employee is provided with information about
acute and chronic HCV infection, the possible long-term sequelae and available
treatment. Lifestyle counselling must address alcohol intake, sexual activity,
household contacts of all ages and the potential for vertical transmission. With
respect to work in a health-care setting the employee must be reminded of the
need to observe universal precautions and safe systems of work to prevent
nosocomial transmission.

Human Immunodeficiency Virus

The risk of seroconversion following a needlestick injury with HIV positive blood is < 0.3% compared with the risk of acquiring HBV which is 5–43%, the higher estimate referring to an HBeAg-positive source (Beckmann and Hendeson, 1992). As with HCV there is no specific active or passive immunization for the prevention of HIV infection. Prevention of occupational infection relies on adherence to universal precautions and safe systems of work. Following any contamination accident the same procedures must be followed as described for HCV. If consent for testing the baseline sample for HIV is withheld it should be stored for two years (Heptonstall *et al.*, 1993). Where consent is given, samples should be periodically retested for a minimum period of six months. In both instances the employee should be provided with information about the symptoms associated with seroconversion (recognizing that seroconversion may be subclinical), the clinical manifestations of HIV infection and address the same lifestyle issues as noted for HCV infection including safe sexual behaviour.

It is known that the progression of clinical HIV infection can be delayed by the administration of anti-retrovirus therapy. There is conflicting evidence, however, about the effectiveness of such treatments, for example using Zidovudine, in preventing seroconversion following occupational exposure (Go *et al.*, 1991). Current guidance from the DH asks that health-care workers are offered counselling and information on the possible benefits from taking prophylactic Zidovudine particularly after deep penetrating inoculum accidents from a source known to be positive for HIV or where there is a high index of suspicion regarding positivity. The success of Zidovudine as a prophylactic measure depends on it being given within hours of exposure. A recent report from CDC (1995) suggests that if Zidovudine is administered immediately after exposure and continued in full dosage (200 mg every 4 hours for 3 days followed by 100–200 mg every 4 hours – 5 times a day – for 25 days) the chances of seroconversion are reduced by 80%. Other research has shown that health-care workers with percutaneous exposure to HIV-positive blood have seoconverted despite the use of Zidovudine. Concerns still exist about drug-related toxicity and, overall, the data from CDC add little to existing evidence. It is likely that Zidovudine will continue to be offered where assessment shows there to be a high degree of risk of occupational transmission rather than a routine prophylactic treatment. Best practice dictates that such information should be given in person and not merely available from an information sheet and the counsellor should be able to discuss the merits and side-effects of Zidovudine so that the health-care worker can make an informed decision.

For other views on AZT see Morgan, Chapter 8 and Waldron, Chapter 15, this volume.

SUMMARY

Despite the recognition of HBV infection as the most important occupational infectious hazard of health-care workers and despite the availability of safe and effective vaccines for over ten years a significant number of health-care workers remain unvaccinated. Risk models for health-care workers at risk from occupationally acquired HBV infection led to the introduction of legislation both in Europe and in the United States. The European Council of Ministers, in passing the Biological Agents Directive and its amending Directive, require all member countries to assess the risk to employees from biological agents in the workplace, to ensure that there are adequate control and preventative measures in place including vaccination where safe and effective vaccines exist.

REFERENCES

Abiteboul, D, Gouaille, B. and Proteau, J. (1990) Prévention de l'hepatite B virale â l'Assistance Publique-Hôpitaux de Paris – Bilan de 7 ans de vaccination par les Medicins du Travail. *Archives Maladies Professionales* 51, 405–412.

Beckmann, S.E. and Hendeson, D.K. (1992) Healthcare workers and hepatitis: risk for infection and management of exposures. *Infectious Disease in Clinical Practice* 1, 424–428.

Bowden, F.J., Pollett, B., Birrell, F. *et al.* (1993) Occupational exposure to the human immunodeficiency virus and other blood-borne pathogens. A six-year prospective study. *Medical Journal of Australia* 158, 810–812.

Burden, A.D. and Whorwell, P.J. (1991) Poor uptake of hepatitis B immunization amongst hospital-based health care staff. *Postgraduate Medical Journal* 67, 256–258.

Campello, C., Majori, S., Poli, A. *et al.* (1992) Prevalence of HCV antibodies in health care workers from northern Italy. *Infection* 20, 224–226.

Carruthers, J. (1990) Hepatitis B prevention policies in the UK. *Proceedings of the European Conference on Hepatitis B as Occupational Hazard*, Geneva, pp. 21–24.

CDC (1994) General recommendations on immunisation. *Morbidity and Mortality Weekly Report* 43, 6–7.

CDC (1995) Case-control study of HIV seroconversion in healthier workers after percutaneous exposure to HIV-infected blood – France, United Kingdom and United States, January 1988 – August 1994. *Morbidity and Mortality Weekly Report* 44, 929–933.

Commission of the Eurpean Communities (1990) *Council Directive on the Protection of Workers from Risks Related to Exposure to Biological Agents at Work.* 90/679/EEC. Brussels.

Commission of the European Communities (1993) *Annex 4: Council Directive on the Classification of Biological Agents.* 93/88/EEC. Brussels.

Control of Substances Hazardous to Health Regulations 1994. SI 3246. HMSO, London.

DH (1990) Guidance for Clinical Health Care Workers: *Protection against Infection with HIV and Hepatitis Viruses. Recommendations of the Expert Advisory Group on AIDS.* Department of Health. HMSO, London.

DH (1993) *Protecting Health Care Workers and Patients from Hepatitis B.* NHS Management Executive; (HSG(93)40): Department of Health, London.

DH (1996) Hepatitis B. In: *Immunisation against Infectious Disease* (Chapter 18, 95–108). Department of Health 1996. London, HMSO.

Forseter, G., Wormser, G.P., Adler, S. *et al.* (1993) Hepatitis C in the health care setting. II. Seroprevalence among haemodialysis staff and patients in suburban New York City. *American Journal of Infection Control* 21, 5–8.

Frisch-Niggemeyer, W., Ambrosch, F. and Hofmann H. (1986) The assessment of immunity against hepatitis B after vaccination. *Journal of Biological Standards* 14, 255–258.

Go, G.W., Baraff, L.J. and Schriger, D.L. (1991) Management guidelines for health care workers exposed to blood and body fluids. *Annals of Emergency Medicine* 20, 1341–1350.

Gouaille, B. (1992) Vaccination contre l'hépatite B des professionnelles du milieu hospitalier. *Immunology and Medicine* 9, 51–56.

Grosheide, P. and Van Damme, P. (1996) *Prevention and Control of Hepatitis B in the Community.* WHO Communicable Diseases Series, No 1. Viral Hepatitis Prevention Board, University of Antwerp, Belgium. p.31.

Heptonstall, J., Gill, O.N., Porter, K. *et al.* (1993) Health care workers and HIV: surveillance of occupational acquired infection in the United Kingdom. *Communicable Disease Report Review* 3, R147–R153.

Hollinger, F.B. (1990) Hepatitis B virus. In: Fields, B.N., Knipe, D.M. *et al.* (eds) *Virology.* Raven Press, New York, pp. 2171–2236.

Journal Officiel (1991) Loi No. 91.73 du 18 Janvier 1991. Parue au du 20 Janvier 1991 modifiant l'article L.10 du code de la Santé Publique.

Kane, M.A., Ghendon, Y. and Lambert, P.H. (1991) Where are we?: the WHO programme for control of viral hepatitis. In: Hollinger, F.B., Lemon, S.M. and Margolis, H.S. (eds) *Viral Hepatitis and Liver Disease.* Williams and Wilkins, Baltimore, Maryland, pp. 706–708.

Leroux, G., Van Hecke, E., Michielson, W. *et al.* (1994) Correlation between in vivo humoral and in vitro cellular immune responses following immunisation with hepatitis B surface antigen(HBsAg) vaccines. *Vaccine* 12, 812–818.

Martos, J.S. (1990) Prevention of hepatitis B in Spain. In: *Proceedings of the European Conference on Hepatitis B as Occupational Hazard.* Geneva. pp. 30–33.

Perez-Trallero, E., Cilla, G., Alcorta, M. *et al.* (1992) Low risk of acquiring the hepatitis C virus for the health personnel. *Medical Clinics of Barcelona* 99, 609–11.

PHLS (1992) Hepatitis Subcommittee. Exposure to hepatitis B virus: guidance on post-exposure prophylaxis. *Communicable Disease Report Review* 2, R97–R101.

PHLS (1993) Hepatitis Subcommittee. Hepatitis C virus: guidance on the risks and current management of occupational exposure. *Communicable Disease Report Review* 3, R135–9.

Polish, L.B., Tong, M.J., Co, R.L. *et al.* (1993) Risk factors for hepatitis C virus infection among health care personnel in a community hospital. *American Journal of Infection Control* 21, 196–200.

Shaw, F.E., Jr, Guess, H.A., Roets, J.M. *et al.* (1989) Effect of anatomic injection site, age and smoking on the immune response to hepatitis B vaccination. *Vaccine* 7, 425–430.

Sodeyama, T., Kiyosawa, K., Urushihara, A. *et al.* (1993) Detection of hepatitis C virus markers and hepatitis C virus genomic-RNA after needlestick accidents. *Archives of Internal Medicine* 153, 1565–1572.

Thomas, D.L., Factor, S.H., Kelen, G.D. *et al.* (1993) Viral hepatitis in health care personnel at The Johns Hopkins Hospital. The seroprevalence of and risk factors for hepatitis B virus and hepatitis C virus infection. *Archives of Internal Medicine* 153, 1705–1712.

Van Damme, P., De Cock G., Cramm, M. *et al.* (1990) Precautions taken by orthopaedic surgeons to avoid infection with HIV and hepatitis B virus. *British Medical Journal* 30, 611.

Van Damme, P. and Tormans, G. (1993) A European risk model. In: *Proceedings of the International Congress on Hepatitis B as an Occupational Hazard,* 10–12 March 1993, Vienna, Austria.

Van Damme, P., Tormans, G., Van Doorslaer, E. *et al.* (1995) European risk model for hepatitis B among health care workers. *European Journal of Public Health* 5, 245–252.

Vogt-Versloot, G. (1990) Hepatitis B: an occupational hazard among nurses in Germany. *Proceedings of the European Conference on Hepatitis B as Occupational Hazard.* Geneva. pp. 25–29.

Westmoreland, D. Player, V., Heap, D.C. *et al.* (1990) Immunisation against hepatitis B – what can we expect? *Epidemiology and Infection* 104, 499-509

CHAPTER 17
Treatment and decontamination of blood spills

P.N. Hoffman and D.A. Kennedy

There is always a risk of blood spill wherever humans are present, either directly from people, e.g. as a consequence of accidental injury at work, haemorrhage in health-care settings, deliberate wounding, or when containers of blood are being handled, e.g. within or during transit to medical laboratories. These events can result in exposure to any infectious agent that may be present in the blood: a wide range of such agents have been recorded (see Jeffries, Chapter 1 and Collins, Chapter 2 this volume), the most important of which are the human immunodeficiency virus (HIV) and the hepatitis B and C viruses (HBV and HCV). Observation and analysis of reports of occupational transmission of HIV – the most acutely observed – has yielded information which puts transmission risks from blood spills into an informed context (see Hunt, Chapter 3 this volume) and it seems that the mere presence of a viable infectious agent is not in itself associated with substantial risk of transmission of that agent. Whilst both HBV and HIV can survive in blood and plasma for more than seven days (Bond et al., 1983; Mougdil and Daar, 1993), they do not appear to transmit readily from spills.

Reviews of worldwide occupational HIV acquisition among health-care workers (Heptonstall et al., 1993) suggested that the predominant mode of transmission is *percutaneous*, mostly associated with fresh blood and hollow-bore needles. Where, less frequently, *mucocutaneous*, i.e. involving exposure of either mucous membrane or non-intact skin, exposure is implicated, it is usually related to large amounts of blood or repeated exposure.

This evidence implies that the real risk from blood-borne viruses in blood spills is when there is contact between the spill and the already-broken skin, or, more likely and posing a greater hazard, when an individual receives a cut from a sharp object within the spill. It is not difficult to visualize unexpected shards of glass or sharp plastic in such situations as a broken specimen container in health-care settings or sharp metal cutting emergency service personnel attending a road traffic accident. Hands may become contaminated if they come into contact with blood spills and hand-to-eye and hand-to-mouth contacts will carry a risk of infection transmission (see Kennedy, Chapter 6 this volume).

There is strong motivation, particularly in health-care settings, to clean up blood spills rapidly as they may be offensive to the public and are not perceived as a sign of high professional standards. Against this background it is prudent to deal with a blood spill in a workplace as soon as possible after it occurs. From a risk perspective, however, it must be accepted that the individual most at risk from pathogens present in the spill is the person who clears it up. This person will also be exposed to any toxic effects of chemicals used in clearing the spill.

Both procedure and disinfection are involved in safe clearance of blood spills, ensuring that the clearer does not come into contact with any infectious agent in the spill; the procedural aspects are the more important. Each will be considered in turn.

Disinfection of Blood Spills

Any disinfectant chosen for the treatment of blood spills must be active against the range of infectious agents which may be present, the blood-borne viruses in particular. This determination is a technically and interpretationally complex area.

Although HIV can be grown in tissue culture, both technical difficulties and the need for a high level of laboratory containment mean that it is not particularly amenable to inactivation studies. Early studies used viral reverse transcriptase activity as a marker of viral infectivity (Spire *et al.*, 1985), but more direct methods, using tissue culture infectivity, soon followed (Resnick *et al.*, 1986). Since then there have been many studies on the chemical inactivation of HIV which have shown it to have innate susceptibility to a very wide variety of inactivating agents. These have been fully reviewed by Sattar and Springthorpe (1991, 1994).

Although HBV has been characterized for decades it cannot be grown in tissue culture to any useful extent. This severely limits our knowledge of its inactivation. Observations resulting in human infection ceased in the late 1960s, leaving the estimation of inactivation to the observance of morphological disruption by electron microscopy, or loss of antigenic markers. As these methods can give false negative results HBV acquired a reputation for a high level of resistance to inactivation. A model using chimpanzee infection showed that HBV was readily inactivated by a range of chemicals: hypochlorites, iodine and isopropanol (Bond *et al.*, 1983), ethanol (Kobayashi *et al.*, 1984), glutaraldehyde (Bond *et al.*, 1983; Kobayashi *et al.* 1984), quaternary ammonium compounds and mixtures of these and phenolics (Prince *et al.*, 1993).

There is no specific information about the disinfection of HCV but, according to its structure, it should be susceptible to those agents that are active against HBV.

Other Factors Affecting the Efficacy of Disinfectants

Unfortunately the determination of the innate susceptibility of a microorganism to a chemical disinfectant is only a starting point for considering which disinfectants are useful in any particular situation. Other factors may, to varying extents, frustrate the action of disinfectants (Ayliffe *et al.*, 1993). In blood spills, these are mainly the failure to penetrate blood that has dried, clotted or even been coagulated by the disinfectant; and inactivation of the disinfectant by blood proteins. The toxicity and materials' compatibility of any chemical may also be relevant, as may the numbers of any infectious agent present. A reduction in viability of 5 $\log_{10}$ is a generally accepted norm of disinfection, although Sattar and Springthorpe (1991) have suggested that a 3 $\log_{10}$ reduction is more appropriate for HIV, given its low concentration in its naturally occurring state. These reviewers also noted that, whereas HIV in a blood spill may be made safe by a comparatively low reduction, this will make little dent in a spill containing 10^{13} infectious HBV particles per ml.

An idea of how blood will affect disinfection in practice may be gained from work that explored the potential of household disinfectants to inactivate HIV in drug injectors' syringes (Flynn *et al.*, 1994). It demonstrates how, whilst HIV is innately sensitive to a wide variety of agents, it may resist inactivation when the test conditions closely replicate real-life situations. With HIV in 'reconstituted blood' (consisting of packed red cells and plasma, to overcome the problem of clotting) only beer, a cola drink and mineral water were completely inactive against HIV. Substances such as malt vinegar and lemon juice inactivated cell-free virus; cider vinegar, wine, vodka, 70% ethanol and 70% propanol inactivated cell-free virus and low numbers (10^4) of cell-associated virus. Household bleach (presumably sodium hypochlorite), distilled malt vinegar and liquid dish detergent (composition unstated but usually sodium alkyl-benzene sulphonate) were the most effective agents and inactivated cell-free virus and 10^4–10^6 cell-associated virus. Although these results may appear to be surprising they represent a hierarchy that takes into account not only the innate susceptibility of HIV to a substance and its strength, but also how they are able to penetrate proteinaceous matter. Concentrated alcohol solutions coagulate the protein in the outer layer of blood spills, forming an effective barrier to further penetration. This is not a novel observation with either alcohol or HIV; Hanson *et al.* (1989) tested chemicals against HIV suspended in 10% serum and dried on to surfaces and concluded that '70% industrial methylated spirit and ethanol were not suitable for surface disinfection of HIV', glutaraldehyde was effective, but is unacceptably toxic for general use on surfaces.

Choice of Disinfectants

The generally-accepted disinfectant of choice for the treatment of blood spills is hypochlorite, either as sodium hypochlorite (liquid bleach), or as sodium dichloroiso-cyanurate (NaDCC), a solid or powdered formulation. Both give hypochlorous acid (HOCl) in aqueous solution, the chemical species primarily responsible for microbicidal activity (Bloomfield and Miles, 1979). However, both these agents bleach fabrics and corrode metals. The oxidizing, and therefore disinfecting, capacity of hypochlorites is still expressed in the misleading form of parts per million of 'available chlorine' (ppm available chlorine) although this actually refers to oxygen loosely bound to chlorine. Hypochlorites need to be at a high concentration for the disinfection of blood spills – 10,000 ppm available chlorine (Department of Health: DH, 1990). This may create problems if domestic bleach is to be used as there is usually no stated concentration on the container and liquid hypochlorite will decay on storage.

NaDCC, on the other hand, is stable over long periods (Hoffman *et al.*, 1981). The main reason for using a high concentration of hypochlorite is that it reacts with, and is neutralized by, the protein in the blood, so as the chemical penetrates a blood spill it becomes progressively less concentrated and consequently less virucidal (van Bueren *et al.*, 1995). It is unlikely to penetrate clotted blood to any relevant extent. Thus there is a comparatively low quality assurance of complete decontamination by this process, depending on the extent of the spill and its physical state. Hypochlorites will release highly toxic gaseous chlorine if mixed with acids, NaDCC particularly so (sodium hypochlorite is usually in an alkaline solution and can withstand small amounts of acid; NaDCC solutions are at the pH of the water in which they are dissolved). To this has been added the general reluctance in many workplaces to use a chemical that will corrode metals and bleach fabrics. This has led to the view that '...thorough cleaning of the surface and wearing of gloves is more important in preventing infection than the use of disinfectants' (Ayliffe *et al.*, 1993).

PROCEDURE FOR CLEARING UP BLOOD SPILLS

Accidents, by their nature, are unexpected and therefore all staff should be aware of the action to be taken if a spill occurs. In establishments where blood spills are likely to present a particular hazard, e.g. in a clinical laboratory, a written *standard operating procedure* (SOP) should be established. All staff likely to come into contact with blood should be offered immunization against HBV infection (see Waldron, Chapter 15 and McCloy, Chapter 16 this volume) and they should receive training so that they feel confident to deal safely with any blood spill.

Blood spill clean-up kits, containing all necessary materials, should be prepared or purchased and be readily available in all areas where spills might occur (Fig. 17.1).

The worker who deals with the spill must wear gloves, and, if the spill is substantial, a disposable apron and protective footwear, all of which must be donned before any contact is made with the spill. Visors, for face and eye protection, should be worn if the spill is extensive. Respiratory protection from infectious agents should not be necessary.

Broken Glass and Other Sharps

As a penetrating skin injury is the primary mode of transmission of blood-borne viruses, it is imperative to protect the clearer against such injury. It is prudent to assume that there is always a possibility of a sharp object in the spill and to keep hands distanced from it. Ordinary gloves only offer protection against the most glancing of contact with sharp objects. Although latex examination-type gloves are most readily available, they are readily punctured. Thick latex (washing-up-type gloves) are more resistant but may reduce dexterity, increasing the risk of a sharps injury. For these reasons it is safest in all but the

Fig. 17.1. Spillage clean-up equipment (Source: CDC/NIH).

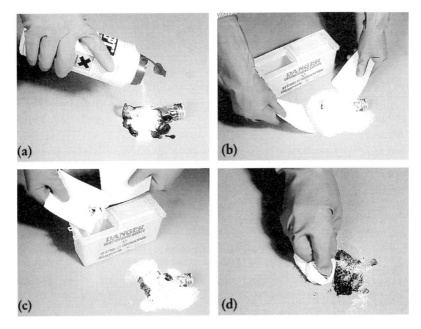

(a)

(b)

(c)

(d)

Fig. 17.2. Treatment of a spillage of body fluid, e.g. a container of blood. (a) Household gloves must be worn and a granular chlorine compound should be sprinkled over the area to inactivate any infectious agent present. (b) Any broken glass or pieces of container must be picked up using forceps or scooped onto strong cardboard material. (c) Inactivated blood and broken glass is then deposited into a sharps bin and the cardboard disposed of as clinical waste. (d) Any residual fluid should be wiped away using tissue, chlorine granules and water. The area should finally be cleaned with detergent solution. (Courtesy of the British Medical Association).

most minor of spills, to keep the gloved hands distanced from the spill at all times. There are several methods for achieving this. Practical methods are use of two rigid pieces of card (filing cards are generally available and are disposable) to scoop a spill solidified by a powder or a gelling agent, or contained within absorbent paper, for small spills; for larger spills a dustpan and brush, ideally disposable, can be used; or for very extensive spills, a mop and bucket.

The disinfectant should be given up to 30 minutes to act wherever possible and then the absorbed spill removed. This may clear the bulk of the spill but may leave a thin, contaminated residue. The whole of the area previously covered with the spill should be treated again with the disinfectant and then mopped up and dried. All broken glass and materials used to decontaminate the spill should be placed in a sharps container, the opening of which is wide enough to take the devices used to pick up the broken glass, and treated as other sharps waste.

A procedure for dealing with blood spillages is shown in Fig. 17.2.

REFERENCES

Ayliffe, G.A.J., Coates, D. and Hoffman, P.N. (1993) _Chemical Disinfection in Hospitals,_ 2nd edn. Public Health Laboratory Service, London.

Bloomfield, S.F. and Miles, G.A. (1979) The antibacterial properties of sodium dichloro-isocyanurate and sodium hypochlorite formulations. _Journal of Applied Bacteriology_ 46, 65–73.

Bond, W.W., Favero, M.F., Petersen, N.J. _et al._ (1983) Inactivation of hepatitis B virus by intermediate to high level disinfectant chemicals. _Journal of Clinical Microbiology_ 18, 525–528.

DH (1990) _Guidance for Clinical Health Care Workers; Protection against Infection with HIV and Hepatitis Viruses._ Recommendations of the Expert Advisory Group on AIDS. Department of Health. HMSO, London.

Flynn, N., Jain, S., Keddie, E.M. _et al._ (1994) _In vitro_ activity of readily available household materials against HIV-1: is bleach enough? _Journal of Acquired Immunodeficiency Syndrome_ 7, 747–753.

Hanson, P.J., Gor, D., Jeffries, D.I. _et al._ (1989) Chemical inactivation of HIV on surfaces. _British Medical Journal_ 298, 862–864.

Heptonstall, J., Gill, O.N., Porter, K. _et al._ (1993) Health care workers and HIV: surveillance of occupationally acquired infection in the United Kingdom. _Communicable Disease Report Review_ 3, R146–153.

Hoffman, P.N., Coates, D. and Death, J.E. (1981) The stability of sodium hypochlorite solutions. In: Collins, C.H., Allwood, M.C., Bloomfield, S.F. _et al. Disinfectants. Their Use and Evaluation of Effectiveness._ Society for Applied Bacteriology Technical Series 18. Academic Press, London.

Kobayashi, H., Tsuzuki, M., Koshimizu, K. _et al._ (1984) Susceptibility of hepatitis B virus to disinfectants or heat. _Journal of Clinical Microbiology_ 20, 214–216.

Moudgil, Y. and Daar, E.S. (1993) Infectious decay of human immunodeficiency virus type 1 in plasma. _Journal of Infectious Disease_ 167, 210–212.

Prince, D.L., Prince, H.N., Thraenhart, O. *et al.* (1993) Methodological approaches to disinfection of human hepatitis virus. *Journal of Clinical Microbiology* 31, 3296–3304.

Resnick, L., Veren, K., Salahuddin, M.S. *et al.* (1986) Stability and inactivation of HTLV-III/LAV under clinical and laboratory conditions. *Journal of the American Medical Association* 255, 1887–1891.

Sattar, S.A. and Springthorpe, V.S. (1991) Survival and disinfectant inactivation of the human immunodeficiency virus: a critical review. *Reviews of Infectious Diseases* 13, 430–437.

Sattar, S.A. and Springthorpe, V.S. (1994) Inactivation of the human immunodeficiency virus: an update. *Reviews in Medical Microbiology* 5, 139–150.

Spire, B., Dormont, D., Barre-Sinoussi, F. *et al.* (1985) Inactivation of lymphadenopathy-associated virus by heat, gamma rays and ultra-violet light. *Lancet* i, 188–189.

van Bueren, J., Simpson, R.A., Salman, H. *et al.* (1995) Inactivation of HIV-1 by chemical disinfectants: sodium hypochlorite. *Epidemiology and Infection* 115, 567–579.

CHAPTER *18*
Disposal of waste blood and blood-contaminated waste

C.H. Collins and D.A. Kennedy

Waste is any material that is no longer useful to the producer and which will be disposed of, or stored for treatment or eventual disposal. Materials which are intended for re-use or recycling are technically not waste.

Waste blood and any waste materials containing, or which have contained, blood are regarded as hazardous waste and are collectively known as 'clinical waste' in the UK and as 'medical waste' or 'biohazardous waste' in the US, whether or not it is contaminated with pathogens. Handling and disposal of it is subject to legal requirements:

- workers, the community and the environment must be protected from any harm that may result from exposure to it; some pathogens remain viable for long periods;
- its nature must be defined;
- its generation, segregation, storage, transport and disposal are subject to control.

DEFINITIONS

In the UK, under the *Collection and Disposal of Waste Regulations 1988*), made under the *Control of Pollution (Amendment) Act 1989*, clinical waste is Controlled Waste. Most of this legislation re-appears, with minor variations, in the *Controlled Waste Regulations 1992* which also categorizes other wastes which arise in hospitals as Industrial Waste, e.g. laboratory waste that arises in premises occupied by scientific research associations.

Under these regulations clinical waste is defined as:

1. Any waste which consists wholly or partly of human or animal tissue, blood or other body fluids, excretions, drugs or other pharmaceutical products, swabs or dressings or syringes, needles or other sharp instruments being waste which, unless rendered safe, may prove hazardous to any persons coming into contact with it; and
2. Any other waste arising from medical, nursing, dental, veterinary, pharmaceutical or similar practice, investigation, treatment, care, teaching or research

©CAB INTERNATIONAL 1997
Occupational Blood-borne Infections (eds C.H. Collins and D.A. Kennedy)

or the collection of blood for transfusion, being waste which may cause infection of any person coming into contact with it.

The Health and Safety Commission (HSAC, 1992) recognizes five categories – Groups A–E – of clinical waste. Three of these include blood and material containing it:

1. *Group A.* All human tissue including blood (whether infected or not), animal carcases and tissue from veterinary centres, hospitals and laboratories, and all related swabs and dressings.
2. *Group B.* Discarded syringe needles. Broken glass or any other contaminated sharp instruments or items.
3. *Group C.* Microbiological cultures and potentially infected waste from pathology departments (laboratory and post-mortem rooms) and other clinical or research laboratories.

In the United States medical, or biohazardous, waste is Regulated Waste under the *Medical Waste Tracking Act 1988* and the US Occupational Health and Safety Administration rules (OSHA, 1991). Of the seven categories of medical waste defined, four (numbers 2, 3, 4 and 6) are concerned with human waste blood and blood-contaminated materials:

2. Human pathological wastes: including tissues, organs, and body parts and body fluids that are removed during surgery or autopsy, or other medical procedures, and specimens of body fluids and their containers.
3. Human blood and blood products: (a) liquid waste human blood; (b) products of blood; (c) items saturated with and/or dripping with human blood or that are caked with dried human blood; including serum, plasma and other blood components, and their containers, which were used or intended for use in either patient care, testing and laboratory analysis or the development of pharmaceuticals. Intravenous bags are also included in this category.
4. Sharps: that have been used in animal or human patient care or treatment or in medical, research or industrial laboratories, including hypodermic needles, syringes (with or without the attached needle), Pasteur pipettes, scalpel blades, blood vials, needles with attached tubing.
6. Isolation wastes: biological waste and discarded materials contaminated with blood, excretion, exudates, or secretions from humans who are isolated to protect others from certain highly communicable diseases.

There is a difference between the hazards presenting to workers and the general public by blood-stained or blood-contaminated waste that is generated during clinical and emergency activities, and that generated in laboratories. The laboratory investigations designed to identify pathogens derived from blood often increase their numbers considerably. For this reason, clinical waste *sensu stricto* and laboratory waste are considered separately below.

There is also a curious anomaly in respect of used sanitary pads and tampons, very large numbers of which are discarded daily. Although the incidence of, e.g. hepatitis B virus, is much the same in both health-care premises and the community, these articles are usually, and for convenience, treated as clinical waste if they are generated in health-care premises. But outside such premises, e.g. in the home or workplace, they are discarded into the municipal waste stream, although some local authorities do make special provision for collection from public toilets in towns and motorway service stations.

PREMISES GENERATING CLINICAL/BIOHAZARDOUS WASTE

Although the various regulations and guidelines stress the 'clinical' origin of this waste, implying that it arises from health care, a considerable proportion is generated in other areas. Table 18.1 lists some, but probably not all, of these.

WASTE LIQUID BLOOD

Blood not in Containers

Waste liquid blood is generated in health care premises, during, for example, surgery and dialysis, as well as in post-mortem rooms and funeral premises (e.g. in embalming). It is clearly not possible to place this liquid in containers specified by the Health and Safety Commission (HSAC, 1992). There are two options for dealing with it: (i) treating it with a chlorine-releasing chemical such as sodium dichloroisocyanate (NaDCC) followed by absorption by cellulose, cotton wool or similar materials – it may then be treated as solid waste; or (ii) discharge to the public sewer (via the sluice), provided that the sewerage authority gives consent. In the US the Environmental Protection Agency (US EPA, 1990) considers that such waste entering a sanitary sewage system, leading to a public operated pre-treatment works, is domestic waste and is not regulated. No adverse public health incidents due to such bulk blood disposal have been reported (Turnberg, 1996).

Table 18.1. Sources of potentially blood-contaminated clinical waste.

Hospitals	Blood transfusion centres
Clinics	Chiropodists
Health centres	Tattooists
General practitioner centres and clinics	Cosmetic piercers
Dental hospitals and surgeries	Acupuncturists
Biomedical and public health laboratories	Funeral undertakers and morticians
Residential nursing homes	Private dwellings where medical treatment is carried out
Custodial services	Public transport
	'Squats'

Adapted from Collins and Kennedy (1993).

Transfusion Packs

In the UK unused transfusion blood is not waste. Packs are returned to the Blood Transfusion Centre, for processing into blood products, e.g. freeze-dried plasma. Elsewhere they may be incinerated.

Laboratories

Blood samples submitted to laboratories rarely exceed 10 ml in volume. Disposal procedures for sample residues and their containers are described under 'laboratory waste', below.

CLINICAL AND LABORATORY WASTE

The wastes listed by the Health and Safety Commission (HSAC, 1992) are not as detailed as those of the US authorities insofar as blood-stained, etc. waste is concerned. Table 18.2, adapted from Collins and Kennedy (1993), and including items from the US regulations, is offered as a guideline, although it is not all-inclusive.

Segregation for Disposal

Clinical waste must be segregated from other kinds of waste at the point of generation.

Table 18.2.   Potentially blood-contaminated clinical and biomedical laboratory waste.

'Soft waste'
Discarded sheets, pillow cases, towels, washcloths, gowns, drapes, gloves, swabs, sponges, soiled dressings, diapers (napkins), maternity pads, sanitary towels and tampons, placentas and other soft tissues

'Hard waste'
Blood transfusion packs and tubing, drainage tubes and receptacles, instruments other than sharps, amputated limbs, teeth

Sharps
Hypodermic needles and syringes, suture and other needles, scalpel and other blades, scissors, forceps, suture cutters

Laboratory waste
Blood specimen containers and blood residues, cuvettes, test containers and instruments used in blood tests, blood cultures, lancets used in blood sampling, swabs and paper tissues used in instrument clean-up and surface wiping operations

Adapted from Collins and Kennedy (1993).

Hard and soft waste

At present, all this waste should be placed in colour-coded bags or other colour-coded containers. In the UK, for example, yellow-coloured plastic bags are used, as specified by the Health and Safety Commission (HSAC, 1992) printed with the words 'Waste for Incineration Only' (Fig. 18.1). There is no international agreement, however, about colour coding or types of containers. In the US, red coloured plastic bags are used. There is a movement in the European Union for plastic bags to be replaced by rigid disposable containers such as those used in some areas for amputated limbs, etc.

Clinical waste must not be placed in paper bags, the black plastic bags specified for household waste, or, except by prior arrangement, the blue or transparent plastic bags specified for autoclaving. Another type of colour-coded plastic bag, yellow with a black band, is intended for certain wastes which could be disposed of by landfill where permitted by the Waste Authority, but when implemented in the UK, the European Union Landfill Directive (Commission of the European Communities, 1991) will ban the landfilling of clinical waste.

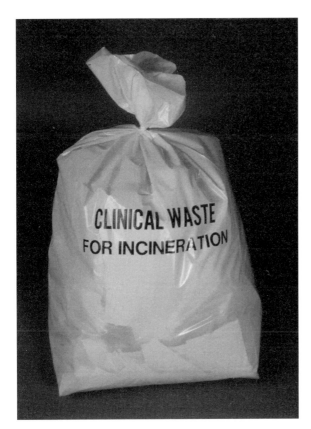

Fig. 18.1. Yellow-coloured plastic bags used for clinical waste in the UK. (Courtesy of the British Medical Association.)

Sharps

These items must be placed in rigid plastic boxes made to the specifications of the appropriate health department. In the UK there is a British Standard (BSI, 1990). Most health authorities purchase sharps containers that conform to this standard, although according to the British Medical Association (BMA, 1990) 'Regrettably, not all containers currently used for sharps disposal meet these requirements. The British Medical Association deplores the continued use of such substandard containers, which pose a real hazard to health care workers'. (See also Waldron, Chapter 15 this volume.) Table 18.3 summarizes appropriate design features of these containers, examples of which are shown in Fig. 18.2. Sharps must *not* be placed in plastic bags of any kind, nor in 'makeshift' containers such as cans or plastic bottles except in cases of dire emergency. Nor should sharps boxes be placed in plastic bags.

Hypodermic needles should not be removed from syringes; the syringe, with needle in place, should be placed in the sharps container. This minimizes the risk of needlestick injuries associated with the removal of needles, and, if the syringe has been used to collect blood, avoids the dispersion of droplets of blood from the syringe nozzle and needle.

The boxes should not be more than three-parts filled. There should be a mark to indicate this level.

Removal and ultimate disposal

Local arrangements for the temporary storage and the removal of yellow bags and sharps containers, by contractors or local waste disposal authorities, are described by Collins and Kennedy (1993). Instructions should be set out in written documents according to the requirements of the Health and Safety Commission (HSAC, 1992) and the Department of the Environment (DE, 1992).

Table 18.3. Design features for sharps disposal containers.

1.	The containers should be puncture-resistant and leakproof, even when knocked over
2.	There should be a handle that does not interfere with the easy insertion of needles and syringes
3.	The predominant colour should be yellow
4.	It should be labelled 'Contaminated Sharps' with instructions that it be destroyed by incineration
5.	The closure should be secure
6.	There should be a prominent horizontal line at the three-quarters full level and instructions that the container should not be filled above that line
7.	It should be possible to dispose of needle and syringe as one unit
8.	There should be no device for the removal of needles from syringes
9.	It should be possible to dispose of needle and syringe with one hand

From Collins and Kennedy (1993); adapted from BSI (1990) and Gwyther (1990).

Fig. 18.2. Various kinds of sharps containers. (Courtesy of M.Y. Cousson, Institut National de Recherche et Sécurité, Paris.)

Regulations in most developed countries require that clinical waste is incinerated. In the UK many hospital incinerators that did not satisfy the current legal requirements have been decommissioned and clinical waste is collected and incinerated by specialist contractors who have the appropriate engineering facilities and expertise.

Documentation, transport and incineration are subject to strict legal requirements. These are too detailed to be given here but are given by Collins and Kennedy (1993) and the former London Waste Regulation Authority (1994–5), now subsumed into the Department of the Environment.

LABORATORY WASTE

The blood-related components of laboratory waste are the containers and their residues after samples have been removed for biochemical, haematological, microbiological and serological examinations, the instruments used for transfer, separation and testing, blood culture bottles, and unfixed tissues that are not required for histological investigations. These are listed in Table 18.4, which, however, is not exhaustive.

Although it is unlikely that viruses and endoparasites will increase in numbers in blood samples during final disposal many bacteria will multiply and increase the risk of infection to those who might be exposed.

Table 18.4.   Disposable laboratory equipment associated with residual blood.

Samples tubes, bottles and other containers of residual blood, CSF, amniotic,
  pericardial, pleural and synovial fluids, and any other residues that are visibly
  contaminated with blood
Cultures made from blood or blood-containing material
Pipettes used for transferring blood, serum, plasma and blood products
Slides with films or smears of blood
Paper towels and swabs used to wipe equipment used in blood examinations and for
  dealing with blood spills
Disposable gloves, aprons and gowns
Syringes and sharps
Commercial and other control sera and quality assurance samples

Adapted from Collins and Kennedy (1993).

Segregation for Disposal

There is no real need to separate these blood sample containers and residues
from other infected laboratory waste (except sharps and certain glass items, see
below). At present the waste should be placed in plastic bags which are light
blue in colour or transparent with a light blue inscription 'Waste for Autoclaving', as specified by the Health and Safety Commission (HSAC, 1992). These
bags should be supported in metal or autoclavable plastic boxes to prevent
dispersal of contents if they burst. Before removal for disposal the bags should
be closed with, for example, Qik Ties.

Sharps, and glass, e.g. slides, glass pasteur pipettes, should be placed in
disinfectant in bench discard jars or in sharps containers, not in plastic bags (see
Collins, 1993).

Making Safe

According to the Health and Safety Commission (HSAC, 1992): 'It is essential
that all infected waste arising from work in clinical laboratories is made safe to
handle by autoclaving before disposal by incineration', and continues: 'If it is
not reasonably practicable to autoclave waste it should be sealed and secured in
yellow, strong, leakproof containers and then transported by special arrangement directly to the incinerator'.

This echoes another publication, that of another Health Services Advisory
Committee, concerning safety in clinical laboratories (HSAC, 1991).

Infected laboratory waste, however, will contain not only blood and any
microorganisms cultured from it, but all other cultures containing large numbers of pathogens. It is clearly inadvisable to transport such infected waste on
the public highway, where it may be dispersed, with a risk to the community,
should the vehicle transporting it be involved in an accident. It is unlikely that
any 'special arrangements' required by the Health and Safety Commission (see
above) will differ from the general arrangements made for other clinical waste.

The Advisory Committee on Dangerous Pathogens (ACDP, 1994), however, simply states, in respect of pathogens that: 'All waste materials must be
made safe before disposal or removal from the laboratory'.

We consider that the '. . . if it is not reasonably practicable . . .' escape
clauses in the Health Services Advisory Commissions booklets (HSAC, 1991,
1992) should be deleted or ignored because it is inconceivable that any prop-
erly-managed laboratory that deals with infectious materials (whether 'accredited' or not) should be without an autoclave and therefore not be capable of
making safe the infected waste that it generates (Collins, 1994).

Fortunately, the latest guidance from the Advisory Committee (ACDP,
1995, 1996) *recommends* that laboratory waste be autoclaved.

Autoclaving procedures and control of sterilization cannot be discussed
here but may be found in most microbiology textbooks (see, for example,
Collins, 1993; Collins *et al.*, 1995). An important feature that must be mentioned here is that the ties on plastic bags must be removed and the bags opened
before they are placed in the autoclave, otherwise steam penetration may be
insufficient to ensure sterilization.

Ultimate Disposal of Infected Laboratory Waste

Once laboratory waste, in its bags, discard jars or sharps containers, has been
made safe it is not longer legally regarded as clinical waste or a biohazard. For
aesthetic reasons, however, and to avoid possible public concern, should it still
be perceived to be a biohazard, it is desirable that it be incinerated, not land-
filled. It is usual to place the blue bags in standard yellow bags and to consign
them and laboratory sharps containers to the clinical waste store to await
collection and incineration.

REFERENCES

ACDP (1994) *Categorization of Pathogens According to Hazard and Categories of
Containment.* Advisory Committee on Dangerous Pathogens, 3rd edn. HMSO,
London.

ACDP (1995) *Protection against Blood-borne Infections in the Work Place: HIV and
Hepatitis.* Advisory Committee on Dangerous Pathogens. HMSO, London.

ACDP (1996) *Categorization of Biological Agents According to Hazard and Categories
of Containment. Advisory Committee on Dangerous Pathogens,* 4th edn. HSE
Books, Sudbury.

BMA (1990) *A Code of Practice for the Safe Use and Disposal of Sharps.* British Medical
Association, London.

BSI (1990) *BS 7320: Specifications for Sharps Containers.* British Standards Institute,
London.

Collection and Disposal of Waste Regulations 1988. SI 819, HMSO, London.

Collins, C.H. (1993) *Laboratory Acquired Infections,* 3rd edn. Butterworth-Heinemann,
Oxford.

Collins, C.H. (1994) Opinion. Infected laboratory waste. *Letters in Applied Microbiology* 19, 61–62.

Collins, C.H. and Kennedy, D.A. (1993) *The Treatment and Disposal of Clinical Waste.* H & H Scientific Ltd., Leeds.

Collins, C.H., Lyne, P.M. and Grange, J.M. (1995) *Collins and Lyne's Microbiological Methods,* 7th edn. Butterworth-Heinemann, Oxford.

Control of Pollution (Amendment) Act 1989. HMSO, London.

Controlled Waste Regulations 1992. SI 588. HMSO, London.

Commission of the European Communities (1991) Proposed Landfill Directive (91/C 190/01). European Commission, Brussels.

DE (1992) *Clinical Wastes: A Technical Memorandum.* Waste Management Paper No. 25. Department of the Environment, HMSO, London.

Gwyther, J. (1990) Sharps disposal containers and their use. *Journal of Hospital Infection* 15, 287–294.

HSAC (1991) *Safe Working and the Prevention of Infection in Clinical Laboratories.* Health Services Advisory Committee, Health and Safety Commission. HMSO, London.

HSAC (1992) *Safe Disposal of Clinical Waste.* Health Services Advisory Committee. Health and Safety Commission. HMSO, London.

London Waste Regulation Authority (1994) *Guidelines for the Segregation, Handling, Transport and Disposal of Clinical Waste,* 2nd edn. LWRA, London (now in Department of the Environment).

Medical Waste Tracking Act 1988. United States Congress, USC 6992. Government Printing Office, Washington.

OSHA (1991) US Occupational Health and Safety Administration. Occupational exposure to blood-borne pathogens. Final rule, *Federal Register* **56**, 64175–64182.

Turnberg, W.L. (1996) *Biohazardous Waste. Risk Assessment, Policy and Management.* John Wiley, New York.

US EPA (1990) *First Interim Report to Congress – Medical Waste Management in the United States.* Environmental Protection Agency, EPA/530-SW-90–051A. Government Printing Office, Washington DC.

CHAPTER 19
Review of the OSHA Regulations on blood-borne infections

D.L. Hunt

The Occupational Safety and Health Administration (OSHA) has been in existence in the United States since 1970 to ensure that employers provide to each of their employees a place of employment which is 'free from recognized hazards that are causing or likely to cause death or serious physical harm'. Historically, the health-care industry was not considered a high risk business, and, thus, a specific standard had not been considered by OSHA for biological safety concerns in this setting prior to 1987.

Although the transmission of blood-borne pathogens (e.g. hepatitis B virus) in the health-care workplace had been well documented, health-care facilities relied on guidelines from the Centers for Disease Control and Prevention (CDC) for safety recommendations specific for their workplace setting. Within one year of the first recognized cases of the newly defined disease, acquired immunodeficiency syndrome (AIDS), the CDC issued guidelines for health-care facilities to prevent occupational transmission of blood-borne pathogens such as the hepatitis B virus (HBV) and the agent causing the new disease, AIDS (CDC, 1982). Updates from the CDC were issued in 1983 and 1985 (CDC, 1983, 1985a, b) that re-emphasized precautions recommended previously for patients known to be infected with hepatitis B. In 1986, the CDC issued its first agent summary statement for work with human T-cell lymphotropic retrovirus III (HTLV-III) in the laboratory setting (CDC, 1986). As information about the prevalence and transmission of the human immunodeficiency virus (HIV) was gathered, guidelines regarding occupational safety precautions evolved. A major recommendation was issued by the CDC in 1987 that is now referred to as Universal Precautions (CDC, 1987a). Occupational infections of health-care workers resulting from percutaneous injuries, nonintact skin, and mucous membrane exposures were reported from patient sources with unknown infection status at the time of the exposures (CDC, 1987b). The major premise of Universal Precautions that was vastly different from previous guidelines involved the use of standard precautions with *all* blood or body fluids, rather than special precautions used only with blood or body fluids from *known* infected patients. This Universal Precautions concept formed the basis for all subsequent guidelines and regulations from other professional organizations.

©CAB INTERNATIONAL 1997
Occupational Blood-borne Infections (eds C.H. Collins and D.A. Kennedy)

In the meantime, OSHA was petitioned by various labour unions representing health-care employees to develop an emergency temporary standard to protect employees from occupational exposure to blood-borne diseases. In response, the US Department of Labor and the US Department of Health and Human Services issued a Joint Advisory Notice in 1987 entitled, *Protection Against Occupational Exposure to Hepatitis B Virus (HBV) and Human Immunodeficiency Virus (HIV)* (US Department of Labor, 1987), which basically reiterated the CDC Universal Precautions guidelines. In 1989, OSHA published a proposed standard, followed by the final standard in December, 1991 entitled, *Occupational Exposure to Bloodborne Pathogens* (OSHA, 1991). In these documents, OSHA issued its conclusion that certain employees face significant health risks as the result of occupational exposure to blood and other potentially infectious materials because of the risk of HBV and HIV occupational transmission. The agency had also concluded that the risk could be significantly decreased with the use of engineering and work practice controls, personal protective clothing and equipment, training, medical follow-up of exposures, vaccinations, and other provisions. It was estimated that compliance with the Standard could prevent approximately 200 deaths and 9000 infections per year from HBV alone. The agency also estimated that compliance with the Standard could reduce needle/sharp exposures by approximately 50% and mucous membrane and open wound exposures by 90% (OSHA, 1991).

OVERVIEW

The provisions of the Blood-borne Pathogen Standard that concern themselves with specific methods for protecting employees from blood-borne pathogen exposures are consistent with the CDC (1987a) Universal Precautions guidelines. Health-care facilities with policies already in place to comply with the CDC Guidelines had little trouble complying with the methods required by OSHA. Some of the bureaucratic aspects of the Standard, such as record keeping and documentation, are more problematic for some institutions. An outline of the requirements of the Standard can be found in Table 19.1, and will be reviewed in this chapter.

OSHA issued the Standard as a 'performance' standard. That is, the employer has a mandate to develop an exposure control plan to provide a safe work environment but is allowed some flexibility in accomplishing this goal. The Standard allows the employer to provide a combination of engineering controls, work practices, and personal protective equipment (PPE) to protect employees from blood-borne pathogen exposure.

OSHA also recognizes that employees in HIV/HBV research laboratories and production facilities have a higher level of risk of infection after an exposure because of concentrated viral preparations. The Standard requires special practices, facility design, and additional training for such laboratories that are consistent with the CDC/NIH laboratory biosafety guidelines for Biosafety Levels

Table 19.1. Major provisons of OSHA blood-borne pathogen standard requirements (OSHA, 1991).

I. Exposure control plan (ECP) the establishment's policy for implementation of procedures relating to control of infectious disease hazards

II. Components of the ECP:
  A. Exposure Risk Determination for All Employees
  B. Control Methods
    1. Universal Precautions
    2. Engineering Controls
    3. Work Practice Controls
    4. Personal Protective Equipment (PPE)
    5. Additional requirements for HIV/HBV research laboratories/production facilities
  C. Housekeeping Practices
  D. Laundry Practices
  E. Regulated Waste Disposal
  F. Tags, Labels and Bags
  G. Training Programmes (additional training for HIV/HBV research/production)
  H. Hepatitis B Vaccination Programme
  I. Post-exposure Evaluation/Follow-up
  J. Record Keeping (Medical, Training)

III. Administrative Controls (Support of ECP, Exposure Determination, Monitoring)

---

2 and 3 (Richmond and McKinney, 1993). Because clinical and diagnostic laboratories may have exposures to human blood, other body fluids, primary human tissue, or cell cultures, they must follow the general requirements of the Standard that are comparable to Biosafety Level 2 practices in the CDC/NIH guidelines.

SCOPE

The Standard applies to all employees designated by the employer as having direct occupational exposure or whose jobs have the likelihood of exposure to blood or other potentially infectious material. Several important definitions contribute to this designation under the Standard:

Occupational exposure

This is 'reasonably anticipated' skin, eye, mucous membrane, or parenteral contact with blood or other potentially infectious material (OPIM) that may result from the performance of an employee's duties. This potential for exposure is evaluated without regard to the use of personal protective equipment such as gloves.

Considerations. The Standard covers employees only with 'reasonably anticipated' exposures in the workplace. Therefore, an employee who may respond to an unanticipated situation may not necessarily be covered. For example, an office clerk who provides cardiopulmonary resuscitation to a client in the hospital lobby would not be covered, unless the clerk is a designated emergency responder to such emergencies, and can reasonably anticipate such actions because of her job responsibilities.

Other potentially infectious materials (OPIM)

Included under this definitional are semen, cerebrospinal fluid (CSF), vaginal secretions, synovial, pleural, pericardial, peritoneal, and amniotic fluids, saliva in dental procedures, or any body fluid visibly contaminated with blood, and any body fluid whose identity is mixed or unknown; any unfixed tissue or organ (other than intact skin) from a human; and HIV-containing cells or tissue cultures, organ cultures, and HIV- or HBV-containing culture medium or other solutions; and blood, organs, or other tissues from experimental animals infected with HIV or HBV.

Considerations. The fluids designated by the Standard are those that have been recognized by the CDC as directly linked to the transmission of HBV and/or HIV. Rather than try to distinguish those fluids for which precautions are necessary versus those for which no requirements exist, institutions may choose to designate *all* body fluids as potentially infectious, such as with the popular 'Body Substance Isolation' practices.

The exposure determination for all employees is one of the most difficult requirements of the Standard. Employers must determine which jobs and/or tasks may expose employees to blood or OPIM. This can be done by either classifying all employees under a job category as having anticipated occupational exposures, or listing the jobs in which some employees have occupational exposures combined with a list of the tasks or procedures in which exposure occurs. For example, all nurses could be included under the provisions of the Standard, or, some nurses with direct patient contact would be included, but those with only administrative responsibilities would not. OSHA allows flexibility with this classification, as long as all affected employees are included. Institutions must keep in mind, however, that once employees are designated as covered by the Standard, they must provide them with hepatitis B vaccination free of charge, and must document annual blood-borne pathogen training. The simplified approach of including all employees in broad job classifications could result in unwarranted costs and record keeping.

Considerations. There may be other workers on the premises who are not considered 'employees', such as volunteers, contract workers, students, etc. However, OSHA states that an employer must provide a safe workplace, and might be cited for 'standards violations to which employees of other employers on their premises are exposed, to the extent that they control the hazard'. In other words, communication of hazards, availability of PPEs, and engineering controls are important for all 'workers' in the institution. An employer is not responsible for vaccinations and required training of employees of another employer. Volunteers and students must be evaluated by the institution for exposure potential, and safety standards equivalent to employees with the same exposure potential should be instituted.

The types of employment potentially covered under the Standard range from the traditional health-care setting to non-health-care such as police, laundry workers, funeral service workers, or housekeepers. Any place of employment that has at least one employee with occupational exposure must establish an Exposure Control Plan (ECP) that contains the exposure determinations discussed above, the methods of compliance with the safety requirements of the Standard, and the post-exposure evaluation procedures. Properly constructed ECPs could be only one page for some facilities with only a few covered employees, or quite voluminous for large medical centres. OSHA requires only that the plan provides the required elements, and that it be accessible to all employees. The fundamental premise of this exercise is to provide the employee with information regarding the hazards in his/her workplace and what the employer is doing to control exposure to them.

METHODS OF CONTROL

A fundamental requirement of the OSHA Blood-borne Pathogen Standard is that the commonsense practices of CDC's Universal Precautions be observed. In conjunction with these guidelines, OSHA approaches, exposure reduction with a hierarchy of control methods listed in order of desirability: engineering controls, work practice controls, and personal protective equipment (PPE). Although this approach is common in hazard abatement in industry, it is a different approach for the health-care setting, which relies more on work practices and PPE as recommended by the CDC guidelines.

Engineering controls eliminate the risk of exposure to the employee without reliance on employee efforts. Examples are microbiological safety cabinets for the containment of splashes, needleless intravenous (IV) ports or self-sheathing needles. Next in the hierarchy are work practice controls to reduce exposures by altering the manner in which a task is performed. Examples are policies to prohibit the re-capping of needles, prompt removal of needle boxes before they are two-thirds full, and decontamination of blood spills. Where occupational exposure remains after the institution of engineering and/ or work practice controls, personal protective equipment is used.

Engineering Controls

In recognizing that human behaviour is inherently less reliable than mechanical controls, OSHA advocates using available technology to remove the potential for employee exposure. This is particularly relevant to the prevention of sharps-related injuries. OSHA admits that 60% of needle injuries would be unaffected by improved work practices and PPE (OSHA, 1991). However, the use of safer needle devices as engineering controls has resulted in a dramatic 93% reduction in sharps-related injuries in one study (Chiarello, 1992). Such devices include recessed needles, needleless IV ports, self-sheathing needles and scalpels, or blunt suture needles. In a Safety Alert, the US Food and Drugs Administration (US FDA, 1992) described the following characteristics of devices that have the potential to reduce the risk of needlestick injuries:

- A fixed safety feature to provide a barrier between the hands and the needle after use; this should allow or require the worker's hands to remain behind the needle at all times.
- The safety feature is an integral part of the device, not an accessory.
- The safety feature is in effect before disassembly and remains in effect after disposal.
- The safety feature is as simple as possible and requires little or no training to use effectively.

> *Considerations.* Protective needle devices are not considered to be mandatory by the OSHA Standard, but need to be evaluated by each institution as an effective means of complying with the intent of the Standard. Many medical centres commenced using needleless or recessed needle systems after the FDA issued the Safety Alert in 1992 and urged that such systems replace hypodermic needles for accessing IV lines.

Other engineering controls that might be used are microbiological safety cabinets to isolate the hazard from the worker, puncture-resistant needle boxes, needle recapping devices and readily accessible hand washing facilities. The Standard requires the routine maintenance and replacement of these controls as is necessary to assure their effectiveness.

Work Practice Controls

Work practice controls reduce the likelihood of exposure by altering the manner in which a task is performed, and is based on employee behaviour rather than on the installation of a physical device such as an engineering control. These two control methods frequently work in tandem because it is often necessary to use appropriate work practices to ensure the correct operation of a piece of equipment. For example, puncture-resistant, leak-proof needle boxes are protective as engineering controls but need to be conveniently located so that employees may immediately dispose of needles rather than recap them for transport to a distant needle (sharps) container.

The OSHA Standard requires many work practices that are considered to be standard safety or infection control practices; these are listed in Box 19.1. Some work practices are notable and are reviewed below.

Sharps handling

In recognizing that unsafe work practices have led to employees suffering sharps injuries, OSHA requires certain practices for handling sharps. For example, recapping needles has resulted in a large proportion of the injuries among health-care workers. The Standard therefore prohibits recapping or removal of needles *unless* there is no alternative in a specific medical procedure. In those cases needles may be recapped by a mechanical device or by an approved single-handed scoop technique.

> *Considerations.* When procedures are identified that require recapping, such as maintaining needle sterility for injecting incremental dosages of medication to one patient, the procedures and the method(s) used to recap safely should be included in the Exposure Control Plan. Note also that this applies only to *contaminated* needles, so those that are used to withdraw sterile fluids for injections are not covered by this requirement.

Box 19.1. Basic safety/infection control practices required by OSHA Blood-borne Pathogen Regulations (OSHA, 1991).

- Employees must wash their hands when contaminated or after removal of gloves and other PPE
- Eating, drinking, smoking, applying cosmetics or lip balm, or handling contact lenses are forbidden in work areas
- No food or drinks are kept in refrigerators, countertops, shelves, etc. where blood or OPIMs are present
- All procedures involving blood or OPIMs are performed in such a manner as to minimize splashing, spraying, spattering and generation of droplets
- Mouth pipetting or suctioning of blood or OPIMs is prohibited
- The worksite is maintained in a clean and sanitary condition

In general, sharps must be handled as little as possible before disposal, i.e. not broken or bent. Since needles must be disposed of as soon as possible after use in a puncture-resistant container there is no need to 'render the needle inoperable' as was the usual practice in past years. Contaminated broken glass-ware must no longer be directly picked up by hand, but by mechanical means, such as tongs, forceps, or dustpan and brush. In HIV/HBV research laborato-ries OSHA restricts the use of needles to parenteral injection and aspiration of fluids from laboratory animals and diaphragm bottles. The Standard also requires that needle-locking syringes or disposable syringe-needle units be used in these situations.

Handwashing
Handwashing has long been one of the basic work practices for preventing transmission of infectious diseases, generally as a means of infection control of nosocomial infections. However, proper attention to handwashing will prevent inadvertent transfer of infectious material from the hands to the eyes, nose or mouth of the employee. Under the Standard, OSHA requires that employers *ensure* that hands are washed when they become contaminated or after removal of protective equipment, i.e. every time gloves are removed. Provisions are made for the use of antiseptic hand cleansers in situations where there are no handwashing facilities, e.g. in emergency vehicles.

Secondary workplace exposures
Many of the other required work practices involve the prevention of environ-mental contamination with subsequent transmission of blood-borne pathogens. Such practices include the transport of infectious materials, decontamination, waste disposal, and basic safety practices such as prohibiting mouth pipetting, eating, drinking, smoking, applying cosmetics or handling contact lenses in workplaces where blood or OPIM are handled.

Safe transport of specimens or infectious materials within the laboratory, the facility, or to other institutions can minimize the potential for accidental spills or injuries. OSHA requires that specimens be packaged in order to pre-vent leakage during transport, i.e. in a closed, leak-proof primary container. In case the primary container becomes contaminated, it should be placed in a secondary container that prevents leakage.

Contaminated equipment that is serviced or shipped from the facility must be decontaminated as much as is feasible. Otherwise it is the employer's respon-sibility to inform the affected employees, the servicing representatives, or man-ufacturers so that they can take appropriate precautions.

Care must also be taken in the handling and transport of contaminated laundry. OSHA requires that such laundry be handled as little as possible with a minimum of agitation, and bagged (not sorted or rinsed) at the source. As with the transport of other potentially infectious material, OSHA requires that laun-dry be placed in a container that prevents leak-through or leakage of fluids to the exterior. Whether the laundry needs to be labelled as biohazardous depends on the work practices of the laundry facility. If the laundry is shipped to an

> *Consideration.* OSHA regulations do not mandate labelling or colour-coding specimens if they are handled only within the facility, a policy implementing Universal Precautions is in effect, and the contents of the containers are recognizable as human material. If the specimens are contained within a secondary container and the specimens are not visible or recognizable as human material the container must be labelled or colour coded. OSHA requires labelling of specimens if they are shipped from one facility to another. Facilities in the US must also comply with other shipping regulations (i.e. those of the Centers for Disease Control and Prevention, the Department of Transportation, the International Air Transport Association) for packaging potentially infectious materials.

outside facility which does not use Universal Precautions to handle it, the facility generating it must place the contaminated laundry in bags or containers which are labelled with the biohazard symbol, or are colour coded.

Besides routine housekeeping standards OSHA requires that surfaces and equipment be decontaminated with 'an appropriate disinfectant' if it is contaminated and at the completion of procedures.

> *Consideration.* An appropriate disinfectant is not defined by OSHA. Facilities need to refer to standard CDC Universal Precautions Guidelines which define adequate characteristics of disinfectants, i.e. one registered by the US Environmental Protection Agency as a 'hospital disinfectant' that is also 'mycobactericidal' for cleaning up blood spills.

The OSHA regulation about 'regulated waste' defines such waste in terms of infectious material that may pose an occupational hazard when not packaged properly for handling and disposal. 'Regulated Waste' as defined by OSHA includes:

- Liquid or semiliquid blood or OPIM.
- Contaminated items that would release blood or OPIM in a liquid or semiliquid state if compressed.
- Items that are caked with dried blood or OPIM and are capable of releasing these materials during handling.
- Contaminated sharps.
- Pathological and microbiological wastes containing blood or OPIM.

The Regulated Waste rules of the Standard emphasize adequate packaging for final transport and disposal. All containers must be closable (and closed prior to removal), constructed to contain the contents and prevent leakage of fluids during handling, storage, transport, or shipping, and labelled according to the specifications of the Standard. Sharps containers must be leakproof, puncture-resistant, easily accessible, and replaced before they are full.

> *Considerations.* OSHA does not define the method of treatment or disposal. Facilities are referred to other US, State, or local medical waste regulations. HIV/HBV research-scale laboratories or production facilities must comply with CDC Biosafety Level 3 practices according to the OSHA Standard. *All* contaminated laboratory waste from such facilities must be decontaminated before disposal.

Personal Protective Equipment

When engineering controls and work practices are instituted, and a potential for exposure continues, the employer must provide PPE to minimize that exposure. The specialized clothing and equipment that constitutes PPE include a variety of gowns, aprons, gloves, shoes and face and head protection. Appropriate use of PPE depends on the kind of procedures and the extent of anticipated exposure. The PPE must prevent blood or OPIM from reaching the employee's work clothes, street clothes, underwear, or body under normal conditions of use. The employer must make sure that the employee has access to appropriate PPE, that it is provided in appropriate sizes, and the employees are trained and monitored on its use. The employer must not only take the responsibility of supplying the PPE, but must arrange for its cleaning, laundering, replacement, or disposal. Employees must not take contaminated PPE home to launder.

> *Considerations.* OSHA recognizes that there may be rare circumstances when the emergent nature of a situation might interfere with an employee's ability to wear appropriate PPE (e.g. emergency response to a gastrointestinal bleed in a hallway). Such circumstances would depend on the employee's professional judgment that PPE use would prevent the delivery of health-care or public safety services, or place an increased hazard to the safety of the worker or co-worker. These instances need to be evaluated and documented to determine whether changes in policies need to be instituted to prevent such occurrences in the future.

Gloves

OSHA requires the use of gloves whenever contact with blood or OPIM is anticipated. OSHA does not require the use of gloves for any specific procedures or situations except in the case of vascular access. Gloves must be worn for phlebotomy or other vascular procedures with an exception for trained phlebotomists in volunteer blood donation centres. OSHA is very concerned about the integrity of gloves, and includes requirements to change gloves 'as soon as feasible' when they are visibly contaminated, torn or defective, or when tasks are completed. Disposable gloves must not be washed or disinfected for re-use because of the enhanced penetration of liquids through undetected holes. Sturdier gloves, such as utility gloves, can be washed or decontaminated for re-use as long as they remain intact.

> *Considerations.* OSHA does not advocate any one type of glove over another. In most cases latex, vinyl, nitrile or other materials may be used as long as they provide adequate protection. If employees develop allergies or sensitivities to a type of glove, however, OSHA requires that the employer finds suitable alternatives. As an example, many health-care workers have developed allergies to latex and must be provided with non-latex gloves that offer adequate protection.

Face protection

Face protection is one of the OSHA requirements that is difficult to enforce. Masks with eye protection, or face shields must be worn when splashes, splatter or sprays are reasonably expected to cause potential face exposures, such as performing arterial punctures, suctioning respiratory secretions or removing cryopreserved samples from liquid nitrogen. Masks with eye protection and face shields also have a passive function in preventing accidental contact of contaminated hands with the eyes, nose and mouth during the course of the work.

> *Considerations.* In laboratory environments face protection may be provided by engineering controls such as a microbiological safety cabinet or a splash shield.

Gowns/other protective body clothing

When soiling of clothing or skin contamination is anticipated. OSHA requires that the employer provide 'appropriate protective clothing'. In keeping with the concept of *performance* standard, OSHA allows the employer to decide the types and characteristics of the protective equipment, depending on the tasks and degree of exposure anticipated. Such protective clothing may include fluid-resistant or fluid-proof gowns, front-buttoned laboratory coats, aprons, elbow-length gauntlets, bootees, knee-thigh boots, caps or hoods.

> *Considerations.* PPE does *not* include scrub suits, uniforms, street clothing or laboratory coats when used as 'identifiers' or decorative wear. Employers are not required to give protection against every conceivable exposure but must protect workers against all 'reasonably anticipated' exposures. As an example, one employer might require only gloves as PPE for a phlebotomist with 15 years phlebotomy experience who works in an out-patient clinical laboratory, while another would require gloves, fluid-proof gowns and face protection for a phlebotomist who might be expected to draw arterial blood, with the possibility of splatter to the body and face. Both employers would be compliant with the OSHA regulations since their decisions are based on anticipated risk of exposure.

MEDICAL CARE

The OSHA Blood-borne Pathogen Standard contains specific requirements for preventive and post-exposure medical care evaluations, mostly taken directly from the US Public Health Service (US PHS) Recommendations. They include requiring the employer:

1. To institute a hepatitis B vaccination programme; and
2. Provide adequate medical follow-up after an exposure incident, i.e. any testing, counselling, or appropriate prophylaxis to reduce the risk of infection or transmission.

The required medical care must be provided to all employees designated by the employer as included under the Standard – that is all those with potential occupational exposure. The care must be provided free of charge to the employee, made available at reasonable times and locations, and provided by or under the supervision of a licensed health-care professional (not necessarily a physician). The medical care must be provided in accordance with the US PHS Recommendations at the time the exposure follow-up takes place. This allows flexibility as new information about appropriate treatment methods for exposed employees becomes available. Any laboratory tests must be performed by an accredited laboratory so that employees can receive reliable results.

Hepatitis B Vaccination

All employees covered by the Standard must be offered hepatitis B immunization after the employee has received training (so that he/she may make an informed decision), but within ten working days of initial assignment, so that employees may receive their three doses as soon as possible. OSHA states that the hepatitis B vaccine must be 'made available' to emphasize the employee's option to participate in the programme. If an employee refuses to accept vaccination he must sign a specifically-worded declination that reiterates the importance of hepatitis B vaccination for the prevention of infection. The refusing employee may later request the vaccine and the employer must provide it at that time if the employee's job responsibilities continue to be designated as a risk. Routine booster doses are not currently required, but would be if subsequently recommended by the US PHS. The vaccine does not have to be offered to previously immunized workers or those with demonstrated immunity, and the employer cannot require serological testing to determine who might already be immune.

Post-exposure Evaluation

The OSHA Standard requires that employers adhere to the post-exposure protocol that requires confidential medical evaluation, follow-up and documentation of an exposure incident. OSHA defines 'exposure incident' as a 'specific eye, mouth, other mucous membrane, non-intact skin or parenteral contact with blood or OPIM that results from the performance of the employee's duties' (OSHA, 1991).

The post-exposure protocol involves the elements listed in Box 19.2.

One of the purposes of having a comprehensive post-exposure protocol is to facilitate the identification of situations, devices or procedures that are sources of exposures. Therefore, documentation of the routes and circumstances of exposure is important in the prevention of future exposures. Institutions can use the evaluation of exposures to determine the need for different or additional PPE, or for justification of engineering controls such as needleless IV systems or self-sheathing devices.

The Standard also requires that the person/source whose blood or OPIM was involved in the exposure be identified and tested as soon as possible for HIV and HBV status unless such identification is infeasible or illegal because of local informed consent laws. The results of all source testing must be made available to the employee, but the employee may be counselled about the laws regarding the confidentiality of this information.

Box 19.2. Elements of a post-exposure evaluation protocol (OSHA, 1991).

- Documentation of route of exposure/circumstances
- Source tested for HIV/HBV/(HCV*)
- Employee counselling/serological testing
- Use of post-exposure measures (US PHS recommendations)
- Health-care professional's written report
- Evaluation of any illness

---

*New recommendation from CDC (CDC, 1996a).

Considerations. Recent recommendations from the CDC request that tests for hepatitis C virus (HCV) also be performed in compliance with local consent laws (CDC, 1996a). Although not specifically written in the text of the Standard, this recommendation becomes mandatory because of the requirement to adhere to US PHS recommendations for post-exposure follow-up.

Exposed employees must also be provided with serological testing as a baseline test and for follow-up. If the employee initially declines HIV testing OSHA requires that employers provide a 90-day 'thinking' period after collecting the blood, to give the employee an opportunity to change his/her mind about testing.

The Standard also requires that the employer provides post-exposure prophylaxis when medically indicated, according to the US PHS recommendations at the time of exposure. When the Standard was issued in 1991 recommendations for prophylaxis for hepatitis B exposures were already in place (CDC, 1989). These included hepatitis B immunoglobulin in combination with hepatitis B vaccine for those exposed employees who were not vaccinated, and a booster dose for those immunized employees whose antibody levels had dropped below a predetermined 'protective' level. In 1996 the CDC issued recommendations for HIV post-exposure prophylaxis after certain high-risk employee exposures (CDC, 1996b). The new recommendation follows the report of a case-control study in which the use of Zidovudine (AZT) prophylaxis following percutaneous exposures was associated with a 79% decrease in the risk of HIV infection among health-care workers (CDC, 1995). It becomes mandatory as a post-exposure prophylaxis method under the OSHA Bloodborne Pathogen Standard.

The health-care professional who is responsible for post-exposure management must inform the employee of the results of the evaluation and of any necessary treatments or follow-up. The health-care professional must also provide the employer with a written opinion within 15 days of the completion of the exposure evaluation to let him know that the employee has been informed of the results and of any further evaluations. In all cases employee confidentiality and consent are of paramount importance.

The CDC and the National Institutes of Health (NIH) recommend that baseline serum samples for all HIV research laboratory personnel be collected and stored (CDC, 1988). They also recommend the establishment of a confidential medical surveillance programme for these facilities and that laboratory workers be tested annually for evidence of seroconversion, with appropriate counselling and medical evaluation for any infected workers.

Record Keeping

Employees' medical records must contain specific information about hepatitis B vaccination status, including dates of immunization or refusal. In addition, the health-care professional's written opinion of post-exposure evaluations and copies of the results of any tests and treatments after exposure must be included. These medical records must be maintained in a confidential manner for the duration of the worker's employment with the institution *plus 30 years*. This is consistent with other record keeping OSHA requirements that would help in the identification of any chronic, long-term effects of workplace exposures.

TRAINING AND INFORMATION

Consistent with OSHA's emphasis of the rights of employees to know about the hazards in their workplace, the Standard requires detailed training of employees at the time of hire (within ten days), annually, and at any time when their occupational risk of exposure changes. The specific elements of training are listed in Box 19.3.

Training must be provided free of charge, during working hours, and be understood by the employees, i.e. appropriate in content, language and literacy level. A person who is 'knowledgeable' about the Standard must be available when the training is given so the employees will have the opportunity to have their questions answered.

Considerations. Commercial packages that provide information about the Standard are available. However, the employer must provide elements of institution-specific information to the employee, such as where to report exposure incidents, and the location of the institution's exposure control plan.

Box 19.3. Elements of a blood-borne pathogen training programme (OSHA, 1991).

- Accessibility of OSHA Blood-borne Pathogen Standard/Institutional Exposure Control Plan
- Blood-borne pathogen information (epidemiology, transmission, symptoms)
- Universal Precautions
- Selection, use and limitations of control methods (engineering, work practices, PPE)
- Emergency and post-exposure management
- HBV vaccination programme
- Hazard communications

HIV/HBV Research Laboratories

Recognizing the unique hazards posed by these research laboratories, OSHA requires additional training for employees in these settings. They must be provided with the general information given to all employees covered under the Standard, *and* they must demonstrate proficiency in standard microbiological techniques. This might involve prior work with human pathogens before work with HIV/HBV, or participation in training programmes with a progression of tasks to demonstrate safe handling techniques.

Record Keeping

Employers must maintain training records on all employees covered by the Standard. In addition to the names, and social security numbers of the employees, these records must include the dates and contents of training sessions and the names and qualifications of the trainers. Training records must be maintained for at least three years.

Other Hazard Communications

OSHA is very much interested in the provision of hazard communications to the unsuspecting 'downstream' worker. For this reason there are many references throughout the Standard about warning labels and colour codings to communicate the potential for hazards. A summary of these requirements is given in Table 19.2. An appropriate label refers to the orange or orange–red universal Biohazard label illustrated in the Standard, or the use of red bags or containers. OSHA does not intend all employers to label *all* containers of blood or OPIM that might be apparent to workers, such as blood specimens (for institutions complying with a Universal Precautions policy), blood products for patient use, or unembalmed cadavers. Rather, the Standard is intended to warn employees of the potential for exposure in secondary storage or transport containers.

COMPLIANCE

After the issue of the Joint Advisory Notice of 1987 (US Departments of Labor/Health and Human Services, 1987), OSHA staff began to visit healthcare facilities, primarily in response to employee complaints. Several citations were issued in the late 1980s under the provision of the General Duty Clause that requires a safe workplace (Bureau of National Affairs, 1988). These citations involved violations of the Personal Protective Equipment Standard (29CFR1910. 132) for failure to provide protection, and failure to tag blood product waste containers under the Hazard Tag Standard (29CFR1910. 145(c)(4)). One citation involved failure to have an effective training programme in place.

In the year following the issue of the OSHA Blood-borne Pathogen Standard (March, 1992 – February, 1993), 1700 violation citations were issued against

Table 19.2. Labelling requirements of OSHA Blood-borne Pathogen Standard (OSHA, 1991).

Item	No label if Universal Precautions used and employees are aware of container use		Biohazard label		Red Bag or container
Regulated waste container (e.g. contaminated sharps)			+	or	+
Re-usable sharps container (e.g. surgical instruments on a tray)			+	or	+
Refrigerator/freezer holding blood or OPIM			+		
Containers used for storage, transport, or shipping of blood			+	or	+
Blood/blood products for clinical use (e.g. transfusions)					
Individual specimen containers of blood or OPIM remaining in facility	+	or	+	or	+
Contaminated equipment needing repair			+		
Specimens or regulated waste shipped off site			+	or	+
Contaminated laundry	a	or	+	or	+
Contaminated laundry sent off site to another facility that does not use Universal Precautions			+	or	+

a – Alternative labelling or colour coding is allowed if all employees recognize the need to use Universal Precautions.

health-care facilities by federal OSHA inspections in the US, resulting in $1.3 million in fines (Pugliese, 1993). Citations began to involve administrative and record keeping issues, such as failure to communicate hazard warnings to workers and to provide hepatitis B vaccination programmes to workers. Likewise, in 1994, the most frequent citations involved the failure to develop an Exposure Control Plan and/or training programmes (American Management Association, 1996).

The OSHA Blood-borne Pathogen Standard requires that institutions monitor and enforce the provisions of the Standard. Although such requirements imply a 'policing' of the workplace, monitoring of health-care facilities has been done in the past for compliance with the requirements of other inspecting groups such as the Joint Commission for Accreditation of Health-care Organizations in hospitals, and the College of American Pathologists in clinical laboratories. Monitoring can be somewhat simplified with regularly scheduled audits, assisted by a check list. As with other rules in the workplace, enforcement is dependent upon administrative support.

The ultimate indication of compliance with the Standard is the reduction in workplace exposures and infection with blood-borne pathogens. The incidence of occupationally-acquired hepatitis B in the United States has declined steadily since the vaccine became available in 1982 (CDC, 1996b). Across the United States, data indicate that reported parenteral injuries in the workplace have declined substantially since the late 1980s. The cause of this trend has not been substantiated, but contributing factors undoubtedly include the institution of Universal Precautions, the availability and use of safety devices such as needleless IV systems, and the compliance with the OSHA Blood-borne Pathogen Standard (Beckman *et al.*, 1994; Jagger and Bentley, 1996).

REFERENCES

American Management Association (1996) *OSHA and the Medical Industry: a Compliance Update*, 3rd edn. Keye Productivity Center, Leawood, KS. 40 pp.

Beckmann, S.E., Vlahov, D., Koziol, D.E. *et al.* (1994) Temporal association between implementation of universal precautions and a sustained, progressive decrease in percutaneous exposures to blood. *Clinical Infectious Diseases* 18, 562–569.

Bureau of National Affairs Inc. (1988) *Occupational Safety and Health Reporter.* 17 February, 1988, pp. 1412–1413.

CDC (1982) Acquired immune deficiency syndrome (AIDS). Precautions for clinical and laboratory staffs. *Morbidity and Mortality Weekly Reports* 31, 577–580.

CDC (1983) AIDS: precautions for health care workers and allied professsionals. *Morbidity and Mortality Weekly Reports* 32N, 450–451.

CDC (1985a) Recommendations for protection against viral hepatitis. *Morbidity and Mortality Weekly Reports* 34, 313–335.

CDC (1985b) Summary: recommendations for preventing transmission of infection with human T-lymphotropic virus III/LAV in the workplace. *Morbidity and Mortality Weekly Reports* 34, 681–695.

CDC (1986) HTLVIII/LAV: agent summary statement. *Morbidity and Mortality Weekly Reports* 35, 540–549.

CDC (1987a) Recommendations for prevention of HIV transmission in health care settings. *Morbidity and Mortality Weekly Reports* 36 (Suppl 2), 3S–18S.

CDC (1987b) Update: human immunodeficiency virus infections in health care workers exposed to blood of infected patients. *Morbidity and Mortality Weekly Reports* 36, 285–289.

CDC (1988) Occupationally-acquired human immunodeficiency virus infections in laboratories producing virus concentrates in large quantities; conclusions and recommendations of an expert team convened by the Director of the National Institutes of Health (NIH). *Morbidity and Mortality Weekly Reports* 37 (Suppl 4), S19–S22.

CDC (1989) Guidelines for the prevention of transmission of human immunodeficiency virus and hepatitis B virus to health care and public safety workers. *Morbidity and Mortality Weekly Reports* 38 (Suppl 6), S3–S31.

CDC (1995) Case control study of HIV seroconversion in health care workers after percutaneous exposure to HIV-infected blood – France, United Kingdom and

United States, January 1988 – August 1994. *Morbidity and Mortality Weekly Reports* 44, 929–933.

CDC (1996a) *Hepatitis surveillance report no. 56.* US Department of Health and Human Services, Atlanta.

CDC (1996b) Update: provisional public health recommendations for chemoprophylaxis after occupational exposure to HIV. *Morbidity and Mortality Weekly Reports* 45, 468–472.

Chiarello, L. (1992) Testimony on needlestick prevention technology. Presented before Congress Committee on Small Business Opportunities and Energy. Washington DC, February 7, 1992.

Jagger, J. and Bentley, M. (1996) Substantial nationwide drop in percutaneous injury rates detected for 1995. *Advances in Exposure Prevention* 2, 1–12.

OSHA (1991) US Occupational Health and Safety Administration. Occupational exposure to blood-borne pathogens, final rule. *Federal Register* 56, 64175–64182.

Pugliese, G. (1993) OSHA fines employers $1.3 million under blood-borne standard. *Infection Control and Hospital Epidemiology* 14, 670.

Richmond, J.Y and McKinney, R.W. (eds) (1993) *Biosafety in Microbiological and Biomedical Laboratories*, 3rd edn. US Department of Health and Human Services Publication No (CDC) 93–8395. Public Health Service, Washington, DC.

US Department of Labor and Department of Health and Human Services (1987) Joint advisory notice. Protection against occupational exposure to hepatitis B virus and human immunodeficiency virus. *Federal Register* 52, 41818–41824.

US FDA (1992) Food and Drug Administration Safety Alert. Needlestick and other risks from hypodermic needles on secondary i.v. administration sets – piggyback and intermittent. Center for Devices and Radiological Health, US Department of Health and Human Services, Rockville, MD.

CHAPTER *20*

Universal Precautions and the advent of Standard Precautions: a review

C.C. Kibbler

INTRODUCTION

The concept of Universal Precautions (UPs) was developed in the USA largely in response to the growing concerns of those managing human immuno-deficiency virus (HIV) infected patients in the health-care setting. Until then the perceived threat of infection from blood and body fluids had been from hepatitis B virus (HBV) and it had been felt that risk groups of patients could be relatively easily identified and specific practices aimed at preventing transmission could be targeted at these groups. It was becoming clear, however, that HIV-infected patients were asymptomatic in the early stages of infection, that antibody tests might be negative whilst a patient was infectious with circulating virus and that risk factors for infection might be unsuspected or undeclared (see Hunt, Chapter 19 this volume).

DEFINITION AND PRINCIPLES

The main principle of UPs is enshrined in the statement that 'all patients should be assumed to be infectious for HIV and other blood-borne pathogens ...' (Centers for Disease Control and Prevention: CDC, 1987). Consequently, pre-cautions to prevent transmission of blood-borne pathogens should be applied '*universally*', i.e. when there is a risk of exposure to the blood or body fluids of any patient. From this followed the basic components of the policy:

1. The risk of exposure of health-care workers or patients to potentially infectious material associated with particular clinical settings or procedures needs to be identified, rather than whether the patient is infected, or not.
2. Appropriate measures should then be applied to prevent exposure to blood and blood-stained body fluids according to the perceived risks.

Clinical settings in which UPs should be applied are those in which there is a risk of exposure to body fluids which may contain HIV, hepatitis B, other hepatitis viruses, or other blood-borne pathogens. Such fluids include:

Occupational Blood-borne Infections (eds C.H. Collins and D.A. Kennedy)

- Blood
- Blood-stained fluids, e.g. urine, faeces, sputum
- Semen
- Human tissues, e.g. placentas
- Cerebrospinal fluid
- Amniotic fluid
- Pericardial fluid
- Pleural fluid
- Vaginal secretions

General measures which should be taken to protect health-care workers are shown in Table 20.1. In addition, the risk of contact with body fluids is minimized by the wearing of personal protective clothing as detailed in Table 20.2 when the clinical situation is appropriate.

PERCEIVED BENEFITS OF UNIVERSAL PRECAUTIONS

The two-tier system which had been in use for many years to prevent the transmission of hepatitis B virus was confusing for health-care workers and, being based on positive identification of a patient's status as a 'high risk' patient and the communication of that status to other staff, is always liable to breakdown. A policy of UPs has the advantage of protecting against all blood-borne infections, including those yet to be discovered. In addition, it protects against transmission of infection from patients in whom the diagnosis of a blood-borne

Table 20.1. General protective measures.

Cover cuts and abrasions with waterproof dressings
Wash hands before and after all procedures and when contaminated with body fluids
Carefully dispose of all sharps
Do not re-sheath needles
Ensure all at-risk staff are vaccinated against hepatitis B

Table 20.2. Appropriate use of personal protective clothing/equipment for Universal Precautions.

Gloves – when there may be direct contact with body fluids, non-intact skin or mucous membranes

Plastic aprons – where contamination of clothing is possible

Impermeable gowns – where there is a likelihood of spillage of large volumes of blood and body fluids

Protective eyewear and masks – where there is a risk of blood or body fluids splashing into the face

infection has not been considered. Finally, by removing the need to identify a patient's infectious status, it serves to improve patient confidentiality.

It was emphasized from the outset that the practice of UPs does not obviate the need for continued adherence to general infection control principles and general hygiene measures, such as hand washing, for preventing transmission of other infectious diseases to both health-care worker and patient.

HISTORICAL BACKGROUND

Universal Precautions were recommended as the means of preventing transmission of HBV and HIV in the USA in 1985 (CDC, 1985). Subsequent updates refined the guidelines (CDC, 1987, 1988, 1991) and outlined their application in different clinical settings. In 1991 the US Occupational Safety and Health Administration (OSHA) embodied them in the Blood-borne Pathogens Standard (OSHA, 1991). This standard applies to all employees who might be exposed to blood or other potentially infectious materials in the workplace. To achieve compliance the employer must have a system for determining and documenting exposure. In addition a written exposure control plan should govern work practices and stipulate the measures, such as the use of personal protective equipment and engineering controls, which must be taken.

Elsewhere in the world UPs have been adopted to a greater or lesser extent: in Canada (Righter, 1991; Osterman, 1995), Europe (Oteo *et al.*, 1991; Nelsing *et al.*, 1993), and the Far East (Anon, 1995). The World Health Organization has also produced guidelines on this issue (Hu *et al.*, 1991).

UNIVERSAL PRECAUTIONS AND OTHER UK GUIDELINES AND LEGISLATION

In the UK guidance incorporating UPs has been issued by the Department of Health (DH, 1990), and various national working parties (Hospital Infection Society Working Party, 1990; Joint Working Party of the Hospital Infection Society and the Surgical Infection Group, 1992; Public Health Laboratory Service Hepatitis Subcommittee: PHLS, 1993). Initially this advice suggested that UPs could be restricted to areas of high prevalence of blood-borne virus infection and that a two-tier approach could be maintained elsewhere, identifying 'high-risk' patients, but latterly the Advisory Committee on Dangerous Pathogens has issued further guidance recommending UPs as the basis for protection in all areas (ACDP, 1995). Although these various sets of UK guidelines have no force of law they enable employers to protect their employees under the terms of the relevant legislation, namely the *Health and Safety at Work Act 1974*, the *Control of Substances Hazardous to Health Regulations 1994*, the *Management of Health and Safety at Work Regulations 1992* and the *Reporting of Incidents, Diseases and Dangerous Occurrences Regulations 1995*.

HOW EFFECTIVE ARE UNIVERSAL PRECAUTIONS?

It is not possible to design a randomized, comparative prospective study to examine the merits of UPs in comparison with patient risk assessment or sero-logical testing, but several important studies have been conducted which shed light on various aspects of the policy.

There is a body of surgical opinion which feels that knowledge of patient status allows individuals to better protect themselves against exposure to the blood and body fluids of that patient. However, a study of surgical procedures in the Operating Department of San Francisco General Hospital, where UPs, together with a safe sharps policy and double gloving, has been practised since 1988, demonstrated no difference in exposure rates between procedures on high- or low-risk patients (Gerberding *et al.*, 1990). The overall exposure rate was low, which was thought to reflect the high standard of infection control practised, and indeed, more than 90% of scrub nurses and house officers and 76% of the attending surgeons wore double gloves.

Several studies have shown that the implementation of UPs, coupled with training, results in a reduction of cutaneous exposures to blood and body fluids (Fahey *et al.*, 1991; Wong *et al.*, 1991) when compared with previous practice. It is possible that such sequential studies might merely be documenting a reduc-tion of potential exposure incidents as health-care workers were made more aware of the risks. However, it has been shown that whilst implementation of UPs may have no effect on the overall rate of exposure incidents, the increase in barrier use prevents direct contact with blood and body fluids and converts actual exposure events into averted ones (Wong *et al.*, 1991).

The efficacy of the personal protective equipment in preventing exposure has been questioned. Studies of the effect of latex gloves upon the volume of blood inoculated during needlestick injury have shown a significant benefit. In an *in vitro* paper prefilter model and an *ex vivo* porcine tissue model the wiping effect which occurs as the needle passes through the glove material has been shown to reduce transferred blood volume by 46–86% (Mast *et al.*, 1993). In addition, in surgical procedures the wearing of two pairs of gloves ('double-gloving') is supported by a study showing that the perforation rate of the inner glove is approximately 2% (Matta *et al.*, 1988) in comparison with perforation rates of 11–54% for single gloves in different types of surgery (Church and Sanderson, 1980; Brough *et al.*, 1988; Matta *et al.*, 1988; Maffulli *et al.*, 1989; Smith and Grant, 1990; Palmer and Rickett, 1992). Hence the *risk* of infection with blood-borne viruses is clearly reduced by the wearing of latex gloves in surgical and other clinical settings, although this has not been correlated with a reduction in *incidence* of infection.

HOW EXPENSIVE ARE UNIVERSAL PRECAUTIONS?

Many people have expressed concerns about the costs of practising UPs. The increased usage of protective clothing, particularly latex gloves and plastic aprons, in centres where the policy has been implemented, inevitably has resource implications. In 1989 OSHA published estimated costs for nationwide operation of the OSHA standard in medical facilities, the police forces, funeral services, fire and rescue services, personnel, and medical equipment repair. The total annual costs were calculated as $852 million, with $195 million per year needed for US hospitals (see Doebbeling and Wenzel, 1990). However, the following year, a study at the University of Iowa Hospitals found that the policy had led to an increase of 92% in the cost of isolation materials at the study hospitals, of which 64% was due to increased glove usage and 25% was due to disposable gowns (Doebbeling and Wenzel, 1990). Applying these data to the USA as a whole meant that UPs would have cost at least $336 million per year at 1990 prices.

In highlighting these costs, some have suggested that resources would be better used by targeting patients shown to be seropositive for blood-borne viruses. They have therefore advocated routine testing for all patients or at least prior to surgery (Ponsford, 1987; Goldman, 1988; Shanson, 1988; Fournier and Zeppa, 1989; Lewis and Montgomery, 1990; Freeman, 1991). However, an economic evaluation comparing routine preoperative HIV testing with UPs has shown that the minimum cost for routine testing is greater than the maximum cost for UPs, assuming that both are equally effective in preventing HIV transmission (Lawrence *et al.*, 1993). It should also be borne in mind that routine testing does not solve the problem of false positive or false negative results and is of little use in emergency surgery.

Attempts to assess the cost effectiveness of UPs are fraught with difficulties and require many assumptions to be made. One such economic evaluation has estimated that the prevention of a single case of HIV seroconversion in a US health-care worker costs $8–129 million and the authors concluded that UPs were not, therefore, cost-effective (Stock *et al.*, 1990). However, this study took no account of the potential for preventing other blood-borne virus infections, such as hepatitis C.

WHAT ARE THE LIMITATIONS OF UNIVERSAL PRECAUTIONS?

Perhaps the most confusing aspect of UPs is the term itself. Staff often fail to realize that the concept should be applied only to the prevention of blood-borne virus infections and not as a *universal* infection control policy to prevent all infections. It would be better if the term 'blood-borne virus precautions' or 'blood and body fluid precautions' had been adopted and many institutions have linked the term with such phrases.

Whilst the basis of a policy of UPs is risk assessment of individual clinical settings, its practice depends upon the availability of personal protective equipment. A policy of identifying 'high-risk' patients allows a pack of personal protective equipment (PPE) to be provided for specific patients, whereas availability of PPE for UPs needs to be ensured by a process of stock control and monitoring/audit. A particular practical problem appears to be with protective eyewear which are frequently missing when needed.

An issue which was raised at the inception of UPs was that of patient alienation. There were concerns that the wearing of gloves and protective eyewear in particular might have a psychological impact on patient care. This has not been studied in detail but widespread adoption of UPs, including in specialties such as dentistry, appears to have been accepted by the majority of patients. Most seem to consider that it is entirely appropriate for health-care workers to protect themselves from occupational exposure.

A number of clinicians have found the adoption of UPs of considerable inconvenience and compliance with the frequent need for gloving and handwashing when performing common tasks such as phlebotomy has been poor in some institutions. However, compliance may be a problem with any infection control policy and can be improved by audit, feedback and education programmes.

The charge of restricting skills has been levelled at UPs. This is chiefly as a consequence of the need to use gloves for minor procedures, such as phlebotomy. It is difficult to sustain such an argument for the use of single gloves, when surgeons have been performing extremely dexterous operations for many years, but some do claim a loss of sensitivity when using two pairs of gloves for surgery, or when using armoured gloves which are designed to reduce the risk of sharps injury (see Scully, Chapter 9, and Collins and Kennedy, Chapter 14, this volume).

THE ADVENT OF STANDARD PRECAUTIONS

A different isolation policy called Body Substance Isolation (BSI) was formulated in 1987 at the Harborview Medical Center in Seattle and the University of California, San Diego (Lynch *et al.*, 1987). This was an attempt to incorporate a policy for the prevention of blood-borne virus infections into an overall policy to prevent nosocomial infection. Health-care workers were instructed to put on clean gloves prior to contact with mucous membranes and non-intact skin and when contact with moist body substances was anticipated. A simple alert sign was used to require individuals to check with ward staff before entering the isolation rooms of patients with respiratory illnesses to determine if masks should be worn.

Body Substance Isolation differed from UPs in extending the need for gloves to be worn when contact with any moist body substance was anticipated and in not requiring hand washing following removal of gloves unless visible soiling of the hands was observed. In addition, BSI appeared to be inadequate for preventing the spread of infections transmitted by the air-borne route and those organisms transmissible from dry skin or the environment by contact.

As a consequence of the need to answer the many criticisms of UPs set out earlier and the failure of BSI to do this, the Centers for Disease Control and Prevention and the Hospital Infection Control Practices Advisory Committee (HIPAC) have recently published the revised *CDC Guideline for Isolation Precautions in Hospitals* (Hospital Infection Control Practices Advisory Committee, 1996a, b). This contains two tiers of precautions. 'Standard Precautions' are designed for the care of all in-patients irrespective of underlying disease or presumed infection status. Transmission-based Precautions are for patients known or suspected to be infected or colonized with transmissible or epidemiologically important pathogens for which additional precautions beyond Standard Precautions are needed to prevent nosocomial transmission. These are based upon knowledge of routes of transmission of these organisms and are classified as Airborne Precautions, Droplet Precautions and Contact Precautions.

Standard Precautions combine the major features of UPs and BSI and apply to blood, all body fluids, secretions and excretions except sweat, regardless of whether they contain visible blood, non-intact skin and mucous membranes. The general measures which form the basis of Standard Precautions are shown in Table 20.3.

THE FUTURE

Standard Precautions seem likely to become adopted elsewhere as national bodies responsible for formulating guidelines on safety and infection control accept the need for clarification of the confusion caused by UPs and the need for unifying infection control policies.

The Australian National Infection Control Working Party has already recommended the adoption of the same precautions for the basic tier of prevention, with 'Additional Precautions' as the second tier (similar to the CDC 'Transmission-based Precautions') and this has been endorsed by the Australian National Health and Medical Research Council (National Health and Medical Research Council, 1996).

However, there may well be further recommendations in the future as the emphasis is placed on newly prominent infectious diseases and the drawbacks and limitations of Standard Precautions become highlighted by investigators. What UPs succeeded in doing was to shift the emphasis away from the concept of the 'high-risk' patient to a strategy of prevention of contamination with fluids which might contain blood-borne viruses. Standard Precautions takes this a step further by integrating it into an overall structure for preventing nosocomial infection.

Table 20.3. Outline of Standard Precautions.

Handwashing
After touching blood, body fluids, secretions, excretions and contaminated items,
    regardless of whether gloves are worn
Immediately after removing gloves and between patients

Gloves
Wear when touching above fluids and items, mucous membranes and non-intact skin
Remove after contact with material that may contain high content of microorganisms
    and between patients

Masks and eyewear
Wear during procedures likely to generate splashes or sprays of above fluids

Gowns
Wear during procedures likely to generate splashes or sprays of above fluids and likely
    to contaminate clothing

Equipment
Ensure appropriate cleaning of re-usable equipment between patients. Handle
    equipment soiled with the above fluids in a safe manner

Environmental control
Ensure adequate routine cleaning procedures

Linen
Handle soiled linen in a safe manner

Occupational health
Carefully dispose of all sharps. Do not recap needles. Take care when handling all
    sharp instruments

Patient placement
Place a patient who contaminates the environment or who does not (or cannot be
    expected to) assist in maintaining appropriate hygiene or environmental control in a
    private room

REFERENCES

ACDP (1995) *Protection Against Blood-borne Viruses in the Workplace: HIV and Hepatitis.* Advisory Committee on Dangerous Pathogens. HMSO, London.

Anon. (1995) Guidelines for implementation of universal precautions. The Hospital Infection Control Group of Thailand. *Journal of the Medical Association of Thailand* 78, Suppl 2, S133–S134.

Brough, S.J., Hunt, T.M. and Barrie, W.M. (1988) Surgical glove perforation. *British Journal of Surgery* 75, 317.

CDC (1985) Recommendations for preventing transmission of infection with human T-lymphotropic virus type III/lymphadenopathy-associated virus in the workplace. *Morbidity and Mortality Weekly Reports* 3, 686, 691–695; 4, 681–686, 691–695.

CDC (1987) Recommendations for prevention of HIV transmission in health care settings. *Morbidity and Mortality Weekly* Reports 36, 1S–18S.

CDC (1988) Update: universal precautions for prevention of transmission of human immunodeficiency virus, hepatitis B virus, and other bloodborne pathogens in health-care settings. *Morbidity and Mortality Weekly Reports* 37, 377–382, 387–388.

CDC (1991) Recommendations for preventing transmission of human immunodeficiency virus and hepatitis B virus to patients during exposure prone invasive procedures. *Morbidity and Mortality Weekly Reports* 40(RR-8), 1–9.

Church J. and Sanderson P. (1980) Surgical glove punctures. *Journal of Hospital Infection* 1, 84.

Control of Substances Hazardous to Health Regulations (1994). HMSO, London.

Doebbeling, B.N. and Wenzel, R.P. (1990) The direct costs of universal precautions in a teaching hospital. *Journal of the American Medical Association* 264, 2083–2087.

DH (1990) Guidance for clinical health care workers: protection against infection with HIV and hepatitis viruses, recommendations of the expert advisory group on AIDS. UK Department of Health. HMSO, London.

Fahey, B.J., Koziol, D.E., Banks, S.M. *et al.* (1991) Frequency of nonparenteral occupational exposures to blood and body fluids before and after universal precautions training. *American Journal of Medicine* 90, 145–153.

Fournier, A.M. and Zeppa, R. (1989) Preoperative screening for HIV infection. A balanced view for the practising surgeon. *Archives of Surgery* 124, 1038–1040.

Freeman, H.E. (1991) HIV testing of asymptomatic patients in U.S. hospitals. *Medical Care* 29, 87–96.

Gerberding, J.L., Littell, C., Tarkington, A. *et al.* (1990) Risk of exposure of surgical personnel to patients' blood during surgery at San Francisco General Hospital. *New England Journal of Medicine* 322, 1788–1793.

Goldman, B. (1988) Doctors divided: AIDS and the physicians at the San Francisco General. *Canadian Medical Association Journal* 138, 736–738.

Health and Safety at Work etc. Act 1974. HMSO, London.

Hospital Infection Control Practices Advisory Committee (1996a) Guideline for isolation precautions in hospitals. Part I. Evolution of isolation practices. *American Journal of Infection Control* 24, 24–31.

Hospital Infection Control Practices Advisory Committee (1996b) Guideline for isolation precautions in hospitals. Part II. Recommendations for isolation precautions in hospitals. *American Journal of Infection Control* 24, 32–45.

Hospital Infection Society Working Party (1990) Recommendation on acquired immune deficiency syndrome. *Journal of Hospital Infection* 15, 7–34.

Hu, D.J., Kane, M.A. and Heymann, D.L. (1991) Transmission of HIV, hepatitis B virus and other blood-borne pathogens in health care settings: a review of risk factors and guidelines for prevention. *Bulletin of the World Health Organization* 69, 623–630.

Joint Working Party of the Hospital Infection Society and the Surgical Infection Study Group (1992) Risks to surgeons and patients from HIV and hepatitis: guidelines on precautions and management of exposure to blood or body fluids. *British Medical Journal* 305, 1337–1343.

Lawrence, V.A., Gafni, A. and Kroenke, K. (1993) Preoperative HIV testing: is it less expensive than universal precautions? *Journal of Clinical Epidemiology* 46, 1219–1227.

Lewis, C.E. and Montgomery, K. (1990) The HIV testing policies of U.S. hospitals. *Journal of the American Medical Association* 264, 2764–2767.

Lynch, P., Jackson, M.M., Cummings, M.J. *et al.* (1987) Rethinking the role of isolation practices in the prevention of nosocomial infections. *Annals of Internal Medicine* 107, 243–246.

Maffulli, N., Capasso, G. and Testa, V. (1989) Glove perforation in elective orthopaedic surgery. *Acta Orthopaedica Scandinavica* 60, 565–566.

Management of Health and Safety at Work Regulations (1992). HMSO, London.

Mast, S.T., Woolwine, J.D. and Gerberding, J.L. (1993) Efficacy of gloves in reducing blood volumes transferred during simulated needlestick injury. *Journal of Infectious Diseases* 168, 1589–1592.

Matta, H., Thompson, A.M. and Rainey, J.B. (1988) Does wearing two pairs of gloves protect operating theatre staff from skin contamination? *British Medical Journal* 297, 597–598.

National Health and Medical Research Council (1996) Infection control in the health care setting. Guidelines for the prevention of transmission of infectious diseases. Australian Government Publishing Service, Canberra.

Nelsing, S., Nielsen, T.L. and Nielsen, J.O. (1993) Occupational blood exposure among health care workers: II. Exposure mechanisms and universal precautions. *Scandinavian Journal of Infectious Diseases*, 25, 199–205.

OSHA (1991) US Occupational Health and Safety Administration. Occupational exposure to blood-borne pathogens. Final rule, *Federal Register* 56, 64175–64182.

Osterman, J.W. (1995) Beyond universal precautions. *Canadian Medical Association Journal* 152, 1051–1055.

Oteo, J.A., Rosino, A., Martinez de Artola, V. *et al.* (1991) Usefulness of warning flags in the handling of biological samples in an emergency room. *Enfermedades Infecciosas y Microbiologia Clinica* 9, 634–636.

Palmer, J.D. and Rickett, J.W.S. (1992) The mechanisms and risk of surgical glove perforation. *Journal of Hospital Infection* 22, 279–286.

PHLS Hepatitis Subcommittee (1993) Hepatitis C virus: guidance on the risks and current management of occupational exposure. *CDR Review* 10, R135–R139.

Ponsford, G. (1987) AIDS in the OR: a surgeon's view. *Canadian Medical Association Journal* 137, 1036–1039.

Reporting of Injuries, Diseases and Dangerous Occurrences Regulations (1995) HMSO, London.

Righter J. (1991) Removal of warning labels from patient specimens. *Canadian Journal of Infection Control* 6, 109.

Shanson, D.C. (1988) Controversies about guidelines to prevent transmission of human immunodeficiency virus in hospitals in Britain. *Journal of Hospital Infection* 11 (Suppl. A), 218–222.

Smith, J.R. and Grant, J.M. (1990) The incidence of glove puncture during Caesarian section. *Journal of Obstetrics and Gynecology* 10, 317–318.

Stock, S.R., Gafni, A. and Bloch, R.F. (1990) Universal precautions to prevent HIV transmission to health care workers: an economic analysis. *Canadian Medical Association Journal* 142, 937–946.

Wong, E.S., Stotka, J.L., Chinchilli, V.M. *et al.* (1991) Are universal precautions effective in reducing the number of occupational exposures among health care workers? *Journal of the American Medical Association* 265, 1123–1128.

GLOSSARY OF ACRONYMS

Glossary of acronyms used in the text. Each is given in full in the appropriate chapter but they are reproduced here for the reader's convenience.

ABE	Accidental blood exposure
ACDP	Advisory Committee on Dangerous Pathogens
ADA	American Dental Association
AIDS	Acquired immunodeficiency syndrome
ALT	Alanine aminotransferase
AZT	Zidovudine
BE	Blood exposure
BMA	British Medical Association
BPL	Beta-propiolactone
BSI	British Standards Institution
BSI	Body substance isolation
BTS	Blood Transfusion Service
CCHF	Congo-Crimea haemorrhagic fever
CDC	Centers for Disease Control and Prevention
CMV	Cytomegalovirus
COSHH	Control of Substances Hazardous to Health Regulations
DE	Department of the Environment
DH	Department of Health (including Department of Health and Social Security and UK Health Departments)
EC	European Community
ECP	Exposure Control Plan
EPA	Environmental Protection Agency
EPINET	Exposure Prevention Information Network
EPIP	Exposure prone invasive procedures
EPP	Exposure prone procedures
FDA	Food and Drugs Administration
FIV	Feline immunodeficiency virus
GDC	General Dental Council
GERES	Groupe d'Étude sur la risque d'exposition au sang
GLP	Good Laboratory Practice
HBcAg	Hepatitis B core antigen
HBeAg	Hepatitis B e antigen
HBIG	Hepatitis B immune globulin
HBsAg	Hepatitis B surface antigen
HBV	Hepatitis B virus
HCC	Hepatocellular carcinoma
HCMV	Human cytomegalovirus

HCV	Hepatitis C virus
HCW	Health-care worker
HDV	Hepatitis D virus
HGV	Hepatitis G virus
HIV	Human immunodeficiency virus
HSC	Health and Safety Commission
HSAC	Health Services Advisory Committee
HTLV	Human T lymphocytic virus
IA	Implantable access
IEC	International Electrotechnical Commission
ISO	International Standards Organization
IV	Intravenous
MCE	Mucocutaneous exposure
MDA	Medical Devices Agency
MRC	Medical Research Council
NCTC	National Collection of Type Cultures
NaDCC	Sodium dichloroisocyanurate
NHS	National Health Service
NANB	(hepatitis) non-A-non-B
NCCLS	National Committee for the Control of Laboratory Standards
NIH	National Institutes of Health
NIR	No identified risk
OPIM	Other potentially infectious material
OSHA	Occupational Safety and Health Agency
PCE	Percutaneous exposure
PCR	Polymerase chain reaction
PDV	Plasma-derived vaccine
PHLS	Public Health Laboratory Service
PPE	Personal protective equipment
SIROH	(Italian Study Group on Occupational Risks of HIV Infection)
SIV	Simian immunodeficiency virus
SOP	Standard operating procedures
UP	Universal Precautions
US PHS	United States Public Health Service
VHF	Viral haemorrhagic fever
WHO	World Health Organization
ZDV	Zidovudine

Index